Process Control

Process Control emphasizes the importance of computers in this modern age of teaching and practicing process control. An introductory textbook, it covers the most essential aspects of process control suitable for a one-semester course.

The text covers classical techniques, but also includes discussion of state-space modeling and control, a modern control topic lacking in most chemical process control introductory texts. MATLAB®, a popular engineering software package, is used as a powerful yet approachable computational tool. Text examples demonstrate how root locus, Bode plots, and time-domain simulations can be integrated to tackle a control problem. Classical control and state-space designs are compared. Despite the reliance on MATLAB, theory and analysis of process control are well presented, creating a well-rounded pedagogical text. Each chapter concludes with problem sets, to which hints or solutions are provided. A Web site provides excellent support in the way of MATLAB outputs of text examples and MATLAB sessions, references, and supplementary notes.

A succinct and readable text, this book will be useful for students studying process control, as well as for professionals undertaking industrial short courses or looking for a brief reference.

Pao C. Chau is Professor of Chemical Engineering at the University of California, San Diego. He also works as a consultant to the biotechnology industry on problems dealing with bioreactor design and control and molecular modeling.

D0140675

CAMBRIDGE SERIES IN CHEMICAL ENGINEERING

Books in the Series:

E. L. Cussler, *Diffusion: Mass Transfer in Fluid Systems, second edition*

Liang-Shih Fan and Chao Zhu, *Principles of Gas–Solid Flows*

Hasan Orbey and Stanley I. Sandler, *Modeling Vapor–Liquid Equilibria: Cubic Equations of State and Their Mixing Rules*

T. Michael Duncan and Jeffrey A. Reimer, *Chemical Engineering Design and Analysis: An Introduction*

John C. Slattery, *Advanced Transport Phenomena*

A. Varma, M. Morbidelli, H. Wu, *Parametric Sensitivity in Chemical Systems*

M. Morbidelli, A. Gavriilidis, and A. Varma, *Catalyst Design: Optimal Distribution of Catalyst in Pellets, Reactors, and Membranes*

E. L. Cussler and G. D. Moggridge, *Chemical Product Design*

Pao C. Chau, *Process Control: A First Course with MATLAB*

Process Control

A First Course with MATLAB

Pao C. Chau

University of California, San Diego

CAMBRIDGE
UNIVERSITY PRESS

CAMBRIDGE UNIVERSITY PRESS
Cambridge, New York, Melbourne, Madrid, Cape Town, Singapore,
São Paulo, Delhi, Dubai, Tokyo, Mexico City

Cambridge University Press
The Edinburgh Building, Cambridge CB2 8RU, UK

Published in the United States of America by Cambridge University Press, New York

www.cambridge.org
Information on this title: www.cambridge.org/9780521002554

First published 2002

A catalogue record for this publication is available from the British Library

Library of Congress Cataloging in Publication Data
Chau, Pao C.
Process control : a first course with MATLAB / Pao C. Chau.
    p. cm. – (Cambridge series in chemical engineering)
Includes bibliographical references and index.
ISBN 0-521-80760-3 – ISBN 0-521-00255-9 (pb.)
1. Chemical process control – Data processing. 2. MATLAB. I. Title. II. Series.
TP155.75 .C42 2002
660.2815 – dc21   2001052567

ISBN 978-0-521-80760-9 Hardback
ISBN 978-0-521-00255-4 Paperback

MATLAB is a registered trademark of The MathWorks, Inc.

Contents

Contents

Contents

Preface

This is an introductory text written from the perspective of a student. The major concern is not of how much material is covered, but rather, how the most important and basic concepts that one should grasp in a first course are presented. If your instructor is using some other text that you are struggling to understand, I hope that I can help you too. The material here is the result of a process of elimination. The writing and the examples are succinct and self-explanatory, and the style is purposely unorthodox and conversational. To a great extent, the style, content, and the extensive use of footnotes are molded heavily by questions raised in class. I left out very few derivation steps. If they are left out, the missing steps are provided as hints in the Review Problems at the back of each chapter. I strive to eliminate those "easily obtained" results that baffle many of us. Most of you should be able to read the material on your own. You just need basic knowledge in differential equations, and it helps if you have taken a course on writing material balances. With the exception of Chapters 4, 9, and 10, which should be skipped in a quarter-long course, it also helps if you proceed chapter by chapter. The presentation of material is not intended for someone to just jump right in the middle of the text. A very strong emphasis is placed on developing analytical skills. To keep pace with the modern computer era, a coherent and integrated approach is taken to using a computational tool. I believe in active learning. When you read the chapters, it is very important that you have MATLAB with its Control Toolbox to experiment and test the examples firsthand.

Notes to Instructors

There are probably more introductory texts on control than on any other engineering disciplines. It is arguable whether we need another control text. As we move into the era of hundred-dollar textbooks, I believe we can lighten the economic burden and, with the Internet, assemble a new generation of modularized texts that soften the printing burden by off-loading selected material to the Web. Still, a key resolve is to scale back on the scope of a text to the most crucial basics. How much students can or be enticed to learn is inversely proportional to the number of pages that they have to read – akin to diminished magnitude and increased lag in frequency response. Therefore, as textbooks become thicker over the years in attempts to reach out to students and are excellent resources from the perspective of instructors, these texts are by no means more effective pedagogical tools. This project was started as a set of review notes when I found that students were having trouble identifying the key concepts in these expansive texts. I also found that these texts in many circumstances deter students from active learning and experimenting on their own.

At this point, the contents are scaled down to fit a one-semester course. On a quarter system, Chapters 4, 9, and 10 can be omitted. With the exception of Chapters 4 and 9, on state-space models, the organization has "evolved" to become very classical. The syllabus is chosen such that students can get to tuning proportional–integral–differential controllers before they lose interest. Furthermore, discrete-time analysis has been discarded. If there is to be one introductory course in the undergraduate curriculum, it is very important to provide an exposure to state-space models as a bridge to a graduate-level course. Chapter 10, on mutliloop systems, is a collection of topics that are usually handled by several chapters in a formal text. This chapter is written such that only the most crucial concepts are illustrated and that it could be incorporated comfortably into a one-semester curriculum. For schools with the luxury of two control courses in the curriculum, this chapter should provide a nice introductory transition. Because the material is so restricted, I emphasize that this is a "first-course" textbook, lest a student mistakenly ignore the immense expanse of the control field. I also have omitted appendices and extensive references. As a modularized tool, I use the *Web Support* to provide references, support material, and detailed MATLAB plots and results.

Homework problems are also handled differently. At the end of each chapter are short, mostly derivation-type problems that are called Review Problems. Hints or solutions are provided for these exercises. To enhance the skill of problem solving, the extreme approach is taken, more so than that of Stephanopoulos (1984), of collecting major homework problems at the back and not at the end of each chapter. My aim is to emphasize the need to understand and integrate knowledge, a virtue that is endearing to ABET, the engineering accreditation body in the United States. These problems do not even specify the associated chapter as many of them involve different techniques. A student has to determine the appropriate route of attack. An instructor may find it aggravating to assign individual parts of a problem, but when all the parts are solved, I hope the exercise will provide a better perspective on how different ideas are integrated.

To be an effective teaching tool, this text is intended for experienced instructors who may have a wealth of their own examples and material, but writing an introductory text is of no interest to them. The concise coverage conveniently provides a vehicle with which they can take a basic, minimalist set of chapters and add supplementary material that they deem appropriate. Even without supplementary material, however, this text contains the most crucial material, and there should not be a need for an additional expensive, formal text.

Although the intended teaching style relies heavily on the use of MATLAB, the presentation is very different from texts that prepare elaborate M-files and even menu-driven interfaces. One of the reasons why MATLAB is such a great tool is that it does not have a steep learning curve. Students can quickly experiment on their own. Spoon-feeding with our misguided intention would only destroy the incentive to explore and learn on one's own. To counter this pitfall, strong emphasis is placed on what students can accomplish easily with only a few MATLAB statements. MATLAB is introduced as walk-through tutorials that encourage students to enter commands on their own. As a strong advocate of active learning, I do not duplicate MATLAB results. Students again are encouraged to execute the commands themselves. In case help is needed, the *Web Support*, however, has the complete set of MATLAB results and plots. This organization provides a more coherent discourse on how one can make use of different features of MATLAB, not to mention save significant printing costs. Finally, the tutorials can easily be revised to keep up with the continual upgrade of MATLAB. At this writing, the tutorials are based on MATLAB Version 6.1 and the object-oriented functions in the Control System Toolbox Version 5.1. Simulink Version 4.1 is also utilized, but its scope is limited to simulating more complex control systems.

As a first-course text, the development of models is limited to stirred tanks, stirred-tank heaters, and a few other examples that are used extensively and repeatedly throughout the chapters. My philosophy is one step back in time. The focus is the theory and the building of a foundation that may help to solve other problems. The design is also formulated to be able to launch into the topic of tuning controllers before students may lose interest. The coverage of Laplace transforms is not entirely a concession to remedial mathematics. The examples are tuned to illustrate immediately how pole positions may relate to time-domain response. Furthermore, students tend to be confused by the many different design methods. As much as I could, especially in the controller design chapters, I used the same examples throughout. The goal is to help a student understand how the same problem can be solved by different techniques.

I have given up the pretense that we can cover controller design and still have time to do all the plots manually. I rely on MATLAB to construct the plots. For example, I take a unique approach to root-locus plots. I do not ignore it as some texts do, but I also do not go into the hand-sketching details. The same can be said of frequency-response analysis. On the whole, I use root-locus and Bode plots as computational and pedagogical tools in ways that can help students to understand the choice of different controller designs. Exercises that may help such thinking are in the MATLAB tutorials and homework problems.

Finally, I have to thank Costas Pozikidris and Florence Padgett for encouragement and support on this project, Raymond de Callafon for revising the chapters on state-space models, and Allan Cruz for proofreading. Last but not least, Henry Lim combed through the manuscript and made numerous insightful comments. His wisdom is sprinkled throughout the text.

Web Support (MATLAB outputs of text examples and MATLAB sessions, references, supplementary notes, and solution manual) is available at http://us.cambridge.org/titles/0521002559.html.

Introduction

C ontrol systems are tightly intertwined in our daily lives so much so that we take them for granted. They may be as low tech and unglamorous as our flush toilet. Or they may be as high tech as electronic fuel injection in our cars. In fact, there is more than a handful of computer control systems in a typical car that we now drive. In everything from the engine to transmission, shock absorber, brakes, pollutant emission, temperature, and so forth, there is an embedded microprocessor controller keeping an eye out for us. The more gadgetry, the more tiny controllers pulling the trick behind our backs.[1] At the lower end of consumer electronic devices, we can bet on finding at least one embedded microcontroller.

In the processing industry, controllers play a crucial role in keeping our plants running – virtually everything from simply filling up a storage tank to complex separation processes and chemical reactors.

As an illustration, let's take a look at a bioreactor (Fig. 1.1). To find out if the bioreactor is operating properly, we monitor variables such as temperature, pH, dissolved oxygen, liquid level, feed flow rate, and the rotation speed of the impeller. In some operations, we may also measure the biomass and the concentration of a specific chemical component in the liquid or the composition of the gas effluent. In addition, we may need to monitor the foam head and make sure it does not become too high. We most likely need to monitor the steam flow and pressure during the sterilization cycles. We should note that the schematic diagram is far from complete. By the time we have added enough details to implement all the controls, we may not recognize the bioreactor. These features are not pointed out to scare anyone; on the other hand, this is what makes control such a stimulating and challenging field.

For each quantity that we want to maintain at some value, we need to ensure that the bioreactor is operating at the desired conditions. Let's use the pH as an example. In control calculations, we commonly use a *block diagram* to represent the problem (Fig. 1.2). We will learn how to use mathematics to describe each of the blocks. For now, the focus is on some common terminology.

To consider pH as a *controlled variable*, we use a pH electrode to measure its value and, with a transmitter, send the signal to a controller, which can be a little black box or a computer. The controller takes in the pH value and compares it with the desired pH, what

---

[1] In the 1999 Mercedes-Benz S-class sedan, there are approximately 40 "electronic control units" that control up to 170 different variables.

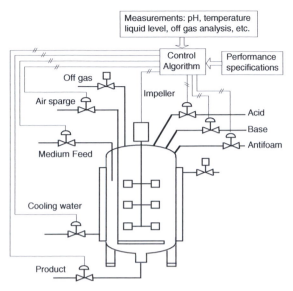

Figure 1.1. Schematic diagram of instrumentation associated with a fermentor. The steam sterilization system and all sensors and transmitters are omitted for clarity. The thick solid lines represent process streams. The thin solid lines represent information flow.

is called the *set point* or the *reference*. If the values are not the same, there is an *error*, and the controller makes proper adjustments by manipulating the acid or the base pump – the *actuator*.[2] The adjustment is based on calculations made with a *control algorithm*, also called the control law. The error is calculated at the summing point, where we take the desired pH minus the measured pH. Because of how we calculate the error, this is a *negative-feedback* mechanism.

This simple pH control example is what we call a *single-input single-output* (SISO) system; the single input is the set point and the output is the pH value.[3] This simple feedback

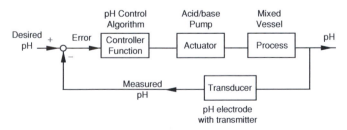

Figure 1.2. Block-diagram representation of a single-input single-output negative-feedback system. Labels within the boxes are general. Labels outside the boxes apply to the simplified pH control discussion.

---

[2] In real life, bioreactors actually use on–off control for pH.

[3] We will learn how to identify input and output variables, how to distinguish among manipulated variables, disturbances, measured variables, and so forth. Do not worry about remembering all the terms here; they will be introduced properly in subsequent chapters.

mechanism is also what we call a *closed loop*. This single-loop system ignores the fact that the dynamics of the bioreactor involves complex interactions among different variables. If we want to take a more comprehensive view, we need to design a *multiple-input multiple-output* (MIMO), or *multivariable*, system. When we invoke the term *system*, we are referring to the *process*[4] (the bioreactor here), the *controller*, and all other instrumentation, such as *sensors*, *transmitters*, and *actuators* (like valves and pumps) that enable us to control the pH.

When we change a specific operating condition, meaning the set point, we would like, for example, the pH of the bioreactor to follow our command. This is what we call *servocontrol*. The pH value of the bioreactor is subjected to external *disturbances* (also called *load changes*), and the task of suppressing or rejecting the effects of disturbances is called *regulatory control*. Implementation of a controller may lead to instability, and the issue of system *stability* is a major concern. The control system also has to be *robust* such that it is not overly sensitive to changes in process parameters.

What are some of the issues when we design a control system? In the first place, we need to identify the role of various variables. We need to determine what we need to control, what we need to manipulate, what the sources of disturbances are, and so forth. We then need to state our design objective and specifications. It may make a difference whether we focus on the servo or on the regulator problem, and we certainly want to make clear, quantitatively, the desired response of the system. To achieve these goals, we have to select the proper control strategy and controller. To implement the strategy, we also need to select the proper sensors, transmitters, and actuators. After all is done, we have to know how to tune the controller. Sounds like we are working with a musical instrument, but that's the jargon.

The design procedures depend heavily on the dynamic model of the process to be controlled. In more advanced model-based control systems, the action taken by the controller actually depends on the model. Under circumstances for which we do not have a precise model, we perform our analysis with approximate models. This is the basis of a field called *system identification and parameter estimation*. Physical insight that we may acquire in the act of model building is invaluable in problem solving.

Although we laud the virtue of dynamic modeling, we will not duplicate the introduction of basic conservation equations. It is important to recognize that all of the processes that we want to control, e.g., bioreactor, distillation column, flow rate in a pipe, drug delivery system, etc., are what we have learned in other engineering classes. The so-called model equations are conservation equations in heat, mass, and momentum. We need force balance in mechanical devices, and, in electrical engineering, we consider circuit analysis. The difference between what we now use in control and what we are more accustomed to is that control problems are *transient* in nature. Accordingly, we include the time-derivative (also called accumulation) term in our balance (model) equations.

What are some of the mathematical tools that we use? In *classical* control, our analysis is based on linear ordinary differential equations with constant coefficients – what is called *linear time invariant* (LTI). Our models are also called *lumped-parameter* models, meaning that variations in space or location are not considered. Time is the only independent variable. Otherwise, we would need partial differential equations in what is called *distributed-parameter* models. To handle our linear differential equations, we rely heavily

---

[4] In most of the control world, a process is referred to as a *plant*. Here "process" is used because, in the process industry, a plant carries the connotation of the entire manufacturing or processing facility.

Introduction

Table 1.1. Examples used in different chapters

Example	Page no.
Example 4.7	72
Example 4.7A	73
Example 4.7B	186
Example 4.7C	187
Example 4.8	75
Example 4.8A	181
Example 5.7	101
Example 5.7A	112
Example 5.7B	123
Example 5.7C	123
Example 5.7D	172
Example 7.2	133
Example 7.2A	135
Example 7.2B	139
Example 7.2C	170
Example 7.2D	171
Example 7.3	133
Example 7.3A	135
Example 7.3B	140
Example 7.4	136
Example 7.4A	172
Example 7.5	137
Example 7.5A	143
Example 7.5B	188

on *Laplace transform*, and we invariably rearrange the resulting algebraic equation into the so-called *transfer functions*. These algebraic relations are presented graphically as block diagrams (as in Fig. 1.2). However, we rarely go as far as solving for the time-domain solutions. Much of our analysis is based on our understanding of the roots of the characteristic polynomial of the differential equation – what we call the *poles*.

At this point, a little secret should be disclosed. Just from the terminology, it may be inferred that control analysis involves quite a bit of mathematics, especially when we go over stability and frequency-response methods. That is one reason why these topics are not immediately introduced. Nonetheless, we have to accept the prospect of working with mathematics. It would be a lie to say that one can be good in process control without sound mathematical skills.

Starting in Chap. 6, a select set of examples is repeated in some subsections and chapters. To reinforce the thinking that different techniques can be used to solve the same problem, these examples retain the same numeric labeling. These examples, which do not follow conventional numbering, are listed in Table 1.1 to help you find them.

It may be useful to point out a few topics that go beyond a first course in control. With certain processes, we cannot take data continuously, but rather in certain selected slow intervals (e.g., titration in freshmen chemistry). These are called *sampled-data* systems. With

4

computers, the analysis evolves into a new area of its own – *discrete-time* or *digital* control systems. Here, differential equations and Laplace transform do not work anymore. The mathematical techniques to handle discrete-time systems are difference equations and *z transforms*. Furthermore, there are *multivariable* and *state-space* controls, which we will encounter in a brief introduction. Beyond the introductory level are optimal control, nonlinear control, adaptive control, stochastic control, and fuzzy-logic control. Do not lose the perspective that control is an immense field. Classical control appears insignificant, but we have to start somewhere, and onward we crawl.

2

Mathematical Preliminaries

lassical process control builds on linear ordinary differential equations (ODEs) and the technique of the Laplace transform. This is a topic that we no doubt have come across in an introductory course on differential equations – like two years ago? Yes, we easily have forgotten the details. Therefore an attempt is made here to refresh the material necessary to solve control problems; other details and steps will be skipped. We can always refer back to our old textbook if we want to answer long-forgotten but not urgent questions.

What Are We Up to?

- The properties of Laplace transform and the transforms of some common functions. We need them to construct a table for doing an **inverse transform**.
- Because we are doing an inverse transform by means of a look-up table, we need to break down any given transfer functions into smaller parts that match what the table has – what are called **partial fractions**. The time-domain function is the sum of the inverse transform of the individual terms, making use of the fact that Laplace transform is a linear operator.
- The time-response characteristics of a model can be inferred from the poles, i.e., the roots of the characteristic polynomial. This observation is independent of the input function and singularly the most important point that we must master before moving onto control analysis.
- After a Laplace transform, a differential equation of deviation variables can be thought of as an input–output model with transfer functions. The causal relationship of changes can be represented by block diagrams.
- In addition to transfer functions, we make extensive use of steady-state gain and time constants in our analysis.
- Laplace transform is applicable to only *linear* systems. Hence we have to **linearize** nonlinear equations before we can go on. The procedure of linearization is based on a first-order Taylor series expansion.

2.1. A Simple Differential Equation Model

First an impetus is provided for solving differential equations in an approach unique to control analysis. The mass balance of a well-mixed tank can be written (see Review Problems) as

$$\tau \frac{dC}{dt} = C_{in} - C, \quad \text{with } C(0) = C_0,$$

where C is the concentration of a component, C_{in} is the inlet concentration, C_0 is the initial concentration, and τ is the space time. In classical control problems, we invariably rearrange the equation as

$$\tau \frac{dC}{dt} + C = C_{in} \tag{2.1}$$

and further redefine variables $C' = C - C_0$ and $C'_{in} = C_{in} - C_0$.[1] We designate C' and C'_{in} as **deviation variables** – they denote how a quantity deviates from the original value at $t = 0$.[2] Because C_0 is a constant, we can rewrite Eq. (2.1) as

$$\tau \frac{dC'}{dt} + C' = C'_{in}, \quad \text{with } C'(0) = 0. \tag{2.2}$$

Note that the equation now has a zero initial condition. For reference, the solution to Eq. (2.2) is[3]

$$C'(t) = \frac{1}{\tau} \int_0^t C'_{in}(z) e^{-(t-z)/\tau} dz. \tag{2.3}$$

If C'_{in} is zero, we have the trivial solution $C' = 0$. It is obvious from Eq. (2.2) immediately. For a more interesting situation in which C' is nonzero or for C to deviate from the initial C_0, C'_{in} must be nonzero, or in other words, C_{in} is different from C_0. In the terminology of differential equations, the right-hand side (RHS) C'_{in} is called the **forcing function**. In control, it is called the *input*. Not only is C'_{in} nonzero, it is, under most circumstances, a function of time as well, $C'_{in} = C'_{in}(t)$.

In addition, the time dependence of the solution, meaning the exponential function, arises from the left-hand side (LHS) of Eq. (2.2), the linear differential operator. In fact, we may recall that the LHS of Eq. (2.2) gives rise to the so-called characteristic equation (or characteristic polynomial).

Do not worry if you have forgotten the significance of the characteristic equation. We will come back to this issue again and again. This example is used just as a prologue. Typically in a class on differential equations, we learn to transform a *linear* ordinary equation into

---

[1] At steady state, $0 = C_{in}^s - C^s$, and if $C_{in}^s = C_0$, we can also define $C'_{in} = C_{in} - C_{in}^s$. We will come back to this when we learn to linearize equations. We will see that we should choose $C_0 = C^s$.

[2] Deviation variables are analogous to *perturbation variables* used in chemical kinetics or in fluid mechanics (linear hydrodynamic stability). We can consider a deviation variable as a measure of how far it is from steady state.

[3] When you come across the term convolution integral later in Eq. (4.10) and wonder how it may come about, take a look at the form of Eq. (2.3) again and think about it. If you wonder where Eq. (2.3) comes from, review your old ODE text on integrating factors. We skip this detail as we will not be using the time-domain solution in Eq. (2.3).

an *algebraic* equation in the *Laplace domain*, solve for the transformed dependent variable, and finally get back the *time-domain* solution with an inverse transformation.

In classical control theory, we make extensive use of a Laplace transform to analyze the dynamics of a system. The key point (and at this moment the trick) is that we will try to predict the time response *without* doing the inverse transformation. Later, we will see that the answer lies in the roots of the characteristic equation. This is the basis of classical control analyses. Hence, in going through Laplace transform again, it is not so much that we need a remedial course. Our old differential equation textbook would do fine. The key task here is to pitch this mathematical technique in light that may help us to apply it to control problems.

2.2. Laplace Transform

Let us first state a few important points about the application of Laplace transform in solving differential equations (Fig. 2.1). After we have formulated a model in terms of a *linear* or a *linearized* differential equation, $dy/dt = f(y)$, we can solve for $y(t)$. Alternatively, we can transform the equation into an algebraic problem as represented by the function $G(s)$ in the Laplace domain and solve for $Y(s)$. The time-domain solution $y(t)$ can be obtained with an inverse transform, but we rarely do so in control analysis.

What we argue (of course it is true) is that the Laplace-domain function $Y(s)$ must contain the same information as $y(t)$. Likewise, the function $G(s)$ must contain the same dynamic information as the original differential equation. We will see that the function $G(s)$ can be "clean looking" if the differential equation has zero initial conditions. That is one of the reasons why we always pitch a control problem in terms of deviation variables.[4] We can now introduce the definition.

The **Laplace transform** of a function $f(t)$ is defined as

$$\mathcal{L}[f(t)] = \int_0^\infty f(t)e^{-st}\,dt, \tag{2.4}$$

where s is the transform variable.[5] To complete our definition, we have the inverse transform,

$$f(t) = \mathcal{L}^{-1}[F(s)] = \frac{1}{2\pi j} \int_{\gamma-j\infty}^{\gamma+j\infty} F(s)e^{st}\,ds, \tag{2.5}$$

where γ is chosen such that the infinite integral can converge.[6] Do not be intimidated by

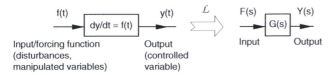

Input/forcing function (disturbances, manipulated variables)   Output (controlled variable)   Input   Output

Figure 2.1. Relationship between time domain and Laplace domain.

[4] But! What we measure in an experiment is the "real" variable. We have to be careful when we solve a problem that provides real data.

[5] There are many acceptable notations for a Laplace transform. Here we use a capital letter, and, if confusion may arise, we further add (s) explicitly to the notation.

[6] If you insist on knowing the details, they can be found on the *Web Support*.

Eq. (2.5). In a control class, we never use the inverse transform definition. Our approach is quite simple. We construct a table of the Laplace transform of some common functions, and we use it to do the inverse transform by means of a look-up table.

An important property of the Laplace transform is that it is a **linear operator**, and the contribution of individual terms can simply be added together (superimposed):

$$\mathcal{L}[af_1(t) + bf_2(t)] = a\mathcal{L}[f_1(t)] + b\mathcal{L}[f_2(t)] = aF_1(s) + bF_2(s). \tag{2.6}$$

Note: The linear property is one very important reason why we can do partial fractions and inverse transforms by means of a look-up table. This is also how we analyze more complex, but linearized, systems. Even though a text may not state this property explicitly, we rely heavily on it in classical control.

We now review the Laplace transforms of some common functions – mainly the ones that we come across frequently in control problems. We do not need to know all possibilities. We can consult a handbook or a mathematics textbook if the need arises. (A summary of the important transforms is in Table 2.1.) Generally, it helps a great deal if you can do the following common ones without having to use a look-up table. The same applies to simple algebra, such as partial fractions, and calculus, such as linearizing a function.

(1) A **constant:**

$$f(t) = a, \qquad F(s) = (a/s). \tag{2.7}$$

The derivation is

$$\mathcal{L}[a] = a \int_0^\infty e^{-st}dt = -\frac{a}{s}e^{-st}\Big|_0^\infty = a\left(0 + \frac{1}{s}\right) = \frac{a}{s}.$$

(2) An **exponential function** (Fig. 2.2):

$$f(t) = e^{-at}, \quad \text{with } a > 0, \qquad F(s) = [1/(s+a)], \tag{2.8}$$

$$\mathcal{L}[e^{-at}] = a \int_0^\infty e^{-at}e^{-st}dt = \frac{-1}{(s+a)}e^{-(a+s)t}\Big|_0^\infty = \frac{1}{(s+a)}.$$

(3) A **ramp function** (Fig. 2.2):

$$f(t) = at \quad \text{for } t \geq 0 \text{ and } a = \text{constant}, \qquad F(s) = (a/s^2), \tag{2.9}$$

$$\mathcal{L}[at] = a \int_0^\infty t e^{-st}dt = a\left(-t\frac{1}{s}e^{-st}\Big|_0^\infty + \int_0^\infty \frac{1}{s}e^{-st}dt\right)$$

$$= \frac{a}{s}\int_0^\infty e^{-st}dt = \frac{a}{s^2}.$$

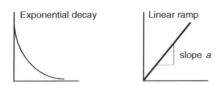

Figure 2.2. Illustration of exponential and ramp functions.

Table 2.1. Summary of a handful of common Laplace transforms

Function	$F(s)$	$f(t)$
The very basic functions	a/s	a or $au(t)$
	a/s^2	at
	$1/(s+a)$	e^{-at}
	$\omega/(s^2+\omega^2)$	$\sin \omega t$
	$s/(s^2+\omega^2)$	$\cos \omega t$
	$\omega/[(s+a)^2+\omega^2]$	$e^{-at}\sin \omega t$
	$(s+a)/[(s+a)^2+\omega^2]$	$e^{-at}\cos \omega t$
	$s^2 F(s) - sf(0) - f'(0)$	$\dfrac{d^2 f}{dt^2}$
	$\dfrac{F(s)}{s}$	$\int_0^t f(t)\,dt$
	$e^{-st_0}F(s)$	$f(t-t_0)$
	A	$A\delta(t)$
Transfer functions in time-constant form	$1/(\tau s+1)$	$(1/\tau)e^{-t/\tau}$
	$\dfrac{1}{(\tau s+1)^n}$	$\dfrac{1}{\tau^n(n-1)!}t^{n-1}e^{-t/\tau}$
	$1/[s(\tau s+1)]$	$1-e^{-t/\tau}$
	$1/[(\tau_1 s+1)(\tau_2 s+1)]$	$\left(e^{-t/\tau_1}-e^{-t/\tau_2}\right)/\tau_1-\tau_2$
	$\dfrac{1}{s(\tau_1 s+1)(\tau_2 s+1)}$	$1+\dfrac{\tau_1 e^{-t/\tau_1}-\tau_2 e^{-t/\tau_2}}{\tau_2-\tau_1}$
	$\dfrac{(\tau_3 s+1)}{(\tau_1 s+1)(\tau_2 s+1)}$	$\dfrac{1}{\tau_1}\dfrac{\tau_1-\tau_3}{\tau_1-\tau_2}e^{-t/\tau_1}+\dfrac{1}{\tau_2}\dfrac{\tau_2-\tau_3}{\tau_2-\tau_1}e^{-t/\tau_2}$
	$\dfrac{(\tau_3 s+1)}{s(\tau_1 s+1)(\tau_2 s+1)}$	$1+\dfrac{\tau_3-\tau_1}{\tau_1-\tau_2}e^{-t/\tau_1}+\dfrac{\tau_3-\tau_2}{\tau_2-\tau_1}e^{-t/\tau_2}$
Transfer functions in pole-zero form	$1/(s+a)$	e^{-at}
	$1/[(s+a)^2]$	$t\,e^{-at}$
	$\dfrac{1}{(s+a)^n}$	$\dfrac{1}{(n-1)!}t^{n-1}e^{-at}$
	$1/[s(s+a)]$	$(1/a)\,(1-e^{-at})$
	$1/[(s+a)(s+b)]$	$[1/(b-a)](e^{-at}-e^{-bt})$
	$s/[(s+a)^2]$	$(1-at)\,e^{-at}$
	$s/[(s+a)(s+b)]$	$[1/(b-a)]\,(be^{-bt}-ae^{-at})$
	$\dfrac{1}{s(s+a)(s+b)}$	$\dfrac{1}{ab}\left[1+\dfrac{1}{a-b}(be^{-at}-ae^{-bt})\right]$

Note: We may find many more Laplace transforms in handbooks or texts, but here we stay with the most basic ones. The more complex ones may actually be a distraction to our objective, which is to understand pole positions.

(4) Sinusoidal functions

$$f(t) = \sin \omega t, \qquad F(s) = [\omega/(s^2+\omega^2)], \tag{2.10}$$

$$f(t) = \cos \omega t, \qquad F(s) = [s/(s^2+\omega^2)]. \tag{2.11}$$

We use the fact that $\sin \omega t = (1/2j)(e^{j\omega t} - e^{-j\omega t})$ and the result with an exponential function to derive

$$\mathcal{L}[\sin \omega t] = \frac{1}{2j} \int_0^\infty (e^{j\omega t} - e^{-j\omega t}) e^{-st} \, dt$$

$$= \frac{1}{2j} \left[\int_0^\infty e^{-(s-j\omega)t} \, dt - \int_0^\infty e^{-(s+j\omega)t} \, dt \right]$$

$$= \frac{1}{2j} \left(\frac{1}{s - j\omega} - \frac{1}{s + j\omega} \right) = \frac{\omega}{s^2 + \omega^2}.$$

The Laplace transform of $\cos \omega t$ is left as an exercise in the Review Problems. If you need a review on complex variables, the *Web Support* has a brief summary.

(5) **Sinusoidal function with exponential decay:**

$$f(t) = e^{-at} \sin \omega t, \qquad F(s) = \frac{\omega}{(s + a)^2 + \omega^2}. \qquad (2.12)$$

Making use of previous results with the exponential and sine functions, we can pretty much do this one by inspection. First, we put the two exponential terms together inside the integral:

$$\int_0^\infty \sin \omega t \, e^{-(s+a)t} \, dt = \frac{1}{2j} \left[\int_0^\infty e^{-(s+a-j\omega)t} \, dt - \int_0^\infty e^{-(s+a+j\omega)t} \, dt \right]$$

$$= \frac{1}{2j} \left[\frac{1}{(s+a) - j\omega} - \frac{1}{(s+a) + j\omega} \right].$$

The similarity to the result of $\sin \omega t$ should be apparent now, if it was not the case with the LHS.

(6) **First-order derivative**, df/dt:

$$\mathcal{L}\left[\frac{df}{dt} \right] = s F(s) - f(0); \qquad (2.13)$$

second-order derivative:

$$\mathcal{L}\left[\frac{d^2 f}{dt^2} \right] = s^2 F(s) - s f(0) - f'(0). \qquad (2.14)$$

We have to use integration by parts here:

$$\mathcal{L}\left[\frac{df}{dt} \right] = \int_0^\infty \frac{df}{dt} e^{-st} \, dt = f(t) e^{-st} \Big|_0^\infty + s \int_0^\infty f(t) e^{-st} \, dt$$

$$= -f(0) + s F(s),$$

$$\mathcal{L}\left[\frac{d^2 f}{dt^2} \right] = \int_0^\infty \frac{d}{dt} \left(\frac{df}{dt} \right) e^{-st} \, dt = \frac{df}{dt} e^{-st} \Big|_0^\infty + s \int_0^\infty \frac{df}{dt} e^{-st} \, dt$$

$$= -\frac{df}{dt} \Big|_0 + s[s F(s) - f(0)].$$

We can extend these results to find the Laplace transform of higher-order derivatives. The key is that, if we use deviation variables in the problem formulation, all the initial-value terms will drop out in Eqs. (2.13) and (2.14). This is how we can get these clean-looking transfer functions in Section 2.6.

(7) An **integral:**

$$\mathcal{L}\left[\int_0^t f(t)\,dt\right] = \frac{F(s)}{s}. \tag{2.15}$$

We also need integration by parts here:

$$\int_0^\infty \left[\int_0^t f(t)\,dt\right] e^{-st}\,dt = -\frac{1}{s}e^{-st}\int_0^t f(t)\,dt\bigg|_0^\infty + \frac{1}{s}\int_0^\infty f(t)e^{-st}\,dt = \frac{F(s)}{s}.$$

2.3. Laplace Transforms Common to Control Problems

We now derive the Laplace transform of functions common in control analysis.

(1) **Step function:**

$$f(t) = Au(t), \qquad F(s) = (A/s). \tag{2.16}$$

We first define the **unit-step function** (also called the Heaviside function in mathematics) and its Laplace transform[7]:

$$u(t) = \begin{cases} 1, & t > 0 \\ 0, & t < 0 \end{cases}; \qquad \mathcal{L}[u(t)] = U(s) = \frac{1}{s}. \tag{2.17}$$

The Laplace transform of the unit-step function (Fig. 2.3) is derived as follows:

$$\mathcal{L}[u(t)] = \lim_{\epsilon \to 0^+} \int_\epsilon^\infty u(t)e^{-st}\,dt = \int_{0^+}^\infty e^{-st}\,dt = \frac{-1}{s}e^{-st}\bigg|_0^\infty = \frac{1}{s}.$$

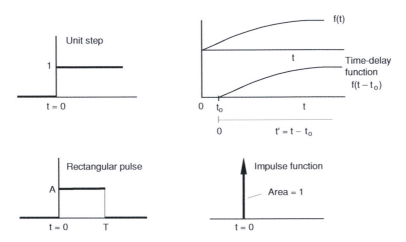

Figure 2.3. Unit-step, time-delay, rectangular, and impulse functions.

[7] Strictly speaking, the step function is discontinuous at $t = 0$, but many engineering texts ignore it and simply write $u(t) = 1$ for $t \geq 0$.

With the result for the unit step, we can see the results of the Laplace transform of any step function $f(t) = Au(t)$:

$$f(t) = Au(t) = \begin{cases} A, & t > 0 \\ 0, & t < 0 \end{cases}; \qquad L[Au(t)] = \frac{A}{s}.$$

The Laplace transform of a step function is essentially the same as that of a constant in Eq. (2.7). When we do the inverse transform of A/s, which function we choose depends on the context of the problem. Generally, a constant is appropriate under most circumstances.

(2) **Dead-time function** (Fig. 2.3):

$$f(t - t_0), \qquad L[f(t - t_0)] = e^{-st_0} F(s). \tag{2.18}$$

The dead-time function is also called the **time-delay**, **transport-lag**, translated, or time-shift function (Fig. 2.3). It is defined such that an original function $f(t)$ is "shifted" in time t_0, and no matter what $f(t)$ is, its value is set to zero for $t < t_0$. This time-delay function can be written as

$$f(t - t_0) = \begin{cases} 0, & t - t_0 < 0 \\ f(t - t_0), & t - t_0 > 0 \end{cases} = f(t - t_0)u(t - t_0).$$

The second form on the far right of the preceding equation is a more concise way of saying that the time-delay function $f(t - t_0)$ is defined such that it is zero for $t < t_0$. We can now derive the Laplace transform,

$$L[f(t - t_0)] = \int_0^\infty f(t - t_0)u(t - t_0)e^{-st}dt = \int_{t_0}^\infty f(t - t_0)e^{-st}dt,$$

and, finally,

$$\int_{t_0}^\infty f(t - t_0)\,e^{-st}dt = e^{-st_0} \int_{t_0}^\infty f(t - t_0)e^{-s(t-t_0)}d(t - t_0)$$

$$= e^{-st_0} \int_0^\infty f(t')e^{-st'}dt' = e^{-st_0} F(s),$$

where the final integration step uses the time-shifted axis $t' = t - t_0$.

(3) **Rectangular-pulse function** (Fig. 2.3):

$$f(t) = \begin{cases} 0, & t < 0 \\ A, & 0 < t < T = A[u(t) - u(t - T)]; \\ 0, & t > T \end{cases} \quad L[f(t)] = \frac{A}{s}(1 - e^{-sT}).$$

$$\tag{2.19}$$

We can generate the rectangular pulse by subtracting a step function with dead time T from a step function. We can derive the Laplace transform by using the formal definition

$$L[f(t)] = \int_0^\infty f(t)e^{-st}dt = A\int_{0+}^T e^{-st}dt = A\frac{-1}{s}e^{-st}\Big|_0^T = \frac{A}{s}(1 - e^{-sT}),$$

or, better yet, by making use of the results of a step function and a dead-time function:

$$\mathcal{L}[f(t)] = \mathcal{L}[A\,u(t) - A\,u(t-T)] = \frac{A}{s} - e^{-sT}\frac{A}{s}.$$

(4) **Unit-rectangular-pulse function:**

$$f(t) = \begin{cases} 0, & t < 0 \\ 1/T, & 0 < t < T = \frac{1}{T}[u(t) - u(t-T)]; \quad \mathcal{L}[f(t)] = \frac{1}{sT}(1 - e^{-sT}). \\ 0, & t > T \end{cases}$$

(2.20)

This is a prelude to the important impulse function. We can define a rectangular pulse such that the area is unity. The Laplace transform follows that of a rectangular pulse function:

$$\mathcal{L}[f(t)] = \mathcal{L}\left[\frac{1}{T}u(t) - \frac{1}{T}u(t-T)\right] = \frac{1}{Ts}(1 - e^{sT}).$$

(5) **Impulse function** (Fig. 2.3):

$$\mathcal{L}[\delta(t)] = 1, \qquad \mathcal{L}[A\delta(t)] = A.$$

(2.21)

The (unit) impulse function is called the **Dirac function** (or simply the delta function) in mathematics.[8] If we suddenly dump a bucket of water into a bigger tank, the impulse function is how we describe the action mathematically. We can consider the impulse function as the unit-rectangular function in Eqs. (2.20) as T shrinks to zero while the height $1/T$ goes to infinity:

$$\delta(t) = \lim_{T \to 0} \frac{1}{T}[u(t) - u(t-T)].$$

The area of this "squeezed rectangle" nevertheless remains at unity:

$$\lim_{T \to 0}\left(T\frac{1}{T}\right) = 1,$$

or, in other words,

$$\int_{-\infty}^{\infty} \delta(t)\,dt = 1.$$

The impulse function is rarely defined in the conventional sense, but rather by its important property in an integral:

$$\int_{-\infty}^{\infty} f(t)\delta(t)\,dt = f(0), \qquad \int_{-\infty}^{\infty} f(t)\delta(t-t_0)\,dt = f(t_0).$$

(2.22)

We easily obtain the Laplace transform of the impulse function by taking the limit of the unit-rectangular-function transform of Eqs. (2.20) with the use of l'Hôpital's rule:

$$\mathcal{L}[\delta(t)] = \lim_{T \to 0}\frac{1 - e^{-sT}}{Ts} = \lim_{T \to 0}\frac{se^{-sT}}{s} = 1.$$

From this result, it is obvious that $\mathcal{L}[A\delta(t)] = A$.

[8] In mathematics, the unit-rectangular function is defined with a height of $1/2T$ and a width of $2T$ from $-T$ to T. We simply begin at $t = 0$ in control problems. Furthermore, the impulse function is the time derivative of the unit-step function.

2.4. Initial- and Final-Value Theorems

Two theorems are now presented that can be used to find the values of the time-domain function at two extremes, $t = 0$ and $t = \infty$, without having to do the inverse transform. In control, we use the final-value theorem quite often. The initial-value theorem is less useful. As we have seen from our first example in Section 2.1, the problems that we solve are defined to have exclusively zero initial conditions.

Initial-Value Theorem:

$$\lim_{s \to \infty} [s F(s)] = \lim_{t \to 0} f(t). \tag{2.23}$$

Final-Value Theorem:

$$\lim_{s \to 0} [s F(s)] = \lim_{t \to \infty} f(t). \tag{2.24}$$

The final-value theorem is valid provided that a final-value exists. The proofs of these theorems are straightforward. We will do the one for the final-value theorem. The proof of the initial-value theorem is in the Review Problems.

Consider the definition of the Laplace transform of a derivative. If we take the limit as s approaches zero, we find

$$\lim_{s \to 0} \int_0^\infty \frac{df(t)}{dt} e^{-st} dt = \lim_{s \to 0} [s F(s) - f(0)].$$

If the infinite integral exists,[9] we can interchange the limit and the integration on the LHS to give

$$\int_0^\infty \lim_{s \to 0} \frac{df(t)}{dt} e^{-st} dt = \int_0^\infty df(t) = f(\infty) - f(0).$$

Now if we equate the RHSs of the previous two steps, we have

$$f(\infty) - f(0) = \lim_{s \to 0} [s F(s) - f(0)].$$

We arrive at the final-value theorem after we cancel the $f(0)$ terms on both sides.

Example 2.1: Consider the Laplace transform $F(s) = \{[6(s - 2)(s + 2)]/[s(s + 1) \times (s + 3)(s + 4)]\}$. What is $f(t = \infty)$?

$$\lim_{s \to 0} s \frac{6(s - 2)(s + 2)}{s(s + 1)(s + 3)(s + 4)} = \frac{6(-2)(2)}{(3)(4)} = -2.$$

Example 2.2: Consider the Laplace transform $F(s) = [1/(s - 2)]$. What is $f(t = \infty)$? Here, $f(t) = e^{2t}$. There is no upper bound for this function, which is in violation of the existence of a final value. The final-value theorem does not apply. If we insist on applying the theorem, we will get a value of zero, which is meaningless.

Example 2.3: Consider the Laplace transform $F(s) = \{[6(s^2 - 4)]/[(s^3 + s^2 - 4s - 4)]\}$. What is $f(t = \infty)$? Yes, another trick question. If we apply the final-value theorem without

---

[9] This is a key assumption and explains why Examples 2.2 and 2.3 do not work. When a function has no bound – what we call unstable – the assumption is invalid.

thinking, we would get a value of 0, but this is meaningless. With MATLAB®, we can use

```
roots([1 1 -4 -4])
```

to find that the polynomial in the denominator has roots -1, -2, and $+2$. This implies that $f(t)$ contains the term e^{2t}, which increases without bound.

As we move on, we will learn to associate the time-exponential terms to the roots of the polynomial in the denominator. From these examples, we can gather that, to have a meaningful, i.e., finite, bounded value, the roots of the polynomial in the denominator must have *negative real parts*. This is the basis of stability, which will formally be defined in Chap. 7.

2.5. Partial-Fraction Expansion

Because we rely on a look-up table to do a reverse Laplace transform, we need the skill to reduce a complex function down to simpler parts that match our table. In theory, we should be able to "break up" a ratio of two polynomials in s into simpler partial fractions. If the polynomial in the denominator, $p(s)$, is of an order higher than the numerator, $q(s)$, we can derive[10]

$$F(s) = \frac{q(s)}{p(s)} = \frac{\alpha_1}{(s + a_1)} + \frac{\alpha_2}{(s + a_2)} + \cdots + \frac{\alpha_i}{(s + a_i)} + \cdots + \frac{\alpha_n}{(s + a_n)}, \qquad (2.25)$$

where the order of $p(s)$ is n and the a_i are the negative values of the roots of the equation $p(s) = 0$. We then perform the inverse transform term by term:

$$f(t) = \mathcal{L}^{-1}[F(s)] = \mathcal{L}^{-1}\left[\frac{\alpha_1}{(s + a_1)}\right] + \mathcal{L}^{-1}\left[\frac{\alpha_2}{(s + a_2)}\right]$$

$$+ \cdots + \mathcal{L}^{-1}\left[\frac{\alpha_i}{(s + a_i)}\right] + \cdots + \mathcal{L}^{-1}\left[\frac{\alpha_n}{(s + a_n)}\right]. \qquad (2.26)$$

This approach works because of the linear property of the Laplace transform.

The next question is how to find the partial fractions in Eq. (2.25). One of the techniques is the so-called Heaviside expansion, a fairly straightforward algebraic method. Three important cases are illustrated with respect to the roots of the polynomial in the denominator: (1) distinct real roots, (2) complex-conjugate roots, and (3) multiple (or repeated) roots. In a given problem, we can have a combination of any of the above. Yes, we need to know how to do them all.

2.5.1. Case 1: $p(s)$ Has Distinct, Real Roots

Example 2.4: Find $f(t)$ of the Laplace transform $F(s) = [(6s^2 - 12)/(s^3 + s^2 - 4s - 4)]$. From Example 2.3, the polynomial in the denominator has roots -1, -2, and $+2$, values

---

[10] If the order of $q(s)$ is higher, we first need to carry out "long division" until we are left with a partial-fraction "residue." Thus the coefficients α_i are also called residues. We then expand this partial fraction. We would encounter such a situation in only a mathematical problem. The models of real physical processes lead to problems with a higher-order denominator.

that will be referred to as poles in Section 2.6. We should be able to write $F(s)$ as

$$\frac{6s^2 - 12}{(s + 1)(s + 2)(s - 2)} = \frac{\alpha_1}{(s + 1)} + \frac{\alpha_2}{(s + 2)} + \frac{\alpha_3}{(s - 2)}.$$

The Heaviside expansion takes the following idea. If we multiply both sides by $(s + 1)$, we obtain

$$\frac{6s^2 - 12}{(s + 2)(s - 2)} = \alpha_1 + \frac{\alpha_2}{(s + 2)}(s + 1) + \frac{\alpha_3}{(s - 2)}(s + 1),$$

which should be satisfied by any value of s. Now if we choose $s = -1$, we should obtain

$$\alpha_1 = \left.\frac{6s^2 - 12}{(s + 2)(s-2)}\right|_{s=-1} = 2.$$

Similarly, we can multiply the original fraction by $(s + 2)$ and $(s - 2)$ to find

$$\alpha_2 = \left.\frac{6s^2 - 12}{(s + 1)(s - 2)}\right|_{s=-2} = 3,$$

$$\alpha_3 = \left.\frac{6s^2 - 12}{(s + 1)(s + 2)}\right|_{s=2} = 1,$$

respectively. Hence,

$$F(s) = \frac{2}{(s + 1)} + \frac{3}{(s + 2)} + \frac{1}{(s - 2)},$$

and the look-up table would give us

$$f(t) = 2e^{-t} + 3e^{-2t} + e^{2t}.$$

When you use MATLAB to solve this problem, be careful when you interpret the results. The computer is useless unless we know what we are doing. We provide only the necessary statements.[11] For this example, all we need is

```
[a,b,k]=residue([6 0 -12],[1 1 -4 -4])
```

Example 2.5: Find $f(t)$ of the Laplace transform $F(s) = [(6s)/(s^3 + s^2 - 4s - 4)]$. Again, the expansion should take the form

$$\frac{6s}{(s + 1)(s + 2)(s - 2)} = \frac{\alpha_1}{(s + 1)} + \frac{\alpha_2}{(s + 2)} + \frac{\alpha_3}{(s - 2)}.$$

One more time, for each term, we multiply the denominators on the RHS and set the resulting equation to its root to obtain

$$\alpha_1 = \left.\frac{6s}{(s + 2)(s - 2)}\right|_{s=-1} = 2, \quad \alpha_2 = \left.\frac{6s}{(s + 1)(s - 2)}\right|_{s=-2} = -3,$$

$$\alpha_3 = \left.\frac{6s}{(s + 1)(s + 2)}\right|_{s=2} = 1.$$

---

[11] From here on, it is important that you go over the MATLAB sessions. Explanation of residue() is in Session 2. Although we do not print the computer results, they can be found on the *Web Support*.

The time-domain function is

$$f(t) = 2e^{-t} - 3e^{-2t} + e^{2t}.$$

Note that $f(t)$ has the identical functional dependence in time as in Example 2.4. Only the coefficients (residues) are different.

The MATLAB statement for this example is

```
[a,b,k]=residue([6 0],[1 1 -4 -4])
```

Example 2.6: Find $f(t)$ of the Laplace transform $F(s) = \{6/[(s + 1)(s + 2)(s + 3)]\}$. This time, we should find

$$\alpha_1 = \frac{6}{(s + 2)(s + 3)}\bigg|_{s=-1} = 3, \quad \alpha_2 = \frac{6}{(s + 1)(s + 3)}\bigg|_{s=-2} = -6,$$

$$\alpha_3 = \frac{6}{(s + 1)(s + 2)}\bigg|_{s=-3} = 3.$$

The time-domain function is

$$f(t) = 3e^{-t} - 6e^{-2t} + 3e^{-3t}.$$

The e^{-2t} and e^{-3t} terms will decay faster than the e^{-t} term. We consider the e^{-t} term, or the pole at $s = -1$, as more dominant. We can confirm the result with the following MATLAB statements:

```
p=poly([-1 -2 -3]);
[a,b,k]=residue(6,p)
```

Note:

(1) The time dependence of the time-domain solution is derived entirely from the roots of the polynomial in the denominator (what we will refer to in Section 2.6 as the poles). The polynomial in the numerator affects only the coefficients α_i. This is one reason why we make qualitative assessment of the dynamic response characteristics entirely based on the poles of the characteristic polynomial.

(2) Poles that are closer to the origin of the complex plane will have corresponding exponential functions that decay more slowly in time. We consider these poles more dominant.

(3) We can generalize the Heaviside expansion into the fancy form for the coefficients

$$\alpha_i = (s + a_i) \frac{q(s)}{p(s)}\bigg|_{s=-a_i},$$

but we should always remember the simple algebra that we have gone through in the preceding examples.

2.5.2. Case 2: $p(s)$ Has Complex Roots

Example 2.7: Find $f(t)$ of the Laplace transform $F(s) = [(s + 5)/(s^2 + 4s + 13)]$. We first take the painful route just so we better understand the results from MATLAB. If we have to do the chore by hand, we much prefer the completing the perfect-square method in Example 2.8. Even without MATLAB, we can easily find that the roots of the polynomial $s^2 + 4s + 13$ are $-2 \pm 3j$ and $F(s)$ can be written as the sum of

$$\frac{s+5}{s^2+4s+13} = \frac{s+5}{[s-(-2+3j)][s-(-2-3j)]} = \frac{\alpha}{s-(-2+3j)} + \frac{\alpha^*}{s-(-2-3j)}.$$

We can apply the same idea as in Example 2.4 to find

$$\alpha = \frac{s+5}{[s-(-2-3j)]}\bigg|_{s=-2+3j} = \frac{(-2+3j)+5}{(-2+3j)+2+3j} = \frac{(j+1)}{2j} = \frac{1}{2}(1-j),$$

and its complex conjugate[12] is

$$\alpha^* = \frac{1}{2}(1+j).$$

The inverse transform is hence

$$f(t) = \frac{1}{2}(1-j)e^{(-2+3j)t} + \frac{1}{2}(1+j)e^{(-2-3j)t}$$

$$= \frac{1}{2}e^{-2t}[(1-j)e^{j3t} + (1+j)e^{-j3t}].$$

We can apply Euler's identity to the result:

$$f(t) = \frac{1}{2}e^{-2t}[(1-j)(\cos 3t + j\sin 3t) + (1+j)(\cos 3t - j\sin 3t)]$$

$$= \frac{1}{2}e^{-2t}[2(\cos 3t + \sin 3t)],$$

which we further rewrite as

$$f(t) = \sqrt{2}e^{-2t}\sin(3t + \phi),$$

where $\phi = \tan^{-1}(1) = \pi/4$ or $45°$. The MATLAB statement for this example is simply

```
[a,b,k]=residue([1 5],[1 4 13])
```

Note:

(1) Again, the time dependence of $f(t)$ is affected by only the roots of $p(s)$. For the general complex-conjugate roots $-a \pm bj$, the time-domain function involves e^{-at} and $(\cos bt + \sin bt)$. The polynomial in the numerator affects only the constant coefficients.

(2) We seldom use the form $(\cos bt + \sin bt)$. Instead, we use the phase-lag form as in the final step of Example 2.7.

Example 2.8: Repeat Example 2.7 with a look-up table. In practice, we seldom do the partial-fraction expansion of a pair of complex roots. Instead, we rearrange the polynomial

---

[12] If you need a review of complex-variable definitions, see the *Web Support*. Many steps in Example 2.7 require these definitions.

$p(s)$ by noting that we can complete the squares:

$$s^2 + 4s + 13 = (s+2)^2 + 9 = (s+2)^2 + 3^2.$$

We then write $F(s)$ as

$$F(s) = \frac{s+5}{s^2+4s+13} = \frac{(s+2)}{(s+2)^2+3^2} + \frac{3}{(s+2)^2+3^2}.$$

With a Laplace-transform table, we find

$$f(t) = e^{-2t}\cos 3t + e^{-2t}\sin 3t,$$

which is the answer with very little work. Compared with how messy the partial fraction was in Example 2.7, this example also suggests that we want to leave terms with conjugate-complex roots as one second-order term.

2.5.3. Case 3: $p(s)$ Has Repeated Roots

Example 2.9: Find $f(t)$ of the Laplace transform $F(s) = \{2/[(s+1)^3(s+2)]\}$. The polynomial $p(s)$ has the roots -1 repeated three times and -2. To keep the numerator of each partial fraction a simple constant, we have to expand to

$$\frac{2}{(s+1)^3(s+2)} = \frac{\alpha_1}{(s+1)} + \frac{\alpha_2}{(s+1)^2} + \frac{\alpha_3}{(s+1)^3} + \frac{\alpha_4}{(s+2)}.$$

To find α_3 and α_4 is routine:

$$\alpha_3 = \left.\frac{2}{(s+2)}\right|_{s=-1} = 2, \quad \alpha_4 = \left.\frac{2}{(s+1)^3}\right|_{s=-2} = -2.$$

The problem is with finding α_1 and α_2. We see that, say, if we multiply the equation with $(s+1)$ to find α_1, we cannot select $s = -1$. What we can try is to multiply the expansion with $(s+1)^3$,

$$\frac{2}{(s+2)} = \alpha_1(s+1)^2 + \alpha_2(s+1) + \alpha_3 + \frac{\alpha_4(s+1)^3}{(s+2)},$$

and then differentiate this equation with respect to s:

$$\frac{-2}{(s+2)^2} = 2\alpha_1(s+1) + \alpha_2 + 0 + [\alpha_4 \text{ terms with } (s+1)].$$

Now we can substitute $s = -1$, which provides $\alpha_2 = -2$.

We can be lazy with the last α_4 term because we know its derivative will contain $(s+1)$ terms and they will drop out as soon as we set $s = -1$. To find α_1, we differentiate the equation one more time to obtain

$$\frac{4}{(s+2)^3} = 2\alpha_1 + 0 + 0 + [\alpha_4 \text{ terms with } (s+1)],$$

which of course will yield $\alpha_1 = 2$ if we select $s = -1$. Hence, we have

$$\frac{2}{(s+1)^3(s+2)} = \frac{2}{(s+1)} + \frac{-2}{(s+1)^2} + \frac{2}{(s+1)^3} + \frac{-2}{(s+2)},$$

and the inverse transform by means of the look-up table is

$$f(t) = 2\left[\left(1 - t + \frac{t^2}{2}\right)e^{-t} - e^{-2t}\right].$$

We can also arrive at the same result by expanding the entire algebraic expression, but that actually takes more work (!), and we leave this exercise to the Review Problems.

The MATLAB command for this example is

```
p=poly([-1 -1 -1 -2]);
[a,b,k]=residue(2,p)
```

Note: In general, the inverse transform of repeated roots takes the form

$$\mathcal{L}^{-1}\left[\frac{\alpha_1}{(s+a)} + \frac{\alpha_2}{(s+a)^2} + \cdots + \frac{\alpha_n}{(s+a)^n}\right]$$

$$= \left[\alpha_1 + \alpha_2 t + \frac{\alpha_3}{2!}t^2 + \cdots + \frac{\alpha_n}{(n-1)!}t^{n-1}\right]e^{-at}.$$

The exponential function is still based on the root $s = -a$, but the actual time dependence will decay slower because of the $(\alpha_2 t + \cdots +)$ terms.

2.6. Transfer Function, Pole, and Zero

Now that we can do Laplace transform, let us return to our first example. The Laplace transform of Eq. (2.2) with its zero initial condition is $(\tau s + 1)C'(s) = C'_{in}(s)$, which we rewrite as

$$\frac{C'(s)}{C'_{in}(s)} = \frac{1}{\tau s + 1} = G(s). \tag{2.27}$$

We define the RHS as $G(s)$, our ubiquitous **transfer function**. It relates an input to the output of a model. Recall that we use deviation variables. The input is the *change* in the inlet concentration, $C'_{in}(t)$. The output, or response, is the resulting *change* in the tank concentration, $C'(t)$.

Example 2.10: What is the time domain response $C'(t)$ in Eq. (2.27) if the change in inlet concentration is (1) a unit-step function and (2) an impulse function?

(1) With a unit-step input, $C'_{in}(t) = u(t)$ and $C'_{in}(s) = 1/s$. Substitution for $C'_{in}(s)$ in Eq. (2.27) leads to

$$C'(s) = \frac{1}{\tau s + 1}\frac{1}{s} = \frac{1}{s} + \frac{-\tau}{\tau s + 1}.$$

After an inverse transform by means of a look-up table, we have $C'(t) = 1 - e^{-t/\tau}$. The change in tank concentration eventually will be identical to the unit-step change in inlet concentration.

(2) With an impulse input, $C'_{in}(s) = 1$, and substitution for $C'_{in}(s)$ in Eq. (2.27) leads to simply

$$C'(s) = \frac{1}{\tau s + 1},$$

and the time-domain solution is $C'(t) = \frac{1}{\tau}e^{-t/\tau}$. The effect of the impulse eventually will decay away.

Finally, you may want to keep in mind that the results of this example can also be obtained by means of the general time-domain solution in Eq. (2.3).

The key of this example is to note that, *irrespective of the input*, the time-domain solution contains the time-dependent function $e^{-t/\tau}$, which is associated with the root of the polynomial in the denominator of the transfer function.

The inherent dynamic properties of a model are embedded in the characteristic polynomial of the differential equation. More specifically, the dynamics is related to the roots of the characteristic polynomial. In Eq. (2.27), the characteristic equation is $\tau s + 1 = 0$, and its root is $-1/\tau$. In a general sense, that is, without specifying what C'_{in} is and without actually solving for $C'(t)$, we can infer that $C'(t)$ must contain a term with $e^{-t/\tau}$. We refer to the root $-1/\tau$ as the **pole** of the transfer function $G(s)$.

We can now state the definitions more generally. For an ODE[13]

$$a_n y^{(n)} + a_{n-1} y^{(n-1)} + \cdots + a_1 y^{(1)} + a_0 y = b_m x^{(m)} + b_{m-1} x^{(m-1)} + \cdots + b_1 x^{(1)} + b_0 x, \tag{2.28}$$

with $n > m$ and zero initial conditions $y^{(n-1)} = \cdots = y = 0$ at $t = 0$, the corresponding Laplace transform is

$$\frac{Y(s)}{X(s)} = \frac{b_m s^m + b_{m-1} s^{m-1} + \cdots + b_1 s + b_0}{a_n s^n + a_{n-1} s^{n-1} + \cdots + a_1 s + a_0} = G(s) = \frac{Q(s)}{P(s)}. \tag{2.29}$$

Generally we can write the transfer function as the ratio of two polynomials in s.[14] When we talk about the mathematical properties, the polynomials are denoted as $Q(s)$ and $P(s)$, but the *same* polynomials are denoted as $Y(s)$ and $X(s)$ when the focus is on control problems or transfer functions. The orders of the polynomials are such that $n \geq m$ for physically realistic processes.[15]

We know that $G(s)$ contains information on the dynamic behavior of a model as represented by the differential equation. We also know that the denominator of $G(s)$ is the

[13] Yes, we try to be general and use an nth-order equation. If you have trouble with the development in this section, think of a second-order equation in all the steps:

$$a_2 y^{(2)} + a_1 y^{(1)} + a_0 y = b_1 x^{(1)} + b_0 x.$$

All the features about poles and zeros can be obtained from this simpler equation.

[14] The exception is when we have dead time. We will come back to this term in Chap. 3.

[15] For real physical processes, the orders of polynomials are such that $n \geq m$. A simple explanation is to look at a so-called lead–lag element when $n = m$ and $y^{(1)} + y = x^{(1)} + x$. The LHS, which is the dynamic model, must have enough complexity to reflect the change of the forcing on the RHS. Thus if the forcing includes a rate of change, the model must have the same capability too.

characteristic polynomial of the differential equation. The roots of the characteristic equation, $P(s) = 0$: $p_1, p_2, \ldots, p_n$, are the poles of $G(s)$. When the poles are real and negative, we also use the time-constant notation:

$$p_1 = -\frac{1}{\tau_1}, \quad p_2 = -\frac{1}{\tau_2}, \ldots, \quad p_n = -\frac{1}{\tau_n}.$$

The poles reveal qualitatively the dynamic behavior of the model differential equation. The term "roots of the characteristic equation" is used interchangeably with "poles of the transfer function."

For the general transfer function in Eq. (2.29), the roots of the polynomial $Q(s)$, i.e., of $Q(s) = 0$, are referred to as the **zeros**. They are denoted by $z_1, z_2, \ldots, z_m$, or, in time-constant notation, by

$$z_1 = -\frac{1}{\tau_a}, \quad z_2 = -\frac{1}{\tau_b}, \ldots, \quad z_m = -\frac{1}{\tau_m}.$$

We can factor Eq. (2.29) into the so-called **pole-zero form:**

$$G(s) = \frac{Q(s)}{P(s)} = \left(\frac{b_m}{a_n}\right) \frac{(s - z_1)(s - z_2)\cdots(s - z_m)}{(s - p_1)(s - p_2)\cdots(s - p_n)}. \tag{2.30}$$

If all the roots of the two polynomials are real, we can factor the polynomials such that the transfer function is in the **time-constant form:**

$$G(s) = \frac{Q(s)}{P(s)} = \left(\frac{b_0}{a_0}\right) \frac{(\tau_a s + 1)(\tau_b s + 1)\cdots(\tau_m s + 1)}{(\tau_1 s + 1)(\tau_2 s + 1)\cdots(\tau_n s + 1)}. \tag{2.31}$$

Equations (2.30) and (2.31) will be a lot less intimidating when we come back to using the examples in Section 2.8. These forms are the mainstays of classical control analysis.

Another important quantity is the **steady-state gain**.[16] With reference to general differential equation model (2.28) and its Laplace transform in Eq. (2.29), the steady-state gain is defined as the final *change* in $y(t)$ relative to a unit *change* in the input $x(t)$. Thus an easy derivation of the steady-state gain is to take a unit-step input in $x(t)$, or $X(s) = 1/s$, and find the final value in $y(t)$:

$$y(\infty) = \lim_{s \to 0}[s\, G(s)\, X(s)] = \lim_{s \to 0}\left[s\, G(s)\frac{1}{s}\right] = \frac{b_0}{a_0}. \tag{2.32}$$

The steady-state gain is the ratio of the two constant coefficients. Take note that the steady-state gain value is based on the transfer function only. From Eqs. (2.31) and (2.32), one easy way to "spot" the steady-state gain is to look at a transfer function in the time-constant form.

Note:

(1) When we talk about the poles of $G(s)$ in Eq. (2.29), the discussion is *regardless* of the input $x(t)$. Of course, the actual response $y(t)$ also depends on $x(t)$ or $X(s)$.

(2) Recall from the examples of partial-fraction expansion that the polynomial $Q(s)$ in the numerator, or the zeros, affects only the coefficients of the solution $y(t)$, but not

---

[16] This quantity is also called the **static gain** or **dc gain** by electrical engineers. When we talk about the model of a process, we also use the term **process gain** quite often to distinguish it from a system gain.

the time-dependent functions. That is why, for qualitative discussions, we focus on only the poles.

(3) For the time-domain function to be made up of only exponential terms that decay in time, *all* the poles of a transfer function must have negative real parts. This is the key to stability analysis, which will be covered in Chap. 7.

2.7. Summary of Pole Characteristics

We now put one and one together. The key is that we can "read" the poles – telling what the *form* of the time-domain function is. We should have a pretty good idea from our exercises in partial fractions. Here, the results are provided one more time in general notation. Suppose we have taken a characteristic polynomial, found its roots, and completed the partial-fraction expansion; this is what we expect in the time domain for each of the terms:

(A) **Real distinct poles:** Terms of the form $c_i/(s - p_i)$, where the pole p_i is a real number, have the time-domain function $c_i e^{p_i t}$. Most often, we have a negative real pole such that $p_i = -a_i$ and the time-domain function is $c_i e^{-a_i t}$.

(B) **Real poles, repeated m times:** Terms of the form

$$\left[\frac{c_{i,1}}{(s - p_i)} + \frac{c_{i,2}}{(s - p_i)^2} + \cdots + \frac{c_{i,m}}{(s - p_i)^m} \right],$$

with the root p_i repeated m times, have the time-domain function

$$\left[c_{i,1} + c_{i,2}t + \frac{c_{i,3}}{2!}t^2 + \cdots + \frac{c_{i,m}}{(m-1)!}t^{m-1} \right] e^{p_i t}.$$

When the pole p_i is negative, the decay in time of the entire response will be slower (with respect to only one single pole) because of the terms involving time in the bracket. This is the reason why we say that the response of models with repeated roots (e.g., tanks-in-series in Section 3.4) tends to be slower or "sluggish."

(C) **Complex-conjugate poles:** Terms of the form $[c_i/(s - p_i)] + [c_i^*/(s - p_i^*)]$, where $p_i = \alpha + j\beta$ and $p_i^* = \alpha - j\beta$ are the complex poles, have the time-domain function $c_i e^{p_i t} + c_i^* e^{p_i^* t}$, which is a form we seldom use. Instead, we rearrange them to give the form (some constant) $e^{\alpha t} \sin(\beta t + \phi)$, where ϕ is the phase lag.

It is cumbersome to write the partial fraction with complex numbers. With complex-conjugate poles, we commonly combine the two first-order terms into a second-order term. With notation that is introduced formally in Chap. 3, we can write the second-order term as

$$\frac{as + b}{\tau^2 s^2 + 2\zeta \tau s + 1},$$

where the coefficient ζ is called the damping ratio. To have complex roots in the denominator, we need $0 < \zeta < 1$. The complex poles p_i and p_i^* are now written as

$$p_i, \ p_i^* = -\frac{\zeta}{\tau} \pm j\frac{\sqrt{1 - \zeta^2}}{\tau}, \quad \text{with } 0 < \zeta < 1,$$

and the time-domain function is usually rearranged to give the form

$$(\text{some constant})e^{(-\zeta t)/\tau}\sin\left(\frac{\sqrt{1-\zeta^2}}{\tau}t+\phi\right),$$

where, again, ϕ is the phase lag.

(D) **Poles on the imaginary axis:** If the real part of a complex pole is zero, then $p = \pm\omega j$. We have a purely sinusoidal behavior with frequency ω. If the pole is zero, it is at the origin and corresponds to the integrator $1/s$. In the time domain, we would have a constant, or a step function.

(E) If a pole has a **negative real part**, it is in the left-hand plane (LHP). Conversely, if a pole has a positive real part, it is in the right-hand plane (RHP) and the time-domain solution is definitely *unstable*.

Note: Not all poles are born equal! *The poles closer to the origin are dominant.*

It is important to understand and be able to identify *dominant poles* if they exist. This is a skill that is used later in what is called model reduction. This is a point that we first observed in Example 2.6. Consider two terms such that $0 < a_1 < a_2$ (Fig. 2.4):

$$Y(s) = \frac{c_1}{(s-p_1)} + \frac{c_2}{(s-p_2)} + \cdots + = \frac{c_1}{(s+a_1)} + \frac{c_2}{(s+a_2)} + \cdots +$$

$$= \frac{c_1/a_1}{(\tau_1 s + 1)} + \frac{c_2/a_2}{(\tau_2 s + 1)} + \cdots.$$

Their corresponding terms in the time domain are

$$y(t) = c_1 e^{-a_1 t} + c_2 e^{-a_2 t} + \cdots + = c_1 e^{-t/\tau_1} + c_2 e^{-t/\tau_2} + \cdots.$$

As time progresses, the term associated with τ_2 (or a_2) will decay away faster. We consider the term with the larger time constant τ_1 as the dominant pole.[17]

Finally, for a complex pole, we can relate the damping ratio ($\zeta < 1$) to the angle that the pole makes with the real axis (Fig. 2.5). Taking the absolute values of the dimensions of the triangle, we can find

$$\theta = \tan^{-1}\left(\frac{\sqrt{1-\zeta^2}}{\zeta}\right), \tag{2.33}$$

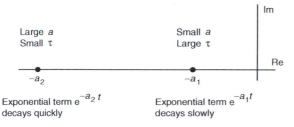

Figure 2.4. Poles with small and large time constants.

[17] Our discussion is valid only if τ_1 is "sufficiently" larger than τ_2. We could establish a criterion, but at the introductory level, we shall keep this as a qualitative observation.

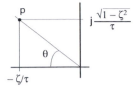

Figure 2.5. Complex pole angular position on the complex plane.

and, more simply,

$$\theta = \cos^{-1} \zeta. \tag{2.34}$$

Equation (2.34) is used in the root-locus method in Chap. 7 when we design controllers.

2.8. Two Transient Model Examples

We now use two examples to review how deviation variables relate to the actual ones, and we can go all the way to find the solutions.

2.8.1. A Transient-Response Example

We routinely test the mixing of **continuous-flow stirred-tanks** (Fig. 2.6) by dumping some kind of inert tracer, say a dye, into the tank and see how they get "mixed up." In more dreamy moments, you can try the same thing with cream in your coffee. However, you are a highly paid engineer, and you must take a more technical approach. The solution is simple. We can add the tracer in a well-defined "manner," monitor the effluent, and analyze how the concentration changes with time. In chemical reaction engineering, you will see that the whole business is elaborated into the study of residence time distributions.

In this example, we have a stirred-tank with a volume V_1 of 4 m^3 being operated with an inlet flow rate Q of 0.02 m^3/s and that contains an inert species at a concentration C_{in} of 1 gmol/m^3. To test the mixing behavior, we purposely turn the knob that doses in the tracer and we jack up its concentration to 6 gmol/m^3 (without increasing the total flow rate) for a duration of 10 s. The effect is a rectangular pulse input (Fig. 2.7).

What is the pulse response in the effluent? If we do not have the patience of 10 s and dump all the extra tracer in at one shot, what is the impulse response?

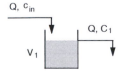

Figure 2.6. A constant-volume continuous-flow well-mixed vessel.

2.8. Two Transient Model Examples

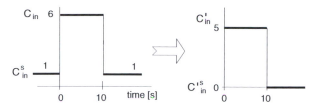

Figure 2.7. A rectangular pulse in real and deviation variables.

The model equation is a continuous-flow stirred-tank without any chemical reaction:

$$V_1 \frac{dC_1}{dt} = Q(C_{in} - C_1).$$

In terms of space time τ_1, it is written as

$$\tau_1 \frac{dC_1}{dt} = C_{in} - C_1, \tag{2.35}$$

where

$$\tau_1 = \frac{V_1}{Q} = \frac{4}{0.02} = 200\,\text{s}.$$

The initial condition is $C(0) = C_1^s$, where C_1^s is the value of the steady-state solution. The inlet concentration is a function of time, $C_{in} = C_{in}(t)$, and will become our input. The analytical results are presented here, and the simulations are done with MATLAB in the Review Problems.

At steady state, Eq. (2.35) is[18]

$$0 = C_{in}^s - C_1^s. \tag{2.36}$$

As suggested in Section 2.1, we define the deviation variables

$$C_1' = C_1 - C_1^s, \qquad C_{in}' = C_{in} - C_{in}^s,$$

and combining Eqs. (2.35) and (2.36) would give us

$$\tau_1 \frac{dC_1'}{dt} = C_{in}' - C_1'$$

with the zero initial condition $C'(0) = 0$. We further rewrite the equation as

$$\tau_1 \frac{dC_1'}{dt} + C_1' = C_{in} \tag{2.37}$$

to emphasize that C_{in}' is the input (or forcing function). The Laplace transform of Eq. (2.37) is

$$\frac{C_1'(s)}{C_{in}'(s)} = \frac{1}{\tau_1 s + 1}, \tag{2.38}$$

---

[18] At steady state, it is obvious from Eq. (2.35) that the solution must be identical to the inlet concentration. You may find it redundant that we add a superscript s to C_{in}^s. The action is taken to highlight the particular value of $C_{in}(t)$ that is needed to maintain the steady state and to make the definitions of deviation variables a bit clearer.

27

where the RHS is the transfer function. Here, it relates changes in the inlet concentration to changes in the tank concentration. This is a convenient form with which we can address different input cases.

Now we have to fit the pieces together for this problem. Before the experiment, that is, at steady state, we must have

$$C_{in}^s = C_1^s = 1. \tag{2.39}$$

Hence the rectangular pulse is really a perturbation in the inlet concentration:

$$C_{in}' = \begin{cases} 0, & t < 0 \\ 5, & 0 < t < 10 \\ 0, & t > 10. \end{cases}$$

This input can be written succinctly as

$$C_{in}' = 5\,[u(t) - u(t - 10)],$$

which then can be applied to Eq. (2.37). Alternatively, we apply the Laplace transform of this input,

$$C_{in}'(s) = \frac{5}{s}(1 - e^{-10s}),$$

and substitute it into Eq. (2.38) to arrive at

$$C_1'(s) = \frac{1}{(\tau_1 s + 1)} \frac{5}{s}(1 - e^{-10s}). \tag{2.40}$$

The inverse transform of Eq. (2.40) gives us the time-domain solution for $C_1'(t)$:

$$C_1'(t) = 5(1 - e^{-t/\tau_1}) - 5\big[1 - e^{-(t-10)/\tau_1}\big]u(t - 10).$$

The most important time dependence of e^{-t/τ_1} arises only from the pole of the transfer function in Eq. (2.38). Again, we can "spell out" the function if we want to: For $t < 10$,

$$C_1'(t) = 5(1 - e^{-t/\tau_1}),$$

and for $t > 10$,

$$C_1'(t) = 5(1 - e^{-t/\tau_1}) - 5\big[1 - e^{-(t-10)/\tau_1}\big] = 5\big[e^{-(t-10)/\tau_1} - e^{-t/\tau_1}\big].$$

In terms of the actual variable, we have, for $t < 10$,

$$C_1(t) = C_1^s + C_1' = 1 + 5\,(1 - e^{-t/\tau_1}),$$

and for $t > 10$,

$$C_1(t) = 1 + 5\big[e^{-(t-10)/\tau_1} - e^{-t/\tau_1}\big].$$

We now want to use an impulse input of equivalent "strength," i.e., the same amount of inert tracer added. The amount of additional tracer in the rectangular pulse is

$$5(\text{gmol/m}^3)\,0.02(\text{m}^3/\text{s})10(\text{s}) = 1\ \text{gmol},$$

which should also be the amount of tracer in the impulse input. Let the impulse input be $C'_{in} = M\delta(t)$. Note that $\delta(t)$ has the unit of inverse time and M has a funny and physically meaningless unit, and we calculate the magnitude of the input by matching the quantities

$$1\,(\text{gmol}) = \int_0^\infty 0.02 \left(\frac{m^3}{s}\right) M \left(\frac{\text{gmol s}}{m^3}\right) \delta(t) \left(\frac{1}{s}\right) dt(s) = 0.02M \quad \text{or}$$

$$M = 50 \left(\frac{\text{gmol s}}{m^3}\right).$$

Thus,

$$C'_{in}(t) = 50\delta(t), \quad C'_{in}(s) = 50,$$

and, for an impulse input, Eq. (2.38) is simply

$$C'_1(s) = \frac{50}{(\tau_1 s + 1)}. \tag{2.41}$$

After inverse transform, the solution is

$$C'_1(t) = \frac{50}{\tau_1} e^{-t/\tau_1},$$

and, in real variables,

$$C_1(t) = 1 + \frac{50}{\tau_1} e^{-t/\tau_1}.$$

We can do a mass balance based on the outlet

$$Q \int_0^\infty C'_1(t)\,dt = 0.02 \frac{50}{\tau_1} \int_0^\infty e^{-t/\tau_1} dt = 1\,(\text{gmol}).$$

Hence mass is conserved and the mathematics is correct.

We now raise a second question. If the outlet of the vessel is fed to a second tank with a volume V_2 of 3 m^3 (Fig. 2.8), what is the time response at the exit of the second tank? With the second tank, the mass balance is

$$\tau_2 \frac{dC_2}{dt} = (C_1 - C_2), \quad \text{where } \tau_2 = \frac{V_2}{Q},$$

or

$$\tau_2 \frac{dC_2}{dt} + C_2 = C_1, \tag{2.42}$$

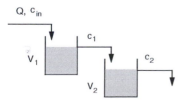

Figure 2.8. Two well-mixed vessels in series.

where C_1 and C_2 are the concentrations in tanks one and two, respectively. The equation analogous to Eq. (2.37) is

$$\tau_2 \frac{dC_2'}{dt} + C_2' = C_1', \tag{2.43}$$

and its Laplace transform is

$$C_2'(s) = \frac{1}{\tau_2 s + 1} C_1'(s). \tag{2.44}$$

With the rectangular pulse to the first vessel, we use the response in Eq. (2.40) and substitute into Eq. (2.44) to give

$$C_2'(s) = \frac{5(1 - e^{-10s})}{s(\tau_1 s + 1)(\tau_2 s + 1)}.$$

With the impulse input, we use the impulse response in Eq. (2.41) instead, and Eq. (2.44) becomes

$$C_2'(s) = \frac{50}{(\tau_1 s + 1)(\tau_2 s + 1)},$$

from which $C_2'(t)$ can be obtained by means of a proper look-up table. The numerical values

$$\tau_1 = \frac{4}{0.02} = 200\,\text{s} \quad \text{and} \quad \tau_2 = \frac{3}{00.2} = 150\,\text{s}$$

can be used. We will skip the inverse transform. It is not always instructive to continue with an algebraic mess. To sketch the time response, we will do so with MATLAB in the Review Problems.

2.8.2. A Stirred-Tank Heater

Temperature control in a **stirred-tank heater** is a common example (Fig. 2.9). We will come across it many times in later chapters. For now, the basic model equation is presented, and we use it as a review of transfer functions.

The heat balance, in standard heat transfer notation, is

$$\rho C_p V \frac{dT}{dt} = \rho C_p Q(T_i - T) + U A (T_H - T), \tag{2.45}$$

where U is the overall heat transfer coefficient, A is the heat transfer area, ρ is the fluid density, C_p is the heat capacity, and V is the volume of the vessel. The inlet temperature $T_i = T_i(t)$ and steam-coil temperature $T_H = T_H(t)$ are functions of time and are presumably given. The initial condition is $T(0) = T^s$, the steady-state temperature.

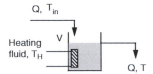

Figure 2.9. A continuous-flow stirred-tank heater.

Before we go on, it should be emphasized that what we subsequently find are nothing but different algebraic manipulations of the same heat balance. First, we rearrange Eq. (2.45) to give

$$\left(\frac{V}{Q}\right)\frac{dT}{dt} = (T_i - T) + \frac{UA}{\rho C_p Q}(T_H - T).$$

The second step is to define

$$\tau = \frac{V}{Q}, \qquad \kappa = \frac{UA}{\rho C_p Q},$$

which leads to

$$\tau\frac{dT}{dt} + (1+\kappa)T = T_i + \kappa T_H. \tag{2.46}$$

At steady state,

$$(1+\kappa)T^s = T_i^s + \kappa T_H^s. \tag{2.47}$$

We now define the deviation variables:

$$T' = T - T^s, \quad T_i' = T_i - T_i^s, \quad T_H' = T_H - T_H^s,$$

$$\frac{dT'}{dt} = \frac{d(T - T^s)}{dt} = \frac{dT}{dt}.$$

Subtracting Eq. (2.47) from transient equation (2.46) would give

$$\tau\frac{dT}{dt} + (1+\kappa)(T - T^s) = (T_i - T_i^s) + \kappa(T_H - T_H^s)$$

or, in deviation variables,

$$\tau\frac{dT'}{dt} + (1+\kappa)T' = T_i' + \kappa T_H'. \tag{2.48}$$

The initial condition is $T'(0) = 0$. Equation (2.48) is identical in form to Eq. (2.46). This is typical of linear equations. Once you understand the steps, you can jump from Eq. (2.46) to (2.48), skipping over the formality.

From here on, we *omit the prime* (') where it would not cause confusion, as it goes without saying that we work with deviation variables. We now further rewrite the same equation as

$$\frac{dT}{dt} + aT = K_i T_i + K_H T_H, \tag{2.48a}$$

where

$$a = \frac{(1+\kappa)}{\tau}, \quad K_i = \frac{1}{\tau}, \quad K_H = \frac{\kappa}{\tau}.$$

A Laplace transform gives us

$$sT(s) + aT(s) = K_i T_i(s) + K_H T_H(s). \tag{2.49}$$

Hence, Eq. (2.48a) becomes

$$T(s) = \left(\frac{K_i}{s+a}\right)T_i(s) + \left(\frac{K_H}{s+a}\right)T_H(s) = G_d(s)T_i(s) + G_p(s)T_H(s), \tag{2.49a}$$

31

where

$$G_d(s) = \frac{K_i}{s+a}, \qquad G_p(s) = \frac{K_H}{s+a}.$$

Of course, $G_d(s)$ and $G_p(s)$ are the transfer functions, and they are in *pole-zero form*. Once again (!), we are working with deviation variables. The interpretation is that *changes* in the inlet temperature and the steam temperature lead to *changes* in the tank temperature. The effects of the inputs are additive and mediated by the two transfer functions.

Are there more manipulated forms of the same old heat balance? You bet. In fact, we very often rearrange Eq. (2.48), without the primes, as

$$\tau_p \frac{dT}{dt} + T = K_d T_i + K_p T_H, \tag{2.48b}$$

where[19]

$$\tau_p = \frac{1}{a} = \frac{\tau}{(1+\kappa)}, \quad K_d = \frac{K_i}{a} = \frac{1}{(1+\kappa)}, \quad K_p = \frac{K_H}{a} = \frac{\kappa}{(1+\kappa)}.$$

After Laplace transform, the model equation is

$$T(s) = G_d(s)T_i(s) + G_p(s)T_H(s), \tag{2.49b}$$

which is identical to Eq. (2.49a) except that the transfer functions are in the *time-constant form*:

$$G_d(s) = \frac{K_d}{\tau_p s + 1}, \qquad G_p(s) = \frac{K_p}{\tau_p s + 1}.$$

In these rearrangements, τ_p is the process time constant and K_d and K_p are the steady-state gains.[20] The denominators of the transfer functions are *identical*; they both are from the LHS of the differential equation – the characteristic polynomial that governs the inherent dynamic characteristic of the process.

Let us try one simple example. If we keep the inlet temperature constant at our desired steady state, the statements in deviation variables (without the prime) are

$$T_i(t) = 0, \qquad T_i(s) = 0.$$

Now we want to know what happens if the steam temperature increases by $10\,°C$. The changes in the deviation variables are

$$T_H = Mu(t), \qquad T_H(s) = (M/s),$$

where $M = 10\,°C$. We can write

$$T(s) = \left(\frac{K_p}{\tau_p s + 1}\right)\frac{M}{s}. \tag{2.50}$$

[19] If the heater is well designed, $\kappa\ (= UA/\rho C_p Q)$ should be much larger than 1. The steady-state gain K_p approaches unity, which means that changing the steam temperature is an effective means of changing the tank temperature. In contrast, K_d is very small, and the tank temperature is insensitive to changes in the inlet temperature. At first reading, you may find the notation confusing – and in some ways this was done on purpose. This is as bad as it gets once you understand the different rearrangements. So go through each step slowly.

[20] K_i and K_H in Eq. (2.49a) are referred to as gains, but *not* the steady-state gains. The process time constant is also called a *first-order lag* or *linear lag*.

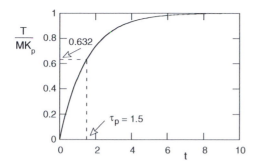

Figure 2.10. Illustration of a first-order response [Eq. (2.51)] normalized by MK_p. The curve is plotted with $\tau_p = 1.5$ (arbitrary time unit). At $t = \tau_p$, the normalized response is 63.2%.

After partial fraction expansion,

$$T(s) = MK_p \left(\frac{1}{s} - \frac{\tau_p}{\tau_p s + 1} \right).$$

An inverse transform by means of a look-up table gives our time-domain solution for the deviation in T[21]:

$$T(t) = MK_p \left(1 - e^{-t/\tau_p} \right). \tag{2.51}$$

Keep a mental imprint of the shape of this first-order step response as shown in Fig. 2.10. As time progresses, the exponential term decays away, and the temperature approaches the new value MK_p. Also illustrated in the figure is the much-used property that, at $t = \tau_p$, the normalized response is 63.2%.

After this exercise, let's hope that we have a better appreciation of the different forms of a transfer function. With one, it is easier to identify the pole positions. With the other, it is easier to extract the steady-state gain and time constants. It is very important for us to learn how to interpret qualitatively the dynamic response from the pole positions and to make a physical interpretation with the help of quantities like steady-state gains and time constants.

2.9.   Linearization of Nonlinear Equations

Because a Laplace transform can be applied to only a linear differential equation, we must "fix" a nonlinear equation. The goal of control is to keep a process running at a specified condition (the steady state). For the most part, if we do a good job, the system should be only slightly perturbed from the steady state such that the dynamics of returning to the steady state is a first-order decay, i.e., a linear process. This is the cornerstone of classical control theory.

What we do is a freshmen calculus exercise in first-order Taylor series expansion about the steady state and reformulate the problem in terms of deviation variables. This is illustrated

---

[21] Note that if we had chosen also $T_H = 0$, $T(t) = 0$ for all t, i.e., nothing happens. Recall once again from Section 2.1 that this is a result of how we define a problem by using deviation variables.

with one simple example. Consider the differential equation that models the liquid level h in a tank with cross-sectional area A:

$$A\frac{dh}{dt} = Q_{in}(t) - \beta h^{1/2}. \tag{2.52}$$

The initial condition is $h(0) = h_s$, the steady-state value. The inlet flow rate Q_{in} is a function of time. The outlet is modeled as a nonlinear function of the liquid level. Both the tank cross-section A and the coefficient β are constants.

We next expand the nonlinear term about the steady-state value h_s (also our initial condition by choice) to provide[22]

$$A\frac{dh}{dt} = Q_{in} - \beta\left[h_s^{1/2} + \frac{1}{2}h_s^{-1/2}(h - h_s)\right]. \tag{2.53}$$

At steady state, we can write Eq. (2.52) as

$$0 = Q_{in}^s - \beta h_s^{1/2}, \tag{2.54}$$

where h_s is the steady-state solution and Q_{in}^s is the particular value of Q_{in} to maintain steady state. If we subtract the steady-state equation from the linearized differential equation, we have

$$A\frac{dh}{dt} = (Q_{in} - Q_{in}^s) - \beta\left[\frac{1}{2}h_s^{-1/2}(h - h_s)\right]. \tag{2.55}$$

We now define the deviation variables:

$$h' = h - h_s, \qquad Q_{in}' = Q_{in} - Q_{in}^s.$$

Substituting them into the linearized equation and moving the h' term to the left should give

$$A\frac{dh'}{dt} + \left(\frac{\beta}{2}h_s^{-1/2}\right)h' = Q_{in}'(t), \tag{2.56}$$

with the zero initial condition $h'(0) = 0$.

It is important to note that the initial condition in Eq. (2.52) has to be h_s, the original steady-state level. Otherwise, we will not obtain a zero initial condition in Eq. (2.56). On the other hand, because of the zero initial condition, the forcing function Q_{in}' must be finite to have a nontrivial solution. Repeating our mantra for the umpteenth time, we find that the LHS of Eq. (2.56) gives rise to the characteristic polynomial and describes the inherent dynamics. The actual response is subject to the inherent dynamics and the input that we impose on the RHS.

---

[22] We casually ignore the possibility of a more accurate second-order expansion. That is because the higher-order terms are nonlinear, and we need a linear approximation. Needless to say that, with a first-order expansion, it is acceptable only if h is sufficiently close to h_s. In case you forgot, the first-order Taylor series expansion can be written as

$$f(x_1, x_2) \approx f(x_{1s}, x_{2s}) + \partial f/\partial x_1|_{x_{1s}, x_{2s}}(x_1 - x_{1s}) + \partial f/\partial x_2|_{x_{1s}, x_{2s}}(x_2 - x_{2s}).$$

Note:

- Always do the linearization before you introduce the deviation variables.
- As soon as we finish the first-order Taylor series expansion, the equation is linearized. All steps that follow are to clean up the algebra with the understanding that terms of the steady-state equation should cancel out and to change the equation to deviation variables with zero initial condition.

A more general description is now provided. Consider an ODE,

$$\frac{dy}{dt} = f(y; \mathbf{u}), \quad \text{with } y(0) = y_s, \tag{2.57}$$

where $\mathbf{u} = \mathbf{u}(t)$ contains other parameters that may vary with time. If $f(y; \mathbf{u})$ is nonlinear, we approximate with Taylor's expansion:

$$\frac{dy}{dt} \approx f(y_s; \mathbf{u}_s) + \left.\frac{\partial f}{\partial y}\right|_{y_s, \mathbf{u}_s} (y - y_s) + \nabla^T f(y_s; \mathbf{u}_s)(\mathbf{u} - \mathbf{u}_s), \tag{2.58}$$

where $\nabla f(y_s; \mathbf{u}_s)$ is a column vector of the partial derivatives of the function with respect to elements in $\mathbf{u}$, $\partial f/\partial u_i$, and evaluated at y_s and $\mathbf{u}_s$. At steady state, Eq. (2.57) is

$$0 = f(y_s; \mathbf{u}_s), \tag{2.59}$$

where y_s is the steady-state solution and $\mathbf{u}_s$ are the values of parameters needed to maintain the steady state. We again define the deviation variables,

$$y' = y - y_s, \quad \mathbf{u}' = \mathbf{u} - \mathbf{u}_s,$$

and subtract Eq. (2.59) from Eq. (2.58) to obtain the linearized equation with zero initial condition:

$$\frac{dy'}{dt} + \left[-\left.\frac{\partial f}{\partial y}\right|_{y_s, \mathbf{u}_s} \right] y' = [\nabla^T f(y_s; \mathbf{u}_s)]\mathbf{u}', \tag{2.60}$$

where we have put quantities being evaluated at steady-state conditions in brackets. When we solve a particular problem, they are just constant coefficients after the substitution of numerical values.

Example 2.11: Linearize the differential equation for the concentration in a mixed vessel; $[V(dC/dt)] = Q_{in}(t)C_{in}(t) - Q_{in}(t)C$, where the flow rate and the inlet concentration are functions of time.

A first-term Taylor expansion of the RHS leads to the approximation

$$V\frac{dC}{dt} \approx [Q_{in,s}C_{in,s} + C_{in,s}(Q_{in} - Q_{in,s}) + Q_{in,s}(C_{in} - C_{in,s})]$$
$$- [Q_{in,s}C_s + C_s(Q_{in} - Q_{in,s}) + Q_{in,s}(C - C_s)],$$

and the steady-state equation, without canceling the flow variable, is

$$0 = Q_{in,s}C_{in,s} - Q_{in,s}C_s.$$

We subtract the two equations and at the same time introduce deviation variables for the dependent variable C and all the parametric variables to obtain

$$V\frac{dC'}{dt} \approx [C_{in,s}Q'_{in} + Q_{in,s}C'_{in}] - [C_s Q'_{in} + Q_{in,s}C'],$$

and, after moving the C' term to the LHS, we obtain

$$V\frac{dC'}{dt} + [Q_{in,s}]C' = [C_{in,s} - C_s]Q'_{in} + [Q_{in,s}]C'_{in}.$$

The final result can be interpreted as stating how changes in the flow rate and the inlet concentration lead to changes in the tank concentration, as modeled by the dynamics on the LHS. Again, we put the constant coefficients evaluated at steady-state conditions in brackets. We can arrive at this result quickly if we understand Eq. (2.60) and apply it carefully.

The final step also has zero initial condition $C'(0) = 0$, and we can take the Laplace transform to obtain the transfer functions if they are requested. As a habit, we can define $\tau = V/Q_{in,s}$ and the transfer functions will be in the time-constant form.

Example 2.12: Linearize the differential equation $(dy/dt) = -xy - \beta y^2 - \gamma^{y-1}$, where $x = x(t)$. Each nonlinear term can be approximated as

$$xy \approx x_s y_s + y_s(x - x_s) + x_s(y - y_s) = x_s y_s + y_s x' + x_s y',$$

$$y^2 \approx y_s^2 + 2y_s(y - y_s) = y_s^2 + 2y_s y',$$

$$\gamma^{y-1} \approx \gamma^{y_s-1} + (\ln\gamma)\gamma^{y_s-1}y'.$$

With the steady-state equation,

$$0 = x_s y_s + \beta y_s + \gamma^{y_s-1},$$

and the usual algebraic work, we arrive at

$$(dy'/dt) + [x_s + 2\beta y_s + (\ln\gamma)\gamma^{y_s-1}]y' = -y_s x'.$$

Example 2.13: What is the linearized form of the reaction rate term $r_A = -k(T)C_A = -k_0 e^{-E/RT}C_A$, where both temperature T and concentration C_A are functions of time?

$$r_A \approx -\left[k_0 e^{-E/RT_s}C_{A,s} + k_0 e^{-E/RT_s}(C_A - C_{A,s}) + \left(\frac{E}{RT_s^2}\right)k_0 e^{-E/RT_s}C_{A,s}(T - T_s)\right].$$

In terms of deviation variables, the linearized approximation is

$$r_A \approx r_{A,s}\left[1 + \frac{1}{C_{A,s}}C'_A + \frac{E}{RT_s^2}T'\right],$$

where $r_{A,s} = -k_0 e^{-E/RT_s}C_{A,s}$.

Note: Although our analyses use deviation variables and not the real variables, examples and homework problems can keep bouncing back and forth. The reason is that, when we do

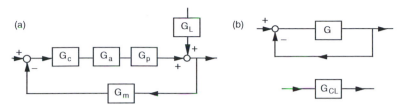

Figure 2.11. (a) Example of a feedback-system block diagram, (b) typical reduced block diagrams.

an experiment, we measure the actual variable, not the deviation variable. You may find this really confusing. All we can do is to be extra careful when we solve a problem.

2.10. Block-Diagram Reduction

The use of **block diagrams** to illustrate a cause-and-effect relationship is prevalent in control. We use operational blocks to represent transfer functions and lines for unidirectional information transmission. It is a nice way to visualize the interrelationships of various components. They will be crucial in helping us identify manipulated and controlled variables and input(s) and output(s) of a system.

Many control systems are complicated-looking networks of blocks. The simplest control system looks like Fig. 2.11(a). The problem is that many theories in control are based on a simple closed-loop or single-block structure [Fig. 2.11(b)].

Hence we must learn how to read a block diagram and *reduce* it to the simplest possible form. We will learn in later chapters how the diagram is related to an actual physical system. First, we do some simple algebraic manipulation and, better yet, do it graphically. It is important to remember that all (graphical) block-diagram reduction is a result of formal algebraic manipulation of transfer functions. When all imagination fails, *always refer back to the actual algebraic equations.*[23]

Of all manipulations, the most important one is the reduction of a feedback loop. Here is the so-called block-diagram reduction and corresponding algebra.

For a negative-feedback system (Fig. 2.12), we have

$$E = R - HY, \tag{2.61}$$

$$Y = GE. \tag{2.62}$$

Using Eq. (2.61) to substitute for E in Eq. (2.62) leads to

$$Y = G[R - HY],$$

which can be rearranged to give, for a **negative-feedback** loop,[24]

$$\frac{Y}{R} = \frac{G}{1 + GH}. \tag{2.63}$$

---

[23] See the *Web Support* for a comment on Mason's gain formula.
[24] Similarly, we can write for the case of *positive* feedback that $E = R + HY$ and $Y = G(R + HY)$, and we have instead $(Y/R) = \{[G(s)]/[1 - G(s)H(s)]\}$.

(a)

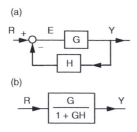

(b)

Figure 2.12. Simple negative-feedback loop and its reduced single closed-loop transfer function form.

The RHS of Eq. (2.63) is what will be referred to as the **closed-loop transfer function** in later chapters.

Note: The important observation is that when we "close" a *negative*-feedback loop, the numerator consists of the product of all the transfer functions along the forward path. The denominator is 1 *plus* the product of all the transfer functions in the entire feedback loop (i.e., both forward and feedback paths). The denominator is also the characteristic polynomial of the closed-loop system. If we have positive feedback, the sign in the denominator is minus.

Here, we try several examples and take the conservative route of writing out the relevant algebraic relations.[25]

Example 2.14: Derive the closed-loop transfer functions C/R and C/L for the system shown in Fig. E2.14. We identify two locations after the summing points with lower case e and a to help us.[26] We can write at the summing point below H,

$$a = -C + KR,$$

and substitute this relation into the equation for the summing point above H to give

$$e = R + Ha = R + H(KR - C).$$

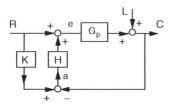

Figure E2.14.

---

[25] A few more simple examples are in the *Web Support*.

[26] How do we decide the proper locations? We do not know for sure, but what should help is after a summing point where information has changed. We may also use the location before a branch-off point, helping us to trace where the information is routed.

We substitute the relation for e into the equation about G_p to obtain

$$C = L + G_p e = L + G_p(R + HKR - HC).$$

The final result is a rearrangement to get C out explicitly:

$$C = \left(\frac{1}{1 + G_p H}\right) L + \left(\frac{G_p(1 + HK)}{1 + G_p H}\right) R.$$

Example 2.15: Derive the closed-loop transfer function C/R for the system with three overlapping negative-feedback loops in Fig. E2.15(a).

The key to this problem is to proceed in steps and "untie" the overlapping loops first. We identify various locations with lowercase a, b, d, f, and g to help us. We first move the branch-off point over to the RHS of G_4 [Fig. E2.15(b)]. We may note that we can write

$$a = H_1 G_3 d = \frac{H_1}{G_4}(G_3 G_4) d,$$

that is, to maintain the same information at the location a, we must divide the branch-off information at C by G_4.

Similarly, we note that at the position g in Fig. E2.15(a),

$$g = G_1(R - f) - bH_1 = G_1 \left(R - f - \frac{bH_1}{G_1}\right),$$

that is, if we move the break-in point from g out to the left of G_1, we need to divide the information by G_1 before breaking in. The block diagram after both the branch-off and the break-in points are moved are shown as Steps 1 and 2 in Fig. E2.15(b). (We could have drawn it such that the loops are flush with one another at R.)

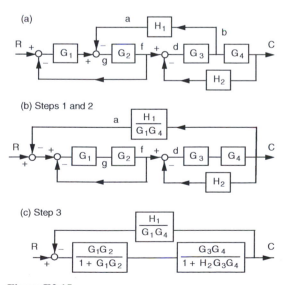

Figure E2.15.

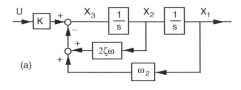

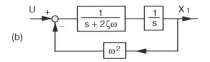

Figure E2.16.

Once the loops are no longer overlapping, the block diagram is easy to handle. We first close the two small loops, shown as Step 3 in Fig. E2.15(c).

The final result is to close the big loop. The resulting closed-loop transfer function is

$$\frac{C}{R} = \frac{G_1 G_2 G_3 G_4}{(1 + G_1 G_2)(1 + H_2 G_3 G_4) + H_1 G_2 G_3}.$$

Example 2.16: Derive the closed-loop transfer function X_1/U for the block diagram in Fig. E2.16(a). We will see this one again in Chap. 4 on state-space models. With the integrator $1/s$, X_2 is the Laplace transform of the time derivative of $x_1(t)$, and X_3 is the second-order derivative of $x_1(t)$.

We can write the algebraic equations about the summing point (the comparator) for X_3 and the two equations about the two integrators $1/s$. We should arrive at the result after eliminating X_2 and X_3.

However, we can also obtain the result quickly by recognizing the two feedback loops. We first "close" the inner loop to arrive at Fig. E2.16(b). With that, we can see the answer. We "close" this loop to obtain

$$\frac{X_1}{U} = \frac{1}{s^2 + 2\zeta\omega s + \omega^2}.$$

Review Problems

(1) Derive Eq. (2.1).
(2) (a) Check that when the RHS is zero, the solution to Eq. (2.2) is zero.
    (b) Derive Eq. (2.3) by using the method of integrating factor.
    (c) Derive the solution $c(t)$ in Eq. (2.3) with an impulse input, $C'_{in} = \delta(t)$, and a unit-step input, $C'_{in} = u(t)$. Show that they are identical when we use the Laplace transform technique, as in Example 2.10.
(3) Prove the linear property of the Laplace transform in Eq. (2.6).
(4) Derive the Laplace transforms of
    (a) $1/(\tau s + 1)$,   (b) $\cos \omega t$,   (c) $e^{-at} \cos \omega t$.
(5) Prove the initial-value theorem.
(6) Show that the inverse transform of $F(s) = [6/(s^3 + s^2 - 4s - 4)]$ is $f(t) = -2e^{-t} + \frac{3}{2}e^{-2t} + \frac{1}{2}e^{2t}$.

(7) Double check α^* in the complex root of Example 2.7 with the Heaviside expansion.

(8) Find the inverse Laplace transform of the general expression $Y(s) = \{[c/(s-p)] + [c^*/(s-p^*)]\}$, where $c = a - bj$ and $p = \alpha + \omega j$. Rearrange the result to a sine function with time lag.

(9) With respect to the repeated root of Example 2.9, show that we could have written

$$2 = \alpha_1(s+1)^2(s+2) + \alpha_2(s+1)(s+2) + \alpha_3(s+2) + \alpha_4(s+1)^3,$$

and that after expanding and collecting terms of the same power in s, we can form the matrix equation

$$\begin{bmatrix} 1 & 0 & 0 & 1 \\ 4 & 1 & 0 & 3 \\ 5 & 3 & 1 & 3 \\ 2 & 2 & 2 & 1 \end{bmatrix} \begin{bmatrix} \alpha_1 \\ \alpha_2 \\ \alpha_3 \\ \alpha_4 \end{bmatrix} = \begin{bmatrix} 0 \\ 0 \\ 0 \\ 2 \end{bmatrix}$$

from which we can solve for the coefficients. Yes, this is a chore not worth doing even with MATLAB. The route that we take in the example is far quicker.

(10) For a general transfer function $G(s) = Y(s)/X(s)$, how can we find the steady-state gain?

(11) Do the partial fractions of $\{(s+1)/[s^2(s^2+4s-5)]\}$.

(12) Find the transfer functions as suggested in Example 2.11.

(13) Derive Eqs. (3.33) and (3.34).

(14) Do the numerical simulation for Subsection 2.8.1.

(15) Regarding Eqs. (2.50) and (2.51) in Subsection 2.8.2:

(a) What is $T(t)$ as $t \to \infty$? What is the actual temperature that we measure?

(b) What are the effects of K_p on the final tank temperature? What is the significance if K_p approaches unity?

(c) What is the time constant for the process?

Hints:

(1) Equation (2.1) is straight from material balance. With Q denoting the volumetric flow rate and V the volume, the balance equation of some species A as denoted by C with reference to Fig. 2.6 is

$$V\frac{dC}{dt} = QC_{in} - QC.$$

Physically, each term in the equation represents

$$\{\text{Accumulation}\} = \{\text{Flow in}\} - \{\text{Flow out}\}.$$

Equation (2.1) is immediately obvious with the definition of space time $\tau = V/Q$.

(2) (c) With the impulse input, $C'(t) = \frac{1}{\tau}e^{-t/\tau}$, and with the unit-step input, $C'(t) = 1 - e^{-t/\tau}$.

(3) This is just a simple matter of substituting the definition of Laplace transform.

(4) Answers are in the Laplace transform summary table. The key to (a) is to rearrange the function as $(1/\tau)/(s + 1/\tau)$, and the result should then be immediately obvious

with Eq. (2.9). The derivations of (b) and (c) are very similar to the case involving $\sin \omega t$.

(5) The starting point is again the Laplace transform of a derivative, but this time we take the limit as $s \to \infty$ and the entire LHS with the integral becomes zero.

(6) This follows Examples 2.4 and 2.5.

(7) $\alpha^* = \dfrac{s+5}{[s-(-2+3j)]}\bigg|_{s=-2-3j} = \dfrac{(-2-3j)+5}{(-2-3j)+2-3j} = \dfrac{(1-j)}{-2j} = \dfrac{1}{2}(1+j).$

(8) We should have

$$y(t) = c\, e^{pt} + c^* e^{p^*t} = (a - bj)e^{(\alpha+\omega j)t} + (a + bj)e^{(\alpha-\omega j)t}.$$

We now do the expansion:

$$y(t) = e^{\alpha t}[(a - bj)(\cos \omega t + j \sin \omega t) + (a + bj)(\cos \omega t - j \sin \omega t)].$$

After cancellation,

$$y(t) = 2e^{\alpha t}(a \cos \omega t + b \sin \omega t),$$

which of course is

$$y(t) = 2e^{\alpha t} A \sin(\omega t + \phi),$$

where $A = (a^2 + b^2)^{1/2}$, and $\phi = \tan^{-1}(a/b)$

(10) Use $X = 1/s$ and the final-value theorem. This is what we did in Eq. (2.32). Note that $y(\infty)$ is not the same as $g(\infty)$.

(11) Use MATLAB to check your algebra.

(12) With the time constant defined as $\tau = V/Q_{\text{in},s}$, the steady-state gain for the transfer function for the inlet flow rate is rate is $(C_{\text{in},s} - C_s)/Q_{\text{in},s}$, and it is 1 for the inlet concentration transfer function.

(13) To find $\tan \theta$ is easy. To find $\cos \theta$, we need to show that the distance from the pole p to the origin (the hypotenuse) is $1/\tau$.

(14) These are the statements that you can use:

```
tau1=200;
G1=tf(1,[tau1 1]);    % Transfer function of the first vessel

pulselength=10;       % Generate a vector to represent the
                      % rectangular pulse
delt=5;               % (We are jumping ahead. These steps
                      % are explained
t=0:delt:1000;        % in MATLAB Session 3.)
u=zeros(size(t));

u(1:pulselength/delt+1)=5;

lsim(G1,u,t);         % The response of the rectangular pulse

hold

y=50*impulse(G1,t);   % Add on the impulse response
plot(t,y)
```

```
tau2=150;                   % Generate the transfer function for
                            % both vessels
G2=tf(1,[tau2 1]);
G=G1*G2;
lsim(G,u,t)                 % Add the responses
y=50*impulse(G,t);
plot(t,y)
```

(15) (a) As $t \rightarrow \infty$, the deviation in $T(t)$ should be $10K_p\,°$C. We have to know the original steady-state temperature in order to calculate the actual temperature.

(b) Because $K_p < 1$, the final change in temperature will not be $10\,°$C. However, if $K_p = 1$, the final change will be $10\,°$C. We can make this statement without doing the inverse transform.

(c) The time constant of the stirred-tank heater is τ_p, not the space time.

3

Dynamic Response

We now derive the time-domain solutions of first- and second-order differential equations. It is not that we want to do the inverse transform, but comparing the time-domain solution with its Laplace transform helps our learning process. What we hope to establish is a better feel between pole positions and dynamic characteristics. We also want to see how different parameters affect the time-domain solution. The results are useful in control analysis and in measuring model parameters. At the end of the chapter, dead time, the reduced-order model, and the effect of zeros are discussed.

What Are We Up to?

- Even as we speak of a time-domain analysis, we invariably still work with a Laplace transform. Time domain and Laplace domain are inseparable in classical control.
- In establishing the relationship between time domain and Laplace domain, we use only first- and second-order differential equations. That is because we are working strictly with linearized systems. As we have seen in partial-fraction expansion, any function can be "broken up" into first-order terms. Terms of complex roots can be combined together to form a second-order term.
- Repeated roots (of multicapacity processes) lead to a sluggish response. Tanks-in-series is a good example in this respect.
- With higher-order models, we can construct approximate reduced-order models based on the identification of dominant poles. This approach is used in empirical controller tuning relations in Chap. 6.
- The dead-time transfer function has to be handled differently in classical control, and we use the Padé approximation for this purpose.

A brief review is in order: Recall that the Laplace transform is a linear operator. The effects of individual inputs can be superimposed to form the output. In other words, an observed output change can be attributed to the individual effects of the inputs. From the stirred-tank heater example in Subsection 2.8.2 we found

$$T(s) = G_d(s)T_i(s) + G_p(s)T_H(s).$$

We can analyze the *change* in tank temperature as a result of individual *changes* in either inlet or steam temperatures without doing the inverse transform. The compromise is that we do not have the time-domain analytical solution, $T(t)$, and cannot as easily think of time.

We can put the example in more general terms. Let's consider an nth-order differential equation and two forcing functions, $x_1(t)$ and $x_2(t)$,

$$a_n \frac{d^n y}{dt^n} + a_{n-1} \frac{d^{n-1} y}{dt^{n-1}} + \cdots + a_1 \frac{dy}{dt} + a_0 y = b_1 x_1(t) + b_2 x_2(t), \tag{3.1}$$

where y is the output deviation variable. We also have the zero initial conditions,

$$y(0) = y'(0) = y''(0) = \cdots = y^{(n-1)}(0) = 0. \tag{3.2}$$

The Laplace transform of Eq. (3.1) leads to

$$Y(s) = G_1(s) X_1(s) + G_2(s) X_2(s), \tag{3.3}$$

where

$$G_1(s) = \frac{b_1}{a_n s^n + a_{n-1} s^{n-1} + \cdots + a_1 s + a_0},$$

$$G_2(s) = \frac{b_2}{a_n s^n + a_{n-1} s^{n-1} + \cdots + a_1 s + a_0} \tag{3.4}$$

are the two transfer functions for the two inputs $X_1(s)$ and $X_2(s)$, respectively.

Take note (again!) that the characteristic polynomials in the denominators of both transfer functions are *identical*. The roots of the characteristic polynomial (the poles) are *independent* of the inputs. It is obvious, because they come from the same differential equation (same process or system). The poles tell us what the time-domain solution, $y(t)$, generally would "look" like. A final reminder: No matter how high the order of n may be in Eqs. (3.4), we can always use partial fractions to break up the transfer functions into first- and second-order terms.

3.1. First-Order Differential Equation Models

This section is a review of the properties of a **first-order differential equation model**. Our Chap. 2 examples of mixed vessels, stirred-tank heater, and homework problems of isothermal stirred-tank chemical reactors all fall into this category. Furthermore, the differential equation may represent either a process or a control system. What we cover here applies to any problem or situation as long as it can be described by a linear first-order differential equation.

We usually try to identify features that are characteristic of a model. Using the examples in Section 2.8 as a guide, we can write a first-order model by using deviation variables with one input and with constant coefficients a_1, a_0, and b in general notation as[1]

$$a_1 \frac{dy}{dt} + a_0 y = bx(t) \tag{3.5}$$

with $a_1 \neq 0$ and $y(0) = 0$. The model, as in Eq. (2.2), is rearranged as

$$\tau \frac{dy}{dt} + y = K x(t), \tag{3.6}$$

where τ is the time constant and K is the steady-state gain.

---

[1] Whether the notation is y or y' is immaterial. The key is to find the initial condition of the problem statement. If the initial condition is zero, the notation must refer to a deviation variable.

In the event that we are modeling a process, we would use a subscript p ($\tau = \tau_p$, $K = K_p$). Similarly, the parameters would be the system time constant and system steady-state gain when we analyze a control system. To avoid confusion, we may use a different subscript for a system.

The Laplace transform of Eq. (3.6) is

$$\frac{Y(s)}{X(s)} = G(s) = \frac{K}{\tau s + 1}, \tag{3.7}$$

where $G(s)$ denotes the transfer function. There is one real pole at $-1/\tau$. (What does it imply in terms of the time-domain function? If you are stuck, refer back to Example 2.10.)

3.1.1. Step Response of a First-Order Model

Consider a step input, $x(t) = Mu(t)$, and $X(s) = M/s$; the output is

$$Y(s) = \frac{K}{(\tau s + 1)} \frac{M}{s} = MK \left[\frac{1}{s} - \frac{\tau}{(\tau s + 1)} \right], \tag{3.8}$$

and the inverse transform gives the solution

$$y(t) = MK \left(1 - e^{-t/\tau} \right). \tag{3.9}$$

We first saw a plot of this function in Fig. 2.10. The output $y(t)$ starts at zero and increases exponentially to a new steady-state MK. A process with this property is called **self-regulating**. The larger the time constant, the slower the response [Fig. 3.1(a)].

We can check the result with the final-value theorem:

$$\lim_{s \to 0} [sY(s)] = \lim_{s \to 0} \left[s \frac{MK}{s(\tau s + 1)} \right] = MK.$$

The new steady state is not changed by a magnitude of M, but is scaled by the gain K [Fig. 3.1(b)]. Consequently we can consider the steady-state gain as the ratio of the observed change in output in response to a unit change in an input, y/M. In fact, this is how we measure K. The larger the steady-state gain, the more sensitive the output to changes in the input. As noted in Fig. 2.10, at $t = \tau$, $y(t) = 0.632\, MK$. This is a result that we often use to estimate the time constant from experimental data.

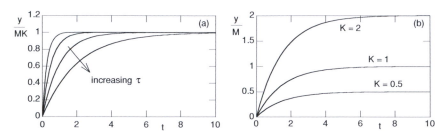

Figure 3.1. Properties of a first-order transfer function in time domain. (a) y/MK: effect of changing the time constant, plotted with $\tau = 0.25, 0.5, 1$, and 2 (arbitrary time unit); (b) y/M: effect of changing the steady-state gain; all curves have $\tau = 1.5$.

3.1.2. Impulse Response of a First-Order Model

Consider an impulse input, $x(t) = M\delta(t)$, and $X(s) = M$; the output is now

$$Y(s) = \frac{MK}{(\tau s + 1)}. \tag{3.10}$$

The time-domain solution, as in Example 2.10, is

$$y(t) = \frac{MK}{\tau} e^{-t/\tau}, \tag{3.11}$$

which implies that the output rises instantaneously to some value at $t = 0$ and then decays exponentially to zero.

3.1.3. Integrating Process

When the coefficient $a_0 = 0$ in differential equation (3.5), we have

$$\frac{dy}{dt} = \left(\frac{b}{a_1}\right) x(t), \tag{3.12}$$

$$\frac{Y(s)}{X(s)} = G(s) = \frac{K_1}{s}, \tag{3.13}$$

where $K_1 = (b/a_1)$. Here, the pole of the transfer function $G(s)$ is at the origin, $s = 0$. The solution of Eq. (3.12), which we could have written immediately without any transform, is

$$y(t) = K_1 \int_0^t x(t)\, dt. \tag{3.14}$$

This is called an **integrating** (also capacitive or non-self-regulating) **process**. We can associate the name with charging a capacitor or filling up a tank.

We can show that, with a step input, the output is a ramp function. When we have an impulse input, the output will not return to the original steady-state value, but will accumulate whatever we have added. (Both items are exercises in the Review Problems.)

Example 3.1: Show that a storage tank with pumps at its inlet and outlet (Fig. E3.1) is an integrating process.

At constant density, the mass balance of a continuous-flow mixed tank is simply

$$A\frac{dh}{dt} = q_{in} - q, \quad \text{with } h(0) = h_s,$$

where A is the cross section and h is the liquid level. The inlet and the outlet flow rates, q_{in} and q, respectively, as dictated by the pumps, are functions of time but not of the liquid

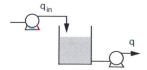

Figure E3.1.

level. At steady state, the flow rates must be the same. Thus we can define deviation variables $h' = h - h_s$, $q'_{in} = q_{in} - q_s$, and $q' = q - q_s$, and the mass balance becomes

$$A\frac{dh'}{dt} = q'_{in} - q', \quad \text{with } h'(0) = 0.$$

The general solution is

$$h'(t) = \frac{1}{A}\int_0^t (q'_{in} - q')\,dt,$$

where the change in the liquid level is a simple time integral on the change in flow rates. In terms of a Laplace transform, the differential equation leads to

$$H(s) = \frac{1}{A}\frac{Q_{in}(s) - Q(s)}{s}.$$

The transfer function has the distinct feature that a pole is at the origin. Because a step input in either q'_{in} or q' would lead to a ramp response in h', there is no steady-state gain at all.

To better observe how the tank works like a capacitor, assume that the inlet flow rate is constant and that we have a pump only at the outlet. The transfer function is now just

$$H(s) = \frac{-Q(s)}{As}.$$

If for some reason the outlet pump slows down, the liquid level in the tank will back up until it overflows. Similarly, if the outlet pump speeds up, the tank will be drained. The tank level will not reach a new steady state with respect to a step change in the input.

3.2. Second-Order Differential Equation Models

We have not encountered examples with a second-order equation, especially one that exhibits oscillatory behavior. One reason is that processing equipment tends to be self-regulating. An oscillatory behavior is most often the result of implementing a controller, and we shall see that in the control chapters. For now, this section provides several important definitions.

A model that leads to a second-order differential equation,

$$a_2\frac{d^2y}{dt^2} + a_1\frac{dy}{dt} + a_0y = b\,x(t), \quad a_2 \neq 0, \quad y(0) = y'(0) = 0, \tag{3.15}$$

is usually rearranged to take the forms

$$\tau^2\frac{d^2y}{dt^2} + 2\zeta\tau\frac{dy}{dt} + y = K\,x(t), \quad \text{or} \quad \frac{d^2y}{dt^2} + 2\zeta\omega_n\frac{dy}{dt} + \omega_n^2 y = K\omega_n^2 x(t), \tag{3.16}$$

where

$$\tau^2 = \frac{a_2}{a_0}, \quad 2\zeta\tau = \frac{a_1}{a_0}, \quad K = \frac{b}{a_0}, \quad \omega_n = \frac{1}{\tau}.$$

The corresponding Laplace transforms are

$$G(s) = \frac{Y(s)}{X(s)} = \frac{K}{\tau^2 s^2 + 2\zeta\tau s + 1} \quad \text{or} \quad \frac{X(s)}{X(s)} = \frac{K\omega_n^2}{s^2 + 2\zeta\omega_n s + \omega_n^2}, \tag{3.17}$$

where τ is the **natural period** of oscillation, ω_n is the **natural** (undamped) **frequency**, ζ is the **damping ratio** (also called the damping coefficient or factor), and K is the steady-state gain.

The characteristic polynomial is

$$p(s) = \tau^2 s^2 + 2\zeta\tau s + 1 = (s - p_1)(s - p_2), \tag{3.18}$$

which provides the poles

$$p_{1,2} = \frac{-2\zeta\tau \pm \sqrt{4\zeta^2\tau^2 - 4\tau^2}}{2\tau^2} = -\frac{\zeta}{\tau} \pm \frac{\sqrt{\zeta^2 - 1}}{\tau}. \tag{3.19}$$

A **stable** process (or system) requires $\zeta > 0$ because we need $\tau > 0$ to be physically meaningful. In addition, the two pole positions, and thus time response, take on four possibilities, depending on the value of ζ:

(1) $\zeta > 1$: Two distinct real poles. This case is called **overdamped**. Here, we can factor the polynomial in terms of two time constants, τ_1 and τ_2,

$$G(s) = \frac{K}{\tau^2 s^2 + 2\zeta\tau s + 1} = \frac{K}{(\tau_1 s + 1)(\tau_2 s + 1)}, \tag{3.20}$$

such that the two real poles are at $-1/\tau_1$ and $-1/\tau_2$.[2]

(2) $\zeta = 1$: Two repeating poles at $-1/\tau$. This case is termed **critically damped**. The natural period τ may be considered the "time constant" in the sense that it is associated with the exponential function. In actual fact, the time response is not strictly exponential, as we saw in Example 2.9 and confirmed in the time-domain solution in Eq. (3.22).

(3) $0 < \zeta < 1$: Two complex-conjugate poles. This situation is considered **underdamped**. We also write $\sqrt{\zeta^2 - 1} = j\sqrt{1 - \zeta^2}$. It is very important to note that τ is *not* the time constant here. The real part of the pole in Eq. (3.19) is $-\zeta/\tau$, and this is the value that determines the exponential decay, as in Eq. (3.23). In this sense, the time constant is τ/ζ.

(4) $\zeta = 0$: Two purely imaginary conjugate poles with frequency $\omega_n = 1/\tau$. This is equivalent to an oscillation with no damping and explains why ω_n is referred to as the natural frequency.

3.2.1. Step-Response Time-Domain Solutions

Consider a step input, $x(t) = Mu(t)$, with $X(s) = M/s$, and the different cases with respect to the value of ζ. We can derive the output response $y(t)$ for the different cases. We rarely

[2] Here, we can find that

$$\tau^2 = \tau_1\tau_2, \quad 2\zeta\tau = (\tau_1 + \tau_2)$$

or

$$\tau = \sqrt{\tau_1\tau_2}, \quad \zeta = \frac{\tau_1 + \tau_2}{2\sqrt{\tau_1\tau_2}}.$$

In this case of having real poles, we can also relate

$$\tau_1 = \frac{\tau}{\zeta - \sqrt{\zeta^2 - 1}}, \quad \tau_2 = \frac{\tau}{\zeta + \sqrt{\zeta^2 - 1}}.$$

Dynamic Response

use these results. They are provided for reference. In the case of the underdamped solution, it is used to derive the characteristic features in the next subsection.

(1) $\zeta > 1$, *overdamped:* The response is sluggish compared with that of critically damped or underdamped cases:

$$y(t) = MK \left[1 - e^{-\zeta t/\tau} \left(\cosh \frac{\sqrt{\zeta^2 - 1}}{\tau} t + \frac{\zeta}{\sqrt{\zeta^2 - 1}} \sinh \frac{\sqrt{\zeta^2 - 1}}{\tau} t \right) \right].$$

(3.21)

This form is unnecessarily complicated. When we have an overdamped response, we typically use the simple exponential form with the $\exp(-t/\tau_1)$ and $\exp(-t/\tau_2)$ terms. (You'll get to try this in the Review Problems.)

(2) $\zeta = 1$, *critically damped:* The response is the "fastest" without oscillatory behavior:

$$y(t) = MK \left[1 - \left(1 + \frac{t}{\tau} \right) e^{-t/\tau} \right].$$

(3.22)

(3) $0 \leq \zeta < 1$, *underdamped:* The response is fast initially, then overshoots and decays to steady state with oscillations. The oscillations are more pronounced and persist longer with smaller ζ:

$$y(t) = MK \left\{ 1 - e^{-\zeta t/\tau} \left[\cos \left(\frac{\sqrt{1 - \zeta^2}}{\tau} t \right) + \frac{\zeta}{\sqrt{1 - \zeta^2}} \sin \left(\frac{\sqrt{1 - \zeta^2}}{\tau} t \right) \right] \right\}.$$

(3.23)

This equation can be rearranged as

$$y(t) = MK \left[1 - \frac{e^{-\zeta t/\tau}}{\sqrt{1 - \zeta^2}} \sin \left(\frac{\sqrt{1 - \zeta^2}}{\tau} t + \phi \right) \right],$$

(3.23a)

where $\phi = \tan^{-1} \left(\frac{\sqrt{1-\zeta^2}}{\zeta} \right)$.

3.2.2. Time-Domain Features of Underdamped Step Response

From the solution of the *underdamped step response* $(0 < \zeta < 1)$, we can derive the following characteristics (Fig. 3.2). They are useful in two respects: (1) fitting experimental data in the measurements of natural period and damping factor, and (2) making control system design specifications with respect to the dynamic response.

(1) **Overshoot** (OS):

$$\text{OS} = \left(\frac{A}{B} \right) = \exp \left(\frac{-\pi \zeta}{\sqrt{1 - \zeta^2}} \right),$$

(3.24)

where A and B are shown in Fig. 3.2. We compute only the first or maximum overshoot in the response. The overshoot increases as ζ becomes smaller. The OS becomes zero as ζ approaches 1.

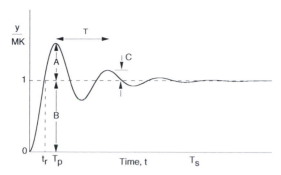

Figure 3.2. Key features in an underdamped response. See text for equations.

The time to reach the peak value is

$$T_p = \frac{\pi \tau}{\sqrt{1 - \zeta^2}} = \frac{\pi}{\omega_n \sqrt{1 - \zeta^2}}. \tag{3.25}$$

This **peak time** is less as ζ becomes smaller and is meaningless when $\zeta = 1$. We can also derive the **rise time** – time for $y(t)$ to cross or hit the final value for the first time – as

$$t_r = \frac{\tau}{\sqrt{1 - \zeta^2}} (\pi - \cos^{-1} \zeta). \tag{3.26}$$

(2) Frequency (or **period of oscillation**) T:

$$\omega = \frac{\sqrt{1 - \zeta^2}}{\tau} \quad \text{or} \quad T = \frac{2\pi \tau}{\sqrt{1 - \zeta^2}} \quad \text{as} \quad \omega = \frac{2\pi}{T}. \tag{3.27}$$

Note that $T = 2T_p$ and the unit of the frequency is radian per time unit.
(3) **Settling time:** The real part of a complex pole in Eq. (3.19) is $-\zeta/\tau$, meaning that the exponential function forcing the oscillation to decay to zero is $e^{-\zeta t/\tau}$, as in Eq. (3.23). If we draw an analogy to a first-order transfer function, the time constant of an underdamped second-order function is τ/ζ. Thus, to settle within $\pm 5\%$ of the final value, we can choose the settling time as[3]

$$T_s = 3\frac{\tau}{\zeta} = \frac{3}{\zeta \omega_n}, \tag{3.28}$$

and if we choose to settle within $\pm 2\%$ of the final value, we can use $T_s = 4\tau/\zeta$.
(4) **Decay ratio** (DR):

$$\text{DR} = \left(\frac{C}{A}\right) = \exp\left(\frac{-2\pi \zeta}{\sqrt{1 - \zeta^2}}\right) = \text{OS}^2. \tag{3.29}$$

The DR is the square of the OS, and both quantities are functions of ζ only. The definitions of C and A are shown in Fig. 3.2.

---

[3] Refer to Review Problem (1) to see why we may pick factors of 3 or 4.

3.3. Processes with Dead Time

Many physiochemical processes involve a time delay between the input and the output. This delay may be due to the time required for a slow chemical sensor to respond or for a fluid to travel down a pipe. A **time delay** is also called **dead time** or **transport lag**. In controller design, the output will not contain the most current information, and systems with dead time can be difficult to control.

From Eq. (2.18) in Section 2.3, the Laplace transform of a time delay is an exponential function. For example, first- and second-order models with dead time will appear as

$$\frac{Y(s)}{X(s)} = \frac{Ke^{-t_d s}}{\tau s + 1}, \quad \frac{Y(s)}{X(s)} = \frac{Ke^{-t_d s}}{\tau^2 s^2 + 2\zeta \tau s + 1}.$$

Many classical control techniques are developed to work only with polynomials in s, and we need some way to tackle the exponential function.

To handle the time delay, we do not simply expand the exponential function as a Taylor series. We use the so-called **Padé approximation**, which puts the function as a ratio of two polynomials. The simplest is the first-order (1/1) Padé approximation:

$$e^{-t_d s} \approx \frac{1 - \frac{t_d}{2} s}{1 + \frac{t_d}{2} s}. \tag{3.30}$$

This is a form that serves many purposes. The term in the denominator introduces a negative pole in the LHP and thus introduces probable dynamic effects to the characteristic polynomial of a problem. The numerator introduces a positive zero in the RHP, which is needed to make a problem to become unstable. (This point will become clear when we cover Chap. 7.) Finally, the approximation is more accurate than a first-order Taylor series expansion.[4]

There are higher-order approximations. For example, the second-order (2/2) Padé approximation is

$$e^{-t_d s} \approx \frac{t_d^2 s^2 - 6t_d s + 12}{t_d^2 s^2 + 6t_d s + 12}. \tag{3.31}$$

Again, this form introduces poles in the LHP and at least one zero is in the RHP. At this point, the important task is to observe the properties of the Padé approximation in numerical simulations.

Example 3.2: Use the first-order Padé approximation to plot the unit-step response of the first order with a dead-time function:

$$\frac{Y}{X} = \frac{e^{-3s}}{10s + 1}.$$

---

[4] We will skip the algebraic details. The simple idea is that we can do long division of a function of the form in approximation (3.30) and match the terms to a Taylor expansion of the exponential function. If we do, we'll find that the (1/1) Padé approximation is equivalent to a third-order Taylor series.

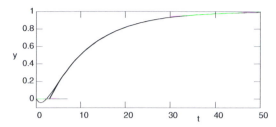

Figure E3.2.

Making use of approximation (3.30), we can construct a plot with the approximation

$$\frac{Y}{X} = \frac{(-1.5s + 1)}{(10s + 1)(1.5s + 1)}.$$

The result is the thin curve in Fig. E3.2. Note how it dips below zero near $t = 0$. This behavior has to do with the first-order Padé approximation, and we can improve the result with a second-order Padé approximation. We will try that in the Review Problems.

Here the observation is that when compared with the original transfer function (the thick solid curve), the approximation is acceptable at larger times. How did we generate the solid curve? We computed the result for the first-order function and then shifted the curve down three time units ($t_d = 3$). The MATLAB statements are

```
td=3;
P1=tf([-td/2 1],[td/2 1]);  % First-order Padé approximation

t=0:0.5:50;
taup=10;
G1=tf(1,[taup 1]);
y1=step(G1*P1,t);  % y1 is first order with Padé approximation of
                   % dead time

y2=step(G1,t);
t2=t+td;           % Shift the time axis for the actual time-delay
                   % function
plot(t,y1,'--',t2,y2);
```

We now move onto a few so-called higher-order or complex processes. We should remind ourselves that all linearized higher-order systems can be broken down into simple first- and second-order units. Other so-called complex processes, like two interacting tanks, are just a math problem in coupled differential equations; these problems are still linear. The following sections serve to underscore these points.

3.4. Higher-Order Processes and Approximations

Staged processes in chemical engineering or compartmental models in bioengineering give rise to higher-order models. The higher-order models are due to a cascade of first-order elements. Numerical calculation is used to illustrate that the resulting response becomes more sluggish, thus confirming our analysis in Example 2.9. We shall also see how the response may be approximated by a lower-order function with dead time. An example of two interacting vessels is given last.

3.4.1. Simple Tanks-in-Series

Consider a series of well-mixed vessels (or compartments) in which the volumetric flow rate and the respective volumes are constant (Fig. 3.3). If we write the mass balances of the first two vessels as in Subection 2.8.1, we have[5]

$$\tau_1 \frac{dc_1}{dt} = c_0 - c_1, \tag{3.32}$$

$$\tau_2 \frac{dc_2}{dt} = c_1 - c_2, \tag{3.33}$$

where $\tau_1 = V_1/q_0$ and $\tau_2 = V_2/q_0$ are the space times of each vessel. Again following Subsection 2.8.1, we find that the Laplace transforms of the mass balance in deviation variables would be

$$\frac{C_1}{C_0} = \frac{1}{\tau_1 s + 1}, \quad \frac{C_2}{C_1} = \frac{1}{\tau_2 s + 1}. \tag{3.34}$$

The effect of changes in $c_0(t)$ on the effluent of the second vessel is evaluated as

$$\frac{C_2}{C_0} = \frac{C_2}{C_1} \frac{C_1}{C_0} = \frac{1}{(\tau_2 s + 1)} \frac{1}{(\tau_1 s + 1)}. \tag{3.35}$$

Obviously, we can generalize to a series of n tanks as in

$$\frac{C_n}{C_0} = \frac{1}{(\tau_1 s + 1) \cdots (\tau_{n-1} s + 1)(\tau_n s + 1)}. \tag{3.36}$$

In this example, the steady-state gain is unity, which is intuitively obvious. If we change the color of the inlet with a food dye, all the mixed tanks will have the same color eventually.

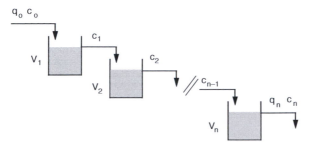

Figure 3.3. Illustration of compartments or tanks-in-series.

[5] Many texts illustrate with a model on the change of inlet flow rate. In such a case, we usually need to assume that the outlet flow rate of each vessel is proportional to the liquid level or the hydrostatic head. The steady-state gains will not be unity.

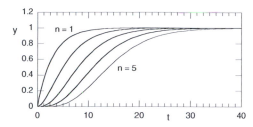

Figure E3.3.

In addition, the more tanks we have in a series, the longer we have to wait until the nth tank "sees" the changes that we have made in the first one. The more tanks in the series, the more sluggish the response of the overall process. Processes that are products of first-order functions are also called **multicapacity** processes.

Finally, if all the tanks have the same space time, $\tau_1 = \tau_2 = \cdots = \tau$, Eq. (3.36) becomes

$$\frac{C_n}{C_0} = \frac{1}{(\tau s + 1)^n}. \tag{3.37}$$

This particular case is not common in reality, but is a useful textbook illustration.

Example 3.3: Making use of Eq. (3.37), show how the unit-step response $C_n(t)$ becomes more sluggish as n increases from 1 to 5.

The exercise is almost trivial with MATLAB. To generate Fig. E3.3, the statements are

```
tau=3;                  % Just an arbitrary time constant
G=tf(1,[tau 1]);
step(G);                % First-order function unit-step response
hold
step(G*G);              % Second-order response
step(G*G*G);            % Third-order response
step(G*G*G*G);          % Fourth-order response
step(G*G*G*G*G);        % Fifth-order response
```

It is clear that, as n increases, the response, as commented in Example 2.9, becomes slower. If we ignore the "data" at small times, it appears that the curves might be approximated with first-order with dead-time functions. We shall do this exercise in the Review Problems.

3.4.2. Approximation with Lower-Order Functions with Dead Time

Following the lead in Example 3.3, we now make use of the result in Example 2.6 and the comments about dominant poles in Section 2.7 to see how we may approximate a transfer function.

Let's say we have a high-order transfer function that has been factored into partial fractions. If there is a large enough difference in the time constants of individual terms, we may try to throw away the small time-scale terms and retain the ones with dominant poles (large time constants). This is our reduced-order model approximation. From Fig. E3.3, we also need to add a time delay in this approximation. The extreme of this idea is to use a **first-order**

with dead-time function. It obviously cannot do an adequate job in many circumstances. Nevertheless, this simple approximation is all we use when we learn to design controllers with empirical tuning relations.

A second-order function with dead time generally provides a better estimate, and this is how we may make a quick approximation. Suppose we have an nth-order process that is broken down into n first-order processes in series with time constants $\tau_1, \tau_2, \ldots, \tau_n$. If we can identify, say, two dominant time constants (or poles) τ_1 and τ_2, we can *approximate* the process as

$$G(s) \approx \frac{Ke^{-t_d s}}{(\tau_1 s + 1)(\tau_2 s + 1)},$$ (3.38)

where

$$t_d \approx \sum_{i \neq 1,2}^{n} \tau_i .$$

The summation to estimate the dead time is over all the other time constants ($i = 3, 4$, etc.). This idea can be extended to the approximation of a first-order with dead-time function.

Example 3.4: Find the simplest lower-order approximation of the following transfer function

$$\frac{Y}{X} = \frac{3}{(0.1s + 1)(0.5s + 1)(s + 1)(3s + 1)}.$$

In this example, the dominant pole is at $-1/3$, corresponding to the largest time constant at 3 (time unit). Accordingly, we may approximate the full-order function as

$$\frac{Y}{X} = \frac{3e^{-1.6s}}{(3s + 1)},$$

where 1.6 is the sum of dead times 0.1, 0.5, and 1. With X representing a unit-step input, the response of the full-order function (solid curve) and that of the first-order with dead-time approximation (dotted curve) are shown in Fig. E3.4. The plotting of the dead-time function is further approximated by the Padé approximation. Even so, the approximation is reasonable when time is large enough. The pole at $-1/3$ can indeed be considered as dominant.

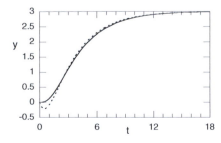

Figure E3.4.

The MATLAB statements are

```
p2=conv([1 1],[3 1]);
p4=conv(conv(p2,[0.5 1]) , [0.1 1]);
G4=tf(3,p4);     % The original full-order function
t=0:0.2:18;
y4=step(G4,t);   % Unit-step response

td=1+0.1+0.5;    % Approximate dead time
P1=tf([-td/2 1],[td/2 1]);
G1=tf(3,[3 1]);
y1=step(P1*G1,t);
plot(t,y1,t,y4)
```

If we follow approximation (3.38), we should approximate the original function with a second-order function with time constants 1 and 3, and dead time 0.6. We will find it to be a much better fit; we will do this in the Review Problems.

3.4.3. Interacting Tanks-in-Series

To complete the discussion, the balance equations are included for the case in which two differential equations may be coupled. The common example is two tanks connected such that there is only a valve between them (Fig. 3.4). Thus the flow between the two tanks depends on the difference in the hydrostatic heads. With constant density, we can write the mass balance of the first tank as

$$A_1 \frac{dh_1}{dt} = q_0 - \left(\frac{h_1 - h_2}{R_1} \right). \tag{3.39}$$

Similarly, for the second vessel we have

$$A_2 \frac{dh_2}{dt} = \left(\frac{h_1 - h_2}{R_1} \right) - \frac{h_2}{R_2}. \tag{3.40}$$

Here we model the flow through the valves with resistances R_1 and R_2, both of which are constants. We rearrange the balance equations a bit, and because both equations are linear, we can quickly rewrite them in deviation variables (without the primes) as

$$\tau_1 \frac{dh_1}{dt} = -h_1 + h_2 + R_1 q_0, \quad h_1(0) = 0, \tag{3.41}$$

$$\tau_2 \frac{dh_2}{dt} = \frac{R_2}{R_1} h_1 - \left(1 + \frac{R_2}{R_1} \right) h_2, \quad h_2(0) = 0, \tag{3.42}$$

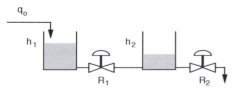

Figure 3.4. Illustration of two tanks interacting in their liquid levels.

where we have defined $\tau_1 = A_1 R_1$ and $\tau_2 = A_2 R_2$. The Laplace transforms of these equations are

$$(\tau_1 s + 1)H_1 - H_2 = R_1 Q_0, \tag{3.43}$$

$$\frac{-R_2}{R_1}H_1 + \left(\tau_2 s + 1 + \frac{R_2}{R_1}\right) H_2 = 0. \tag{3.44}$$

We have written the equations in a form that lets us apply Cramer's rule. The result is

$$H_1 = \frac{R_1 \tau_2 s + (R_1 + R_2)}{p(s)} Q_0, \qquad H_2 = \frac{R_2}{p(s)} Q_0, \tag{3.45}$$

where the characteristic polynomial is

$$p(s) = (\tau_1 s + 1)(\tau_2 s + 1 + R_2/R_1) - R_2/R_1. \tag{3.46}$$

We do not need to carry the algebra further. The points to be made are clear. First, even the first vessel has a second-order transfer function; it arises from the interaction with the second tank. Second, if we expand Eq. (3.46), we should see that the interaction introduces an extra term in the characteristic polynomial, but the poles should remain real and negative.[6] That is, the tank responses remain overdamped. Finally, we may be afraid (!) that the algebra might become hopelessly tangled with more complex models. Indeed, we would prefer to use state-space representation based on Eqs. (3.41) and (3.42). After Chaps. 4 and 9, you can try this problem in Homework Problem II (39).

3.5. Effect of Zeros in Time Response

The inherent dynamics is governed by the poles, but the zeros can impart finer "fingerprint" features by modifying the coefficients of each term in the time-domain solution. That was the point that was attempted to be made with the examples in Section 2.5. Two common illustrations on the effects of zeros are the lead–lag element and the sum of two functions in parallel.

3.5.1. Lead–Lag Element

The so-called lead–lag element is a semiproper function with a first-order lead divided by a first-order lag:

$$\frac{Y}{X} = \frac{\tau_z s + 1}{\tau s + 1}, \tag{3.47}$$

where τ_z and τ are two time constants. We have to wait until the chapters on controllers to see that this function is the basis of a derivative controller and not till the frequency

[6] To see this, you need to go one more step to get

$$p(s) = \tau_1 \tau_2 s^2 + (\tau_1 + \tau_2 + \tau_1 R_2/R_1)s + 1$$

and compare the roots of this polynomial with the case with no interaction:

$$p(s) = (\tau_1 + 1)(\tau_2 + 1).$$

Note how we have an extra term when the tanks interact.

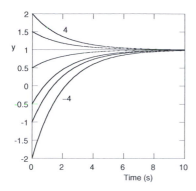

Figure 3.5. Time response of a lead–lag element with $\tau = 2\,$s. The curves from top to bottom are plotted with $\tau_z = $ 4, 3, 2, 1, -1, -2, and $-4\,$s.

response chapter to appreciate the terms lead and lag. For now, we take a quick look at its time response.

For the case with a unit-step input such that $X = 1/s$, we have, after partial-fraction expansion,

$$Y = \frac{1}{s} + \frac{\tau_z - \tau}{\tau s + 1}, \tag{3.48}$$

and the inverse transform by means of a look-up table yields

$$y(t) = 1 - \left(1 - \frac{\tau_z}{\tau}\right) e^{-t/\tau}. \tag{3.49}$$

There are several things that we want to take note of. First, the exponential function is dependent on only τ, or in other words, the pole at $-1/\tau$. Second, with Eq. (3.49), the actual time response depends on whether $\tau < \tau_z$, $\tau > \tau_z$, or $\tau_z < 0$ (Fig. 3.5). Third, when $\tau = \tau_z$, the time response is just a horizontal line at $y = 1$, corresponding to the input $x = u(t)$. This is also obvious from Eq. (3.47), which becomes just $Y = X$. When a zero equals a pole, we have what is called a **pole-zero cancellation**. Finally, the value of y is nonzero at time zero. We may wonder how that could be the case when we use differential equation models that have zero initial conditions. The answer has to do with the need for the response to match the rate of change term in the input. We will get a better picture in Chap. 4 when we cover state-space models.

3.5.2. Transfer Functions in Parallel

There are circumstances in which a complex process may involve two competing (i.e., opposing) dynamic effects that have different time constants. One example is the increase in inlet temperature to a tubular catalytic reactor with exothermic kinetics. The initial effect is that the exit temperature will momentarily decrease as increased conversion near the entrance region depletes reactants at the distal, exit end. Given time, however, higher reaction rates lead to higher exit temperatures.

To model this highly complex and nonlinear dynamics properly, we need the heat and the mass balances. In classical control, however, we would replace them with a linearized model that is the sum of two functions in parallel:

$$\frac{Y}{X} = \frac{K_1}{\tau_1 s + 1} + \frac{K_2}{\tau_2 s + 1}. \tag{3.50}$$

We can combine the two terms to give the second-order function

$$\frac{Y}{X} = \frac{K(\tau_z s + 1)}{(\tau_1 s + 1)(\tau_2 s + 1)}, \tag{3.51}$$

where

$$K = K_1 + K_2, \qquad \tau_z = \frac{K_1 \tau_2 + K_2 \tau_1}{K_1 + K_2}.$$

For circumstances in which the two functions represent opposing effects, one of them has a negative steady-state gain. In the following illustration, we choose to have $K_2 < 0$.

From Eq. (3.51), the time response $y(t)$ should be strictly overdamped. However, this is not necessarily the case if the zero is positive (or $\tau_z < 0$). We can show with algebra how various ranges of K_i and τ_i may lead to different zeros $(-1/\tau_z)$ and time responses. However, we will not do that. (We will use MATLAB to take a closer look in the Review Problems, though.) The key, once again, is to appreciate the principle of superposition with linear models. Thus we should get a rough idea of the time response simply based on the form in Eq. (3.50).

The numerical calculation is illustrated in Fig. 3.6. The input is a unit step, $X = 1/s$, and the two steady-state gains are $K_1 = 3$ and $K_2 = -1$ such that $|K_1| > |K_2|$. We consider the three cases in which τ_1 is roughly the same as τ_2 [Fig. 3.6(a)], $\tau_1 \ll \tau_2$ [Fig. 3.6(b)], and (c) $\tau_1 \gg \tau_2$ [Fig. 3.6(c)]. We should see that the overall response is overdamped in the case of Fig. 3.6(a), but in the case of Fig. 3.6(b) we can have an overshoot and in the case of Fig. 3.6(c) an initial inverse response. Note that all three cases have the same overall steady gain of $K = 2$.[7]

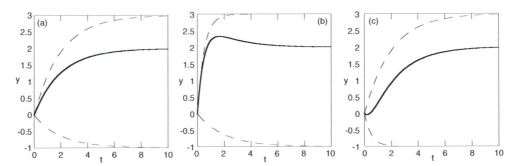

Figure 3.6. Time-response calculations with different time constants. In all cases, $K_1 = 3$, $K_2 = -1$, and the individual terms in Eq. (3.50) are indicated by the dashed curves. Their superimposed responses y are the solid curves. (a) $\tau_1 = \tau_2 = 2$; (b) $\tau_1 = 0.5$, $\tau_2 = 2$; (c) $\tau_1 = 2$, $\tau_2 = 0.5$.

[7] When you repeat this exercise with MATLAB in the Review Problems, check that τ_z is negative in the case of Fig. 3.6(c). More commonly, we say that this is the case with a positive zero. After we have learned frequency response, we will see that this is an example of what is referred to as a nonminimum phase.

Review Problems

(1) With respect to the step response of the first-order model in Eq. (3.9), make a table for $y(t)/MK_p$, with $t/\tau_p = 1, 2, 3, 4$, and 5. It is helpful to remember the results for $t/\tau_p = 1, 3$, and 5.

(2) It is important to understand that the time constant τ_p of a process, say, a stirred-tank, is not the same as the space time τ. Review this point with the stirred-tank heater example in Chap. 2. Further, derive the time constant of a **continuous-flow stirred-tank reactor** (CSTR) with a first-order chemical reaction.

(3) Write the time-response solutions to the integrating process in Eq. (3.14) when the input is (1) a unit step and (2) an impulse. How are they different from the solutions to a self-regulating process?

(4) Derive the time-constant relationships given in footnote 2.

(5) With respect to the overdamped solution of second-order equation (3.21), derive the step response $y(t)$ in terms of the more familiar $\exp(-t/\tau_1)$ and $\exp(-t/\tau_2)$. This is much easier than Eq. (3.21) and more useful too!

(6) Show that when $\zeta = 0$ (natural period of oscillation, no damping), the process (or system) oscillates with a constant amplitude at the natural frequency ω_n. (The poles are at $\pm\omega_n$ and the period is $2\pi\tau$.)

(7) Use MATLAB to make plots of OS and DR as functions of the damping ratio.

(8) What is the expected time response when the *real part* of the pole is zero in a second-order function? The pole can be just zero or have purely imaginary parts.

(9) Plot the unit-step response by using just first- and second-order Padé approximations (3.30) and (3.31). Try also the step response of a first-order function with dead time as in Example 3.2. Note that, although the approximation to the exponential function itself is not that good, the approximation to the entire transfer function is not as bad as long as $t_d \ll \tau$. How do you plot the exact solution in MATLAB?

(10) Use MATLAB to observe the effect of higher-order multicapacity models as in Example 3.3. Try to fit the fifth-order case with a first-order with dead-time function.

(11) With respect to Example 3.4, try also a second-order with dead-time approximation.

(12) We do not have a rigorous criterion to determine when a pole is absolutely dominant. Plot the exponential decay with different time constants to get an idea when the terms associated with smaller time constants can be omitted.

(13) With MATLAB, try to do a unit-step response of a lead–lag element as in Eq. (3.49).

(14) Repeat the time-response simulation of inverse response in Section 3.5.2. Calculate the value of zero in each case.

Hints:

(1) $y(t)/MK_p$ at $t/\tau_p = 1, 2, 3, 4$, and 5 are 0.63, 0.86, 0.95, 0.98, and 0.99.

(2) The mass balance of a CSTR with a first-order chemical reaction is very similar to the problem in Subsection 2.8.1. We just need to add the chemical reaction term. The balance written for reactant A will appear as

$$V\frac{dC_A}{dt} = q(C_0 - C_A) - VkC_A,$$

where C_A is the molar concentration of A, V is the reactor volume, q is the volumetric flow rate, C_0 is the inlet concentration of A, and k is the first-order reaction rate constant. If we define space time $\tau = V/q$, the equation can be rewritten as

$$\tau \frac{dC_A}{dt} + (1 + k\tau)C_A = C_0.$$

This is a linear equation if k and τ are constants. Now if we follow the logic in Subsection 2.8.2, we should find that the time constant of a CSTR with a first-order reaction is $\tau/(1 + k\tau)$.

(3) Part of the answer is already in Example 3.1.

(5) This is really an algebraic exercise in partial fractions. The answer hides in Table 2.1.

(6) This is obvious from Eq. (3.17) or Eq. (3.19).

(7) Plot Eq. (3.29) with $0 < \zeta < 1$. You can write a small M-file to do the plotting too.

(8) See Review Problem (6).

(9) Follow Example 3.2. For the first order approximation, we can try, for example,

```
td=3;                    % Use an M-file to rerun with different
                         % values
P1=tf([-td/2 1],[td/2 1]);
step(P1);                % Note how the response starts from
                         % negative values
t=0:0.1:50;
taup=10;
G1=tf(1,[taup 1]);
y1=step(G1*P1,t);  % y1 is first order with Padé approx of
                         % dead time
y2=step(G1,t);     % y2 has no time delay
t2=t+td;
plot(t,y1, t2,y2,'-.');
% Note how the Padé approx has a dip at the beginning
```

(10) The MATLAB statements to do the unit-step response are already in Example 3.4. You may repeat the computation with a different time constant. The statements to attempt fitting the five tanks-in-series responses are

```
tau=3;
G=tf(1,[tau 1]);
[y5,t]=step(G*G*G*G*G);    % The fifth-order calculation
G1=tf(1,[12 1]);
y1=step(G1,t);             % Using a time shift to do the
t1=t+3;                    % first-order with dead-time plot
plot(t,y5, t1,y1)
```

The choice of the time constant and dead time is meant as an illustration. The fit will not be particularly good in this example because there is no one single dominant pole in the fifth-order function with a pole repeated five times. A first-order with dead-time function will never provide a perfect fit.

(11) Both first- and second-order approximation statements are here:

```
q=3;
p2=conv([1 1],[3 1]);  % Second-order reduced model
```

```
p4=conv(conv(p2,[0.5 1]),[0.1 1]);
roots(p4)                % check
G2=tf(q,p2);
G4=tf(q,p4);
step(G4)
hold
td=0.1+0.5;
P1=tf([-td/2 1],[td/2 1]);
step(P1*G2);             % Not bad!
td=1+0.1+0.5;
G1=tf(q,[3 1]);          % First-order approximation
step(P1*G1);             % is not that good in this case
hold off
```

(12) Below are the MATLAB statements that we may use for a visual comparison of exponential decay with different time constants. In rough engineering calculations, a pole may already exhibit acceptable dominant behavior if other time constants are 1/3 or less:

```
tau=1;
t=0:0.1:5;
f=exp(-t/tau);
plot(t,f)
hold
          % Now add curves with smaller time constants
frac=0.3;
f=exp(-t/(frac*tau));
plot(t,f)
frac=0.2;
f=exp(-t/(frac*tau));
plot(t,f)
frac=0.1;
f=exp(-t/(frac*tau));
plot(t,f)
```

(13) Try to vary the zero as in

```
tz=3;             % Try to vary tz, zero is -1/tz
G=tf([tz 1],[5 1]);
step(G);
```

(14) Try to vary the values of τ_1 and τ_2. To display the value of zero is trivial:

```
k1=3; k2=-1;
tau1=2; tau2=0.5;
k=k1+k2;
tz=(k1*tau2+k2*tau1)/k;
G=tf(k*[tz 1], conv([tau1 1],[tau2 1]));
step(G);
```

4

State-Space Representation

The limitation of transfer function representation becomes obvious as we tackle more complex problems. For complex systems with multiple inputs and outputs, transfer function matrices can become very clumsy. In the so-called modern control, the method of choice is state-space or state variables in the time domain – essentially a matrix representation of the model equations. The formulation allows us to make use of theories in linear algebra and differential equations. It is always a mistake to tackle modern control without a firm background in these mathematical topics. For this reason, we will not overreach by doing both the mathematical background and the control together. Without a formal mathematical framework, the explanation is made by means of examples as much as possible. The actual state-space control has to be delayed until after we tackle classical transfer function feedback systems.

What Are We Up to?

- Learning how to write the state-space representation of a model.
- Understanding the how a state-space representation is related to the transfer function representation.

4.1. State-Space Models

Just as we are feeling comfortable with transfer functions, we now switch gears totally. Nevertheless, we are still working with *linearized differential equation* models in this chapter. Whether we have a high-order differential equation or multiple equations, we can always rearrange them into a set of first-order differential equations. Bold statements indeed! We will see that when we go over the examples.

With state-space models, a set of differential equations is put in standard matrix form,

$$\dot{x} = Ax + Bu, \tag{4.1}$$

$$y = Cx + Du, \tag{4.2}$$

where **x** is the **state-variable** vector, **u** is the input, and **y** is the output. The time derivative is denoted by the overdot. In addition, **A** is the process (plant) matrix, **B** is the input matrix, **C** is the output matrix, and **D** is the feedthrough matrix. Very few processes (and systems) have an input that has a direct influence on the output. Hence **D** is usually zero.

When we discuss single-input single-output (SISO) models, scalar variables should be used for the input, output, and feedthrough: u, y, and D. For convenience, we keep the notation for **B** and **C**, but keep in mind that, in this case, **B** is a column vector and **C** is a row vector. If **x** is of the order of n, then **A** is $n \times n$, **B** is $n \times 1$, and **C** is $1 \times n$.[1]

The idea behind the use of Eqs. (4.1) and (4.2) is that we can make use of linear system theories, and we can analyze complex systems much more effectively. There is no unique way to define the state variables. What will be shown is just one of many possibilities.

Example 4.1: Derive the state-space representation of a *second-order differential equation* in a form similar to that of Eq. (3.16):

$$\frac{d^2 y}{dt^2} + 2\zeta \omega_n \frac{dy}{dt} + \omega_n^2 y = K u(t). \tag{E4.1}$$

We can do blindfolded the transfer function of this equation with zero initial conditions:

$$G_p(s) = \frac{Y(s)}{U(s)} = \frac{K}{s^2 + 2\zeta \omega_n s + \omega_n^2}. \tag{E4.2}$$

Now let's do something different. First, we rewrite the differential equation as

$$\frac{d^2 y}{dt^2} = -2\zeta \omega_n \frac{dy}{dt} - \omega_n^2 y + K u(t),$$

and define the state variables,[2]

$$x_1 = y, \qquad x_2 = \frac{dx_1}{dt}, \tag{E4.3}$$

which allow us to redefine the second-order equation as a set of two coupled first-order equations. The first differential equation is the definition of the state variable x_2 in Eq. (E4.3); the second is based on the differential equation

$$\frac{dx_2}{dt} = -2\zeta \omega_n x_2 - \omega_n^2 x_1 + K u(t). \tag{E4.4}$$

We now put the result in a matrix equation:

$$\begin{bmatrix} \dot{x}_1 \\ \dot{x}_2 \end{bmatrix} = \begin{bmatrix} 0 & 1 \\ -\omega_n^2 & -2\zeta \omega_n \end{bmatrix} \begin{bmatrix} x_1 \\ x_2 \end{bmatrix} + \begin{bmatrix} 0 \\ K \end{bmatrix} u(t). \tag{E4.5}$$

We further write

$$y = [1 \quad 0] \begin{bmatrix} x_1 \\ x_2 \end{bmatrix} \tag{E4.6}$$

---

[1] If you are working with only SISO problems, it would be more appropriate to replace the notation **B** with **b** and **C** with **c**^{T}, and write d for **D**.

[2] This exercise is identical to how higher-order equations are handled in numerical analysis and would come as no surprise if you have taken a course on numerical methods.

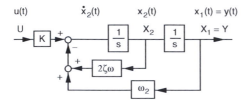

Figure E4.2.

as a statement that x_1 is our output variable. Compare the results with Eqs. (4.1) and (4.2), and we see that, in this case,

$$\mathbf{A} = \begin{bmatrix} 0 & 1 \\ -\omega_n^2 & -2\zeta\omega_n \end{bmatrix}, \quad \mathbf{B} = \begin{bmatrix} 0 \\ K \end{bmatrix}, \quad \mathbf{C} = [1 \quad 0], \quad D = 0.$$

To find the eigenvalues of $\mathbf{A}$, we solve its characteristic equation:

$$|\lambda\mathbf{I} - \mathbf{A}| = \lambda(\lambda + 2\zeta\omega_n) + \omega_n^2 = 0. \tag{E4.7}$$

We can use the MATLAB function `tf2ss()` to convert the transfer function in Eq. (E4.2) to state-space form:

```
z=0.5; wn=1.5; % Pick two sample numbers for ζ and ωn
p=[1 2*z*wn wn*wn];
[a,b,c,d]=tf2ss(wn*wn,p)
```

However, you will find that the MATLAB result is not identical to Eq. (E4.5). It has to do with the fact that there is no unique representation of a state-space model. To avoid unnecessary confusion, the differences with MATLAB are explained in MATLAB Session 4.

One important observation that we should make immediately: The characteristic polynomial of the matrix $\mathbf{A}$ [Eq. (E4.7)] is *identical* to that of the transfer function [Eq. (E4.2)]. Needless to say, the eigenvalues of $\mathbf{A}$ are the poles of the transfer function. It is a reassuring thought that different mathematical techniques provide the same information. It should come as no surprise if we remember our linear algebra.

Example 4.2: Draw the block diagram of the state-space representation of the second-order differential equation in Example 4.1.
The result is in Fig. E4.2. It is quite easy to understand if we note that the transfer function of an integrator is $1/s$. Thus the second-order derivative is located before the two integrations are made. The information at the summing point also adds up to the terms of the second-order differential equation. The resulting transfer function is identical to Eq. (E4.2). The reduction to a closed-loop transfer function was done in Example 2.16.

Example 4.3: Let's try another model with a slightly more complex input. Derive the state-space representation of the differential equation

$$\frac{d^2 y}{dt^2} + 0.4\frac{dy}{dt} + y = \frac{du}{dt} + 3u, \quad y(0) = dy/dt(0) = 0, \quad u(0) = 0,$$

which has the transfer function

$$\frac{Y}{U} = \frac{s + 3}{s^2 + 0.4s + 1}.$$

The method that we will follow is more for illustration than for its generality. Let's introduce a variable X_1 between Y and U:

$$\frac{Y}{U} = \frac{X_1}{U}\frac{Y}{X_1} = \left(\frac{1}{s^2 + 0.4s + 1}\right)(s + 3).$$

The first part, X_1/U, is a simple problem itself:

$$\frac{X_1}{U} = \left(\frac{1}{s^2 + 0.4s + 1}\right)$$

is the Laplace transform of

$$\frac{d^2 x_1}{dt^2} + 0.4\frac{dx_1}{dt} + x_1 = u.$$

With Example 4.1 as the hint, we define the state variables $x_1 = x_1$ (i.e., the same), and $x_2 = dx_1/dt$. Using steps similar to those of Example 4.1, we find that the result, as equivalent to Eq. (4.1), is

$$\begin{bmatrix} \dot{x}_1 \\ \dot{x}_2 \end{bmatrix} = \begin{bmatrix} 0 & 1 \\ -1 & -0.4 \end{bmatrix}\begin{bmatrix} x_1 \\ x_2 \end{bmatrix} + \begin{bmatrix} 0 \\ 1 \end{bmatrix} u. \tag{E4.8}$$

As for the second part $Y/X_1 = (s + 3)$, it is the Laplace transform of $y = (dx_1/dt) + 3x_1$. We can use the state variables defined above to rewrite as

$$y = x_2 + 3x_1, \tag{E4.9}$$

or, in matrix form,

$$y = [3 \quad 1]\begin{bmatrix} x_1 \\ x_2 \end{bmatrix},$$

which is the form of Eq. (4.2).

With MATLAB, the statements for this example are

```
q=[1 3];
p=[1 0.4 1];
roots(p)
[a,b,c,d]=tf2ss(q,p)
eig(a)
```

Comments at the end of Example 4.1 also apply here. The result should be correct, and we should find that both the roots of the characteristic polynomial p and the eigenvalues of the matrix a are $-0.2 \pm 0.98j$. We can also check by going backward:

```
[q2,p2]=ss2tf(a,b,c,d,1)
```

and the original transfer function is recovered in q2 and p2.

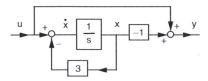

Figure E4.4.

Example 4.4: Derive the state-space representation of the lead–lag transfer function:

$$\frac{Y}{U} = \frac{s+2}{s+3}.$$

We follow the hint in Example 4.3 and write the transfer function as

$$\frac{Y}{U} = \frac{X}{U}\frac{Y}{X} = \frac{1}{s+3}s + 2.$$

From $X/U = 1/(s+3)$, we have $sX = -3X + U$, or in time domain,

$$\frac{dx}{dt} = -3x + u \tag{E4.10}$$

and from $Y/X = s + 2$, we have $Y = sX + 2X$ and substitution for sX leads to

$$Y = (-3X + U) + 2X = -X + U.$$

The corresponding time-domain equation is

$$y = -x + u. \tag{E4.11}$$

Thus all the coefficients in Eqs. (4.1) and (4.2) are scalar, with $A = -3$, $B = 1$, $C = -1$, and $D = 1$. Furthermore, (E4.10) and (E4.11) can be represented by the block diagram in Fig. E4.4.

We may note that the coefficient D is not zero, meaning that, with a lead–lag element, an input can have an instantaneous effect on the output. Thus, although the state variable x has zero initial condition, it is not necessarily so with the output y. This analysis explains the mystery with the inverse transform of this transfer function in Eq. (3.49).

The MATLAB statement for this example is

```
[a,b,c,d]=tf2ss([1 2], [1 3])
```

The next two examples illustrate how state-space models can handle a multiple-input multiple output (MIMO) problem. A simple example shows how to translate information in a block diagram into a state-space model. Some texts rely on signal-flow graphs, but we do not need them with simple systems. Moreover, we can handle complex problems easily with MATLAB. Go over MATLAB Session 4 before reading Example 4.7A.

Example 4.5: Derive the state space representation of two continuous-flow stirred-tank reactors in series **(CSTR-in-series)**. Chemical reaction is first order in both reactors. The reactor volumes are fixed, but the volumetric flow rate and inlet concentration are functions of time.

This example is used to illustrate how state-space representation can handle complex models. First, we use the solution to Review Problem 2 in Chap. 3 and write the mass balances of reactant A in chemical reactors 1 and 2:

$$V_1 \frac{dc_1}{dt} = q(c_0 - c_1) - V_1 k_1 c_1, \tag{E4.12}$$

$$V_2 \frac{dc_2}{dt} = q(c_1 - c_2) - V_2 k_2 c_2. \tag{E4.13}$$

Because q and c_0 are input functions, the linearized equations in deviation variables and with zero initial conditions are (with all primes omitted in the notation)

$$V_1 \frac{dc_1}{dt} = q_s c_0 + (c_{0s} - c_{1s})q - (q_s + V_1 k_1)c_1 \tag{E4.14}$$

$$V_2 \frac{dc_2}{dt} = q_s c_1 + (c_{1s} - c_{2s})q - (q_s + V_2 k_2)c_2. \tag{E4.15}$$

The missing steps are similar to the steps that we did Example 2.11. Dividing the equations by the respective reactor volumes and defining space times $\tau_1 = V_1/q_s$ and $\tau_2 = V_2/q_s$, we obtain

$$\frac{dc_1}{dt} = \frac{1}{\tau_1} c_0 + \left(\frac{c_{0s} - c_{1s}}{V_1} \right) q - \left(\frac{1}{\tau_1} + k_1 \right) c_1, \tag{E4.16}$$

$$\frac{dc_2}{dt} = \frac{1}{\tau_2} c_1 + \left(\frac{c_{1s} - c_{2s}}{V_2} \right) q - \left(\frac{1}{\tau_2} + k_2 \right) c_2. \tag{E4.17}$$

Up to this point, the exercise is identical to what we learned in Chap. 2. In fact, we can now take the Laplace transform of these two equations to derive the transfer functions. In state-space models, however, we would put the two linearized equations in matrix form. Analogous to Eq. (4.1), we now have

$$\frac{d}{dt} \begin{bmatrix} c_1 \\ c_2 \end{bmatrix} = \begin{bmatrix} -\left(\dfrac{1}{\tau_1} + k_1 \right) & 0 \\ \dfrac{1}{\tau_2} & -\left(\dfrac{1}{\tau_2} + k_2 \right) \end{bmatrix} \begin{bmatrix} c_1 \\ c_2 \end{bmatrix} + \begin{bmatrix} \dfrac{1}{\tau_1} & \dfrac{c_{0s} - c_{1s}}{V_1} \\ 0 & \dfrac{c_{1s} - c_{2s}}{V_2} \end{bmatrix} \begin{bmatrix} c_0 \\ q \end{bmatrix}. \tag{E4.18}$$

The output $\mathbf{y}$ in Eq. (4.2) can be defined as

$$\begin{bmatrix} y_1 \\ y_2 \end{bmatrix} = \begin{bmatrix} 1 & 0 \\ 0 & 1 \end{bmatrix} \begin{bmatrix} c_1 \\ c_2 \end{bmatrix} = \begin{bmatrix} c_1 \\ c_2 \end{bmatrix} \tag{E4.19}$$

if we are to use two outputs. In SISO problems, we likely would measure and control only c_2, and hence we would define instead

$$y = [0 \quad 1] \begin{bmatrix} c_1 \\ c_2 \end{bmatrix} \tag{E4.20}$$

with c_2 as the only output variable.

Example 4.6: Derive the transfer function Y/U and the corresponding state-space model of the block diagram in Fig. E4.6.

69

State-Space Representation

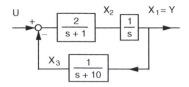

Figure E4.6.

From block-diagram reduction in Chap. 2, we can easily spot that

$$\frac{Y}{U} = \frac{\frac{2}{s(s+1)}}{1 + \frac{2}{s(s+1)(s+10)}},$$

which is reduced to

$$\frac{Y}{U} = \frac{2(s+10)}{s^3 + 11s^2 + 10s + 2}. \tag{E4.21}$$

This transfer function has closed-loop poles at -0.29, -0.69, and -10.02. (Of course, we computed them by using MATLAB.)

To derive the state-space representation, one visual approach is to identify locations in the block diagram where we can assign state variables and write out the individual transfer functions. In this example, we have chosen to use (Fig. E4.6)

$$\frac{X_1}{X_2} = \frac{1}{s}, \quad \frac{X_2}{U - X_3} = \frac{2}{s+1}, \quad \frac{X_3}{X_1} = \frac{1}{s+10}, \quad Y = X_1;$$

the final equation is the output equation. We can now rearrange each of the three transfer functions from which we write their time-domain equivalent:

$$sX_1 = X_2, \qquad \frac{dx_1}{dt} = x_2; \tag{E4.22a}$$

$$sX_2 = -X_2 - 2X_3 + 2U, \qquad \frac{dx_2}{dt} = -x_2 - 2x_3 + 2u; \tag{E4.22b}$$

$$sX_3 = -10X_3 + X_1, \qquad \frac{dx_3}{dt} = x_1 - 10x_3. \tag{E4.22c}$$

The rest is trivial. We rewrite the differential equations in matrix form as

$$\frac{d}{dt} \begin{bmatrix} x_1 \\ x_2 \\ x_3 \end{bmatrix} = \begin{bmatrix} 0 & 1 & 0 \\ 0 & -1 & -2 \\ 1 & 0 & -10 \end{bmatrix} \begin{bmatrix} x_1 \\ x_2 \\ x_3 \end{bmatrix} + \begin{bmatrix} 0 \\ 2 \\ 0 \end{bmatrix} u, \tag{E4.23}$$

$$y = [1 \ 0 \ 0] \begin{bmatrix} x_1 \\ x_2 \\ x_3 \end{bmatrix}. \tag{E4.24}$$

We can check with MATLAB that the model matrix **A** has eigenvalues -0.29, -0.69, and -10.02. They are identical to the closed-loop poles. Given a block diagram, MATLAB can put the state-space model together for us easily. To do that, we need to learn some closed-loop MATLAB functions; we will leave this illustration to MATLAB Session 5.

An important reminder: Eq. (E4.23) has zero initial conditions $\mathbf{x}(0) = 0$. This is a direct consequence of deriving the state-space representation from transfer functions. Otherwise, Eq. (4.1) is not subjected to this restriction.

4.2. Relation of State-Space Models to Transfer Function Models

From Example 4.6, we may see why the primary mathematical tools in modern control are based on linear system theories and time-domain analysis. Part of the confusion in learning these more advanced techniques is that the umbilical cord to Laplace transform is not entirely severed, and we need to appreciate the link between the two approaches. On the bright side, if we can convert a state-space model to transfer function form, we can still make use of classical control techniques. A couple of examples in Chap. 9 will illustrate how classical and state-space techniques can work together.

We can take the Laplace transform of matrix equation (4.1) to give

$$s\mathbf{X}(s) = \mathbf{A}\mathbf{X}(s) + \mathbf{B}U(s), \tag{4.3}$$

where the capital $\mathbf{X}$ does not indicate that it is a matrix, but rather is used in keeping with our notation of Laplace variables. From Eq. (4.3), we can extract $\mathbf{X}$ explicitly as

$$\mathbf{X}(s) = (s\mathbf{I} - \mathbf{A})^{-1}\mathbf{B}U(s) = \mathbf{\Phi}(s)\mathbf{B}U(s), \tag{4.4}$$

where

$$\mathbf{\Phi}(s) = (s\mathbf{I} - \mathbf{A})^{-1} \tag{4.5}$$

is the resolvent matrix. More commonly, we refer to the **state-transition matrix** (also called the fundamental matrix), which is its inverse transform:

$$\mathbf{\Phi}(t) = \mathcal{L}^{-1}[(s\mathbf{I} - \mathbf{A})^{-1}]. \tag{4.6}$$

We use the same notation, $\mathbf{\Phi}$, for the time function and its Laplace transform, and we add the t or s dependence only when it is not clear in which domain the notation is used.

Setting $\mathbf{D} = 0$ and $\mathbf{X}(s)$ as derived in Eq. (4.4), we find that the output $Y(s) = \mathbf{C}\mathbf{X}(s)$ becomes

$$Y(s) = \mathbf{C}\mathbf{\Phi}(s)\mathbf{B}U(s). \tag{4.7}$$

In this case, in which U and Y are scalar quantities, $\mathbf{C}\mathbf{\Phi}\mathbf{B}$ must also be scalar.[3] In fact, if we make an association between Eq. (4.7) and what we have learned in Chap. 2, $\mathbf{C}\mathbf{\Phi}\mathbf{B}$ is our ubiquitous transfer function. We can rewrite Eq. (4.7) as

$$Y(s) = G_p(s)U(s), \tag{4.7a}$$

where

$$G_p(s) = \mathbf{C}\mathbf{\Phi}(s)\mathbf{B}. \tag{4.8}$$

Hence we can view the transfer function as how the Laplace transform of the state-transition matrix $\mathbf{\Phi}$ mediates the input $\mathbf{B}$ and the output $\mathbf{C}$ matrices. We may wonder how this output

---

[3] From Eq. (4.5), we see that $\mathbf{\Phi}$ is an $n \times n$ matrix. Because $\mathbf{B}$ is $n \times 1$ and $\mathbf{C}$ is $1 \times n$, $\mathbf{C}\mathbf{\Phi}\mathbf{B}$ must be 1×1.

equation is tied to the matrix $\mathbf{A}$. With linear algebra, we can rewrite the definition of $\boldsymbol{\Phi}$ in Eq. (4.5) as

$$\boldsymbol{\Phi}(s) = (s\mathbf{I} - \mathbf{A})^{-1} = \frac{\mathrm{adj}(s\mathbf{I} - \mathbf{A})}{\det(s\mathbf{I} - \mathbf{A})}. \tag{4.5a}$$

Substitution of this form in Eq. (4.8) provides a more informative view of the transfer function:

$$G_p(s) = \frac{\mathbf{C}[\mathrm{adj}(s\mathbf{I} - \mathbf{A})]\mathbf{B}}{\det(s\mathbf{I} - \mathbf{A})}. \tag{4.8a}$$

The characteristic polynomial clearly is

$$\det(s\mathbf{I} - \mathbf{A}) = 0. \tag{4.9}$$

This is the result that we have arrived at, albeit less formally, in Example 4.1. Again, the poles of G_p are identical to the eigenvalues of the model matrix $\mathbf{A}$.

Example 4.7: The results in this section are illustrated with a numerical version of Example 4.5. Consider again two CSTR-in-series, with $V_1 = 1$ m^3, $V_2 = 2$ m^3, $k_1 = 1$ min^{-1}, $k_2 = 2$ min^{-1}, and, initially at steady state, $\tau_1 = 0.25$ min, $\tau_2 = 0.5$ min, and inlet concentration $c_{0s} = 1$ kmol/m^3. Derive the transfer functions and the state-transition matrix in which both c_0 and q are input functions.

With the steady-state form of Eqs. (E4.12) and (E4.13), we can calculate

$$c_{1s} = \frac{c_{0s}}{1 + k_1\tau_1} = \frac{1}{1 + 0.25} = 0.8, \qquad c_{2s} = \frac{c_{1s}}{1 + k_2\tau_2} = \frac{0.8}{1 + 2(0.5)} = 0.4.$$

In addition, we find that $1/\tau_1 = 4$ min^{-1}, $1/\tau_2 = 2$ min^{-1}, $(1/\tau_1 + k_1) = 5$ min^{-1}, $(1/\tau_2 + k_2) = 4$ min^{-1}, $(c_{0s} - c_{1s})/V_1 = 0.2$ kmol/m^6, and $(c_{1s} - c_{2s})/V_2 = 0.2$ kmol/m^6. We substitute these numerical values into Eqs. (E4.16) and (E4.17) and take the Laplace transform of these equations to obtain (for more general algebraic result, we should take the transform first)

$$C_1(s) = \frac{4}{s+5}C_0(s) + \frac{0.2}{s+5}Q(s), \tag{E4.25}$$

$$C_2(s) = \frac{2}{s+4}C_1(s) + \frac{0.2}{s+4}Q(s).$$

Further substitution for $C_1(s)$ with Eq. (E4.25) into $C_2(s)$ gives

$$C_2(s) = \frac{8}{(s+4)(s+5)}C_0(s) + \frac{0.2(s+7)}{(s+4)(s+5)}Q(s) \tag{E4.26}$$

Equations (E4.25) and (E4.26) provide the transfer functions relating changes in flow rate Q and inlet concentration C_0 to changes in the two tank concentrations.

With the state-space model, the substitution of numerical values into Eq. (E4.18) leads to the dynamic equation

$$\frac{d}{dt}\begin{bmatrix} c_1 \\ c_2 \end{bmatrix} = \begin{bmatrix} -5 & 0 \\ 2 & -4 \end{bmatrix}\begin{bmatrix} c_1 \\ c_2 \end{bmatrix} + \begin{bmatrix} 4 & 0.2 \\ 0 & 0.2 \end{bmatrix}\begin{bmatrix} c_0 \\ q \end{bmatrix}. \tag{E4.27}$$

With the model matrix $\mathbf{A}$, we can derive

$$(s\mathbf{I} - \mathbf{A}) = \begin{bmatrix} s+5 & 0 \\ -2 & s+4 \end{bmatrix},$$

$$\Phi(s) = (s\mathbf{I} - \mathbf{A})^{-1} = \frac{1}{(s+5)(s+4)} \begin{bmatrix} s+4 & 0 \\ 2 & s+5 \end{bmatrix}. \tag{E4.28}$$

We consider Eq. (E4.19) in which both concentrations c_1 and c_2 are outputs of the model. The transfer function in Eq. (4.7) is now a matrix:

$$\mathbf{G}_p(s) = \mathbf{C}\Phi(s)\mathbf{B} = \frac{1}{(s+5)(s+4)} \begin{bmatrix} s+4 & 0 \\ 2 & s+5 \end{bmatrix} \begin{bmatrix} 4 & 0.2 \\ 0 & 0.2 \end{bmatrix}, \tag{E4.29}$$

where $\mathbf{C}$ is omitted as it is just the identity matrix [Eq. (E4.19)].[4] With input $\mathbf{u}(s) = [C_0(s)\ Q(s)]^T$, we can write output equation (4.6) as

$$\begin{bmatrix} C_1(s) \\ C_2(s) \end{bmatrix} = \mathbf{C}\Phi(s)\mathbf{B}\mathbf{u}(s) = \frac{1}{(s+5)(s+4)} \begin{bmatrix} 4(s+4) & 0.2(s+4) \\ 8 & 0.2(s+7) \end{bmatrix} \begin{bmatrix} C_0(s) \\ Q(s) \end{bmatrix}. \tag{E4.30}$$

This is how MATLAB returns the results. We can, of course, clean up the algebra to give

$$\begin{bmatrix} C_1(s) \\ C_2(s) \end{bmatrix} = \begin{bmatrix} \dfrac{4}{(s+5)} & \dfrac{0.2}{(s+5)} \\ \dfrac{8}{(s+5)(s+4)} & \dfrac{0.2(s+7)}{(s+5)(s+4)} \end{bmatrix} \begin{bmatrix} C_0(s) \\ Q(s) \end{bmatrix}, \tag{4.30a}$$

which is identical to what we have obtained earlier in Eqs. (E4.25) and (E4.26). The case of only one output, as in Eq. (E4.20), is easy, and that is covered in Example 4.7A.

To wrap things up, we can take the inverse transform of Eq. (E4.30a) to get the time-domain solutions:

$$\begin{bmatrix} c_1 \\ c_2 \end{bmatrix} = \begin{bmatrix} 4e^{-5t} & 0.2e^{-5t} \\ 8(e^{-4t} - e^{-5t}) & 0.2(3e^{-4t} - 2e^{-5t}) \end{bmatrix} \begin{bmatrix} c_0 \\ q \end{bmatrix}. \tag{E4.31}$$

Example 4.7A: Repeat Example 4.7 using MATLAB.

If you understand the equations in Example 4.7, you are ready to tackle the same problem with MATLAB.

```
t1=0.25; t2=0.5;        % Define the variables
k1=1; k2=2;
V1=1; V2=2;
cos=1;

% Calculate the steady-state values. MATLAB should return
c1s=cos/(1+k1*t1);      % 0.8
c2s=c1s/(1+k2*t2);      % 0.4
```

[4] Be careful with the notation. Uppercase italic C with subscripts 1 and 2 denotes concentration in the Laplace domain. The boldface uppercase $\mathbf{C}$ is the output matrix.

```
                                  % Coefficients of A and B using (E4.18)
a11=-(1/t1+k1);                   % -5
a12=0;
a21=1/t2;
a22=-(1/t2+k2);                   % -4
b11=1/t1;
b12=(cos-c1s)/V1;                 % 0.2
b21=0;
b22=(c1s-c2s)/V2;                 % 0.2


                                  % Finally build A and B in (E4.27)
a=[a11 a12; a21 a22];             % [-5 0; 2 4]
b=[b11 b12; b21 b22];             % [4 0.2; 0 0.2]
eig(a)                            % Check that they are -4, -5


c=[1 0; 0 1];                     % Define C such that both C₁ and C₂ are
                                  % outputs
d=[0 0; 0 0];
```

With all the coefficient matrices defined, we can do the conversion to transfer function. The function ss2tf() works with only one designated input variable. Thus for the *first* input variable C_0 we use

```
                                  % MATLAB returns, for input no. 1
[q1,p]=ss2tf(a,b,c,d,1)           % q1=[0 4 16; 0 0 8]
                                  % p =[1 9 20] = (s+4)(s+5)
```

The returned vector p is obviously the characteristic polynomial. The matrix q1 is really the first *column* of the transfer function matrix in Eq. (E4.30), denoting the two terms describing the effects of changes in C_0 on C_1 and C_2. Similarly, the second column of the transfer function matrix in Eq. (E4.30) is associated with changes in the *second* input Q and can be obtained with

```
[q2,p]=ss2tf(a,b,c,d,2)           % q2=[0 .2 .8; 0 .2 1.4]
                                  % The first term is 0.2(s+4) because
                                  % MATLAB retains p=(s+4)(s+5)
```

If C_2 is the only output variable, we define **C** according to output equation (E4.20). Matrices **A** and **B** remain the same. The respective effects of changes of C_0 and Q on C_2 can be obtained with

```
c=[0 1]; d=[0 0];                 % C2 is the only output
[q21,p]=ss2tf(a,b,c,d,1)          % Co as input; q21=[0 0 8]
[q22,p]=ss2tf(a,b,c,d,2)          % Q as input; q22=[0 .2 1.4]
```

We should find that the result is the same as the second *row* of Eq. (E4.30), denoting the two terms describing the effects of changes in C_0 and Q on C_2.

Similarly, if C_1 is the only output variable, we use instead

```
c=[1 0]; d=[0 0];          % C1 is the only output
[q11,p]=ss2tf(a,b,c,d,1)   % q11=[0 4 16]
[q12,p]=ss2tf(a,b,c,d,2)   % q12=[0 .2 .8]
```

and the result is the first row of (E4.30).

Example 4.8: Develop a fermentor model that consists of two mass balances, one for the cell mass (or yeast), C_1, and the other for glucose (or substrate), C_2. We have to forget about the alcohol for now. The cell mass balance [Eq. (E4.32)] has two terms on the RHS. The first one describes cell growth by use of the specific growth rate $\mu = \mu(C_2)$. The second term accounts for the loss of cells that is due to outlet flow Q, which in turn is buried inside the notation D, the dilution rate:

$$\frac{dC_1}{dt} = \mu C_1 - DC_1. \tag{E4.32}$$

The specific growth rate and dilution rate are defined as

$$\mu = \mu(C_2) = \mu_m \frac{C_2}{K_m + C_2}, \qquad D = \frac{Q}{V}.$$

The glucose balance has three terms on the RHS. The first accounts for the consumption by the cells. The last two terms account for the flow of glucose into and out of the fermentor:

$$\frac{dC_2}{dt} = -\frac{\mu C_1}{Y} + D(C_{20} - C_2) \tag{E4.33}$$

The maximum specific growth rate μ_m, Monod constant K_m, and cell yield coefficient Y are constants. In Eq. (E4.33), C_{20} is the inlet glucose concentration.

The dilution rate D is dependent on the volumetric flow rate Q and the volume V and really is the reciprocal of the space time of the fermentor. In this problem, the fermentor volume V is fixed, and we vary the flow rate Q. Hence it is more logical to use D (and easier to think) as it is proportional to Q.

Our question is to formulate this model under two circumstances: (1) when we vary only the dilution rate, and (2) when we vary both the dilution rate and the amount of glucose input. We also derive the transfer function model in the second case. In both cases, C_1 and C_2 are the two outputs.

To solve this problem, we obviously have to linearize the equations. In vector form, the nonlinear model is

$$\frac{d\mathbf{x}}{dt} = f(\mathbf{x}, D), \tag{E4.34}$$

where $\mathbf{x} = [C_1 \ C_2]^T$ and

$$f(\mathbf{x}, D) = \begin{bmatrix} f_1(\mathbf{x}, D) \\ f_2(\mathbf{x}, D) \end{bmatrix} = \begin{bmatrix} [\mu(C_2) - D]C_1 \\ -\dfrac{\mu(C_2)\,C_1}{Y} + D(C_{20} - C_2) \end{bmatrix}. \tag{E4.35}$$

75

We first take the inlet glucose, C_{20}, to be a constant (i.e., no disturbance) and vary only the dilution rate D. From the steady-state form of Eqs. (E4.32) and (E4.33), we can derive (without special notation for the steady-state values)

$$D(C_{20} - C_2) = \frac{\mu C_1}{Y}, \qquad C_1 = Y(C_{20} - C_2). \tag{E4.36}$$

Now we linearize the two equations about the steady state. We expect to arrive at (with the primes dropped from all the deviation variables and partial derivatives evaluated at the steady state)

$$\frac{d}{dt} \begin{bmatrix} C_1 \\ C_2 \end{bmatrix} = \begin{bmatrix} \dfrac{\partial f_1}{\partial C_1} & \dfrac{\partial f_1}{\partial C_2} \\[2mm] \dfrac{\partial f_2}{\partial C_1} & \dfrac{\partial f_2}{\partial C_2} \end{bmatrix}_{\text{ss}} \begin{bmatrix} C_1 \\ C_2 \end{bmatrix} + \begin{bmatrix} \dfrac{\partial f_1}{\partial D} \\[2mm] \dfrac{\partial f_2}{\partial D} \end{bmatrix}_{\text{ss}} D. \tag{E4.37}$$

Using Eq. (E4.35) to evaluate the partial derivatives, we should find

$$\frac{d}{dt} \begin{bmatrix} C_1 \\ C_2 \end{bmatrix} = \begin{bmatrix} 0 & C_1 \mu' \\[2mm] -\dfrac{\mu}{Y} & -\dfrac{C_1}{Y}\mu' - \mu \end{bmatrix}_{\text{ss}} \begin{bmatrix} C_1 \\ C_2 \end{bmatrix} + \begin{bmatrix} -C_1 \\[2mm] \dfrac{C_1}{Y} \end{bmatrix}_{\text{ss}} D = \mathbf{Ax} + \mathbf{B}D, \tag{E4.38}$$

where μ' is the derivative with respect to the substrate C_2:

$$\mu' = \frac{d\mu}{dC_2} = \mu_m \frac{K_m}{(K_m + C_2)^2}. \tag{E4.39}$$

All the coefficients in $\mathbf{A}$ and $\mathbf{B}$ are evaluated at steady-state conditions. From Eq. (E4.32), $D = \mu$ at steady state. Hence the coefficient a_{11} in $\mathbf{A}$ is zero.

To complete the state-space model, we find that the output equation is

$$\begin{bmatrix} C_1 \\ C_2 \end{bmatrix} = \begin{bmatrix} 1 & 0 \\ 0 & 1 \end{bmatrix} \begin{bmatrix} C_1 \\ C_2 \end{bmatrix}. \tag{E4.40}$$

In this case, the output matrix $\mathbf{C}$ is just an identity matrix.

Now, we will see what happens with *two inputs*. In practice, we most likely would make a highly concentrated glucose stock and dose it into a main feed stream that contains the other ingredients. What we manipulate is the dosage rate. Consider that the glucose feedstock has a fixed concentration C_{2f} and an adjustable feed rate q_f, and the other nutrients are being fed at a rate of q_0. The effective glucose feed concentration is

$$C_{20} = \frac{q_f C_{2f}}{q_f + q_0} = \frac{q_f C_{2f}}{Q}, \tag{E4.41}$$

where $Q = q_f + q_0$ is the total inlet flow rate and the dilution rate is

$$D = \frac{Q}{V} = \frac{q_f + q_0}{V}. \tag{E4.42}$$

The general fermentor model equation as equivalent to Eq. (E4.34) is

$$\frac{d\mathbf{x}}{dt} = f(\mathbf{x}, \mathbf{u}), \tag{E4.43}$$

where the state-space remains $\mathbf{x} = [C_1 \ C_2]^T$, but the input is the vector $\mathbf{u} = [D_0 \ D_f]^T$.

Here, $D_0 = q_0/V$ and $D_f = q_f/V$ are the respective dilution rates associated with the two inlet streams. That is, we vary the main nutrient feed and glucose dosage flow rates to manipulate this system. The function f is

$$f(\mathbf{x}, \mathbf{u}) = \begin{bmatrix} f_1(\mathbf{x}, \mathbf{u}) \\ f_2(\mathbf{x}, \mathbf{u}) \end{bmatrix} = \begin{bmatrix} \mu(C_2)C_1 - (D_0 + D_f)C_1 \\ -\dfrac{\mu(C_2)\,C_1}{Y} + D_f C_{2f} - (D_0 + D_f)C_2 \end{bmatrix}. \qquad \text{(E4.44)}$$

At steady state,

$$\mu = (D_0 + D_f) = D, \qquad \text{(E4.45)}$$

$$C_1 = Y(C_{20}^* - C_2), \qquad \text{(E4.46)}$$

where

$$C_{20}^* = \frac{D_f C_{2f}}{D_0 + D_f}.$$

The equations linearized about the steady state [with the primes dropped from the deviation variables as in Eq. (E4.38)] are

$$\frac{d}{dt}\begin{bmatrix} C_1 \\ C_2 \end{bmatrix} = \begin{bmatrix} 0 & C_1\mu' \\ -\dfrac{\mu}{Y} & -\dfrac{C_1}{Y}\mu' - \mu \end{bmatrix}_{ss} \begin{bmatrix} C_1 \\ C_2 \end{bmatrix} + \begin{bmatrix} -C_1 & -C_1 \\ -C_2 & (C_{2f} - C_2) \end{bmatrix}_{ss} \begin{bmatrix} D_0 \\ D_f \end{bmatrix} = \mathbf{Ax} + \mathbf{Bu}.$$

$$\text{(E4.47)}$$

The output equation remains the same as in Eq. (E4.40). Laplace transform of the model equations and rearrangement lead us to

$$\begin{bmatrix} C_1 \\ C_2 \end{bmatrix} = \begin{bmatrix} G_{11} & G_{12} \\ G_{21} & G_{22} \end{bmatrix}\begin{bmatrix} D_0 \\ D_f \end{bmatrix}, \qquad \text{(E4.48)}$$

where the four open-loop plant transfer functions are

$$G_{11} = \frac{-\left(\frac{C_1}{Y}\right)_{ss} s - \left[C_1\left(\frac{C_1\mu'}{Y} + \mu\right) + C_1\mu' C_2\right]_{ss}}{p(s)}, \qquad \text{(E4.49)}$$

$$G_{12} = \frac{-\left(\frac{C_1}{Y}\right)_{ss} s + \left[C_1\mu'(C_{2f} - C_2) - C_1\left(\frac{C_1\mu'}{Y} + \mu\right)\right]_{ss}}{p(s)}, \qquad \text{(E4.50)}$$

$$G_{21} = \frac{-(C_2)_{ss}s + \left(\frac{C_1\mu}{Y}\right)_{ss}}{p(s)}, \qquad \text{(E4.51)}$$

$$G_{22} = \frac{(C_{2f} - C_2)_{ss}s + \left(\frac{C_1\mu}{Y}\right)_{ss}}{p(s)}, \qquad \text{(E4.52)}$$

and the characteristic polynomial is

$$p(s) = s^2 + \left(\frac{C_1\mu'}{Y} + \mu\right)_{ss} s + \left(\frac{C_1\mu\,\mu'}{Y}\right)_{ss}. \tag{E4.53}$$

Until we can substitute numerical values and turn the problem over to a computer, we have to admit that the state-space form of Eq. (E4.47) is much cleaner to work with.

This completes our "feel-good" examples. It may not be too obvious, but the hint is that linear system theory can help us analyze complex problems. We should recognize that state-space representation can do everything in classical control and more, and we should feel at ease with the language of state-space representation.

4.3. Properties of State-Space Models

This section contains brief remarks on some transformations and the state-transition matrix. The scope is limited to materials that we may draw on from introductory linear algebra.

4.3.1. Time-Domain Solution

We can find the solution to Eq. (4.1), which is simply a set of first-order differential equations. As analogous to how Eq. (2.3) was obtained, we now use the matrix exponential function as the integration factor, and the result is (see the hints in the Review Problems)

$$\mathbf{x}(t) = e^{\mathbf{A}t}\mathbf{x}(0) + \int_0^t e^{-\mathbf{A}(t-\tau)}\mathbf{B}u(\tau)\,d\tau, \tag{4.10}$$

where the first term on the RHS evaluates the effect of the initial condition and the second term is the so-called convolution integral that computes the effect of the input $u(t)$.

The point is that state-space representation is general and is not restricted to problems with zero initial conditions. When Eq. (4.1) is homogeneous (i.e., $\mathbf{B}u = 0$), the solution is simply

$$\mathbf{x}(t) = e^{\mathbf{A}t}\mathbf{x}(0). \tag{4.11}$$

We can also solve the equation by using a Laplace transform. Starting again from Eq. (4.1), we can find (see Review Problems)

$$\mathbf{x}(t) = \mathbf{\Phi}(t)\mathbf{x}(0) + \int_0^t \mathbf{\Phi}(t-\tau)\mathbf{B}u(\tau)\,d\tau, \tag{4.12}$$

where $\mathbf{\Phi}(t)$ is the state-transition matrix defined in Eq. (4.6). Comparing Eq. (4.10) with Eq. (4.12), we can see that

$$\mathbf{\Phi}(t) = e^{\mathbf{A}t}. \tag{4.13}$$

We have shown how the state-transition matrix can be derived in a relatively simple problem in Example 4.7. For complex problems, there are numerical techniques that we can use to compute $\mathbf{\Phi}(t)$, or even the Laplace transform $\mathbf{\Phi}(s)$, but, of course, we shall skip them.

One idea (not that we will really do this) is to apply the Taylor series expansion on the exponential function of $\mathbf{A}$ and evaluate the state-transition matrix with

$$\Phi(t) = e^{\mathbf{A}t} = \mathbf{I} + \mathbf{A}t + \frac{1}{2!}\mathbf{A}^2 t^2 + \frac{1}{3!}\mathbf{A}^3 t^3 + \cdots. \tag{4.14}$$

We can derive, instead of an infinite series, a closed-form expression for the exponential function. For an $n \times n$ matrix $\mathbf{A}$, we have

$$e^{\mathbf{A}t} = \alpha_0(t)\mathbf{I} + \alpha_1(t)\mathbf{A} + \alpha_2(t)\mathbf{A}^2 + \cdots + \alpha_{n-1}(t)\mathbf{A}^{n-1}. \tag{4.15}$$

The challenge is to find those coefficients $\alpha_i(t)$, which we shall skip.[5]

4.3.2. Controllable Canonical Form

Although there is no unique state-space representation of a system, there are "standard" ones that control techniques make use of. Given any state equations (and if some conditions are met), it is possible to convert them to these standard forms. A couple of important canonical forms are covered in this subsection.

A tool that we should be familiar with from introductory linear algebra is the similarity transform, which allows us to transform a matrix into another one but retains the same eigenvalues. If a state $\mathbf{x}$ and another $\bar{\mathbf{x}}$ are related by a so-called **similarity transform**, the state-space representations constructed with $\mathbf{x}$ and $\bar{\mathbf{x}}$ are considered to be equivalent.[6]

For the nth-order differential equation[7]

$$y^{(n)} + a_{n-1}y^{(n-1)} + \cdots + a_1 y^{(1)} + a_0 y = u(t), \tag{4.16}$$

we define

$$x_1 = y, \quad x_2 = y^{(1)}, \quad x_3 = y^{(2)}, \ldots, \quad \text{and } x_n = y^{(n-1)}. \tag{4.17}$$

The original equation in Eq.(4.16) can now be reformulated as a set of first-order differential equations:

$$\begin{aligned}
\dot{x}_1 &= x_2, \\
\dot{x}_2 &= x_3, \\
&\vdots \\
\dot{x}_{n-1} &= x_n,
\end{aligned} \tag{4.18}$$

---

[5] We need only the general form of Eq. (4.15) later in Chap. 9. There are other properties of the state–transition matrix that have been skipped, but the text has been structured such that they are not needed here.

[6] That includes transforming a given system to the controllable canonical form. We can say that state-space representations are unique up to a similarity transform. As for transfer functions, we can say that they are unique up to scaling in the coefficients in the numerator and the denominator. However, the derivation of canonical transforms requires material from Chap. 9 and is not crucial for the discussion here. These details are provided on the *Web Support*.

[7] Be careful when you read the MATLAB manual; it inverts the index of coefficients as in $y^{(n)} + a_1 y^{(n-1)} + \cdots + a_{n-1}y^{(1)} + a_n y$. Furthermore, here we use a simple RHS in the ODE. You find more general, and thus messier, derivations in other texts.

and finally

$$\dot{x}_n = -a_0 x_1 - a_1 x_2 - \cdots - a_{n-1} x_n + u(t).$$

This set of equations, of course, can be put in matrix form as in Eq. (4.1):

$$\dot{\mathbf{x}} = \begin{bmatrix} 0 & 1 & 0 & \cdots & 0 \\ 0 & 0 & 1 & \cdots & 0 \\ \vdots & & & & \vdots \\ 0 & 0 & 0 & \cdots & 1 \\ -a_0 & -a_1 & -a_2 & \cdots & -a_{n-1} \end{bmatrix} \mathbf{x} + \begin{bmatrix} 0 \\ 0 \\ \vdots \\ 0 \\ 1 \end{bmatrix} u = \mathbf{A}\mathbf{x} + \mathbf{B}u. \tag{4.19}$$

The output equation equivalent to Eq. (4.2) is

$$\mathbf{y} = [1 \ 0 \ 0 \ \cdots \ 0]\mathbf{x} = \mathbf{C}\mathbf{x}. \tag{4.20}$$

The system of equations in Eqs. (4.19) and (4.20) is called the **controllable canonical form** (also phase-variable canonical form). As the name implies, this form is useful in doing controllability analysis and in doing pole-placement system design – topics that are covered in Chap. 9.

With all the zeros along the leading diagonal, we can find relatively easily that the characteristic equation of $\mathbf{A}$, $|s\mathbf{I} - \mathbf{A}| = 0$, is

$$s^n + a_{n-1}s^{n-1} + \cdots + a_1 s + a_0 = 0, \tag{4.21}$$

which is immediately obvious from Eq. (4.16) itself. We may note that the coefficients of the characteristic polynomial are contained in matrix $\mathbf{A}$ in Eq. (4.19). Matrices with this property are called the **companion form**. When we use MATLAB, its `canon()` function returns a companion matrix that is the transpose of $\mathbf{A}$ in Eq. (4.19); this form is called the **observable canonical form**. We shall see that in MATLAB Session 4.

4.3.3. Diagonal Canonical Form

Here, we want to transform a system matrix $\mathbf{A}$ into a diagonal matrix $\mathbf{\Lambda}$ that is made up of the eigenvalues of $\mathbf{A}$. In other words, all the differential equations are decoupled after the transformation.

For a given system of equations in Eq. (4.1), in which $\mathbf{A}$ has *distinct eigenvalues*, we should find a transformation with a matrix $\mathbf{P}$,

$$\bar{\mathbf{x}} = \mathbf{P}^{-1}\mathbf{x} \quad \text{or} \quad \mathbf{x} = \mathbf{P}\bar{\mathbf{x}}, \tag{4.22}$$

such that

$$\dot{\bar{\mathbf{x}}} = \mathbf{\Lambda}\mathbf{x} + \bar{\mathbf{B}}u, \tag{4.23}$$

where now $\bar{\mathbf{B}} = \mathbf{P}^{-1}\mathbf{B}$, and $\mathbf{\Lambda} = \mathbf{P}^{-1}\mathbf{A}\mathbf{P}$ is a diagonal matrix made up of the eigenvalues of $\mathbf{A}$. The transformation matrix (also called the modal matrix) $\mathbf{P}$ is made up of the eigenvectors of $\mathbf{A}$. In control, Eq. (4.23) is called the **diagonal canonical form**.

If $\mathbf{A}$ has repeated eigenvalues (multiple roots of the characteristic polynomial), the result, again from introductory linear algebra, is the **Jordan canonical form**. Briefly, the transformation matrix $\mathbf{P}$ now needs a set of generalized eigenvectors, and the transformed matrix

$\mathbf{J} = \mathbf{P}^{-1}\mathbf{A}\mathbf{P}$ is made of Jordan blocks for each of the repeated eigenvalues. For example, if matrix $\mathbf{A}$ has three repeated eigenvalues λ_1, the transformed matrix should appear as

$$\mathbf{J} = \begin{bmatrix} \mathbf{J}_{11} & \mathbf{0} \\ \mathbf{0} & \mathbf{J}_{22} \end{bmatrix}, \tag{4.24}$$

where

$$\mathbf{J}_{11} = \begin{bmatrix} \lambda_1 & 1 & 0 \\ 0 & \lambda_1 & 1 \\ 0 & 0 & \lambda_1 \end{bmatrix}$$

and $\mathbf{J}_{22}$ is a diagonal matrix made up of eigenvalues $\lambda_4, \ldots, \lambda_n$. Because in Chap. 9 such cases are not used, the details are left to a second course in modern control.

Example 4.9: For a model with the following transfer function,

$$Y/U = \frac{1}{(s+3)(s+2)(s+1)},$$

find the diagonal and observable canonical forms with MATLAB.
The statements to use are

```
G=zpk([],[-1 -2 -3],1);
S=ss(G);                    % S is the state-space system
canon(S)                    % Default is the diagonal form
canon(S,'companion')        % This is the observable companion
```

There is no messy algebra. We can be spoiled! Further details are in MATLAB Session 4.

Review Problems

(1) Fill in the gaps in the derivation of Eqs. (E4.25) and (E4.26) in Example 4.7.
(2) Write the dimensions of all the matrixes in Eq. (4.6) for the general case of MIMO models. Take $\mathbf{x}$ to be $n \times 1$, $\mathbf{y}$ to be $m \times 1$, and $\mathbf{u}$ to be $k \times 1$.
(3) Derive Eq. (4.9) using Eq. (4.3) as the starting point.
(4) For the SISO system shown in Fig. R4.4, derive the state-space representation. Show that the characteristic equation of the model matrix is identical to the closed-loop characteristic polynomial as derived from the transfer functions.
(5) Derive Eq. (4.10).
(6) Derive Eq. (4.12).
(7) Derive Eq. (4.23).

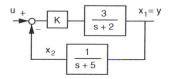

Figure R4.4.

Hints:

(2) $\mathbf{A}$ is $(n \times n)$, $\mathbf{B}$ $(n \times k)$, $\mathbf{C}$ $(m \times n)$, $\mathbf{\Phi}$ $(n \times n)$, and $\mathbf{C\Phi B}$ $(m \times k)$.

(4) Multiply the K to the transfer function to give a gain of $3K$. Then the rest is easier than Example 4.6.

(5) We multiply Eq. (4.1) by $\exp(-\mathbf{A}t)$ to give $e^{-\mathbf{A}t}[\dot{\mathbf{x}} - \mathbf{A}\mathbf{x}] = e^{-\mathbf{A}t}\mathbf{B}u$, which is

$$\frac{d}{dt}(e^{-\mathbf{A}t}\mathbf{x}) = e^{-\mathbf{A}t}\mathbf{B}u.$$

Integration with the initial condition gives

$$e^{-\mathbf{A}t}\mathbf{x}(t) - \mathbf{x}(0) = \int_0^t e^{-\mathbf{A}\tau}\mathbf{B}u(\tau)\,d\tau,$$

which is one step away from Eq. (4.10).

(6) The Laplace transform of Eq. (4.1) with nonzero initial conditions is

$$s\mathbf{X} - \mathbf{x}(0) = \mathbf{A}\mathbf{X} + \mathbf{B}U$$

or

$$\mathbf{X} = (s\mathbf{I} - \mathbf{A})^{-1}\mathbf{x}(0) + (s\mathbf{I} - \mathbf{A})^{-1}\mathbf{B}U.$$

Substituting in the definition $\mathbf{\Phi}(s) = (s\mathbf{I} - \mathbf{A})^{-1}$, we have

$$\mathbf{X} = \mathbf{\Phi}(s)\mathbf{x}(0) + \mathbf{\Phi}(s)\mathbf{B}U.$$

The time-domain solution vector is the inverse transform

$$\mathbf{x}(t) = \mathcal{L}^{-1}[\mathbf{\Phi}(s)]\mathbf{x}(0) + \mathcal{L}^{-1}[\mathbf{\Phi}(s)\mathbf{B}U],$$

and if we invoke the definition of a convolution integral (from calculus), we have Eq. (4.12).

(7) We first substitute $\mathbf{x} = \mathbf{P}\bar{\mathbf{x}}$ into Eq. (4.1) to give

$$\mathbf{P}\frac{d}{dt}\bar{\mathbf{x}} = \mathbf{A}\mathbf{P}\bar{\mathbf{x}} + \mathbf{B}u.$$

Then we multiply the equation by the inverse, $\mathbf{P}^{-1}$,

$$\frac{d}{dt}\bar{\mathbf{x}} = \mathbf{P}^{-1}\mathbf{A}\mathbf{P}\bar{\mathbf{x}} + \mathbf{P}^{-1}\mathbf{B}u,$$

which is Eq. (4.23).

5

Analysis of Single-Loop Control Systems

We now finally launch into the material on controllers. State-space representation is more abstract, and it helps to understand controllers in the classical sense first. We will come back to state-space controller design later. The introduction stays with the basics. Our primary focus is to learn how to design and tune a classical proportional–integra–derivative (PID) controller. Before that, we first need to know how to set up a problem and derive the closed-loop characteristic equation.

What Are We Up to?

- Introducing the basic PID control schemes
- Deriving the closed-loop transfer function of a system and understanding its properties

5.1. PID controllers

We use a simple liquid-level controller to illustrate the concept of a classic feedback control system.[1] In this example (Fig. 5.1), we monitor the liquid level in a vessel and use the information to adjust the opening of an effluent valve to keep the liquid level at some user-specified value (the **set point** or **reference**). In this case, the liquid level is both the **measured variable** and the **controlled variable** – they are the same in a SISO system. In this respect, the controlled variable is also the output variable of the SISO system. A **system** refers to the **process** that we need to control plus the **controller** and accompanying accessories, such as sensors and actuators.[2]

Let's say we want to keep the liquid level at the set point h^s, but a sudden surge in the inlet flow rate q_i (the **disturbance** or **load**) increases h such that there is a deviation $h' = h - h^s > 0$. The deviation can be rectified if we open up the valve (or we can think in

---

[1] In Fig. 5.1, we use the actual variables because they are what we measure. Regardless of the notations in a schematic diagram, the block diagram in Fig. 5.2 is based on deviation variables and their Laplace transform.

[2] Recall footnote 4 in Chap. 1: A process is referred to as a plant in control engineering.

Analysis of Single-Loop Control Systems

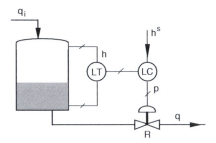

Figure 5.1. Schematic diagram of a liquid-
level control system.

terms of lowering the flow resistance R). Here we assume that the level controller will send
out an appropriate signal to the valve to accomplish the task. It is logical to think that the
signal from the controller, $p(t)$, should be a function of the deviation.

However, because we are to implement *negative* feedback, we base our decision on the
error, which is defined as

$$e(t) = h^s(t) - h(t),$$

which is the negative of the deviation (Fig. 5.2). The actual controller output is

$$p(t) = p^s + f[e(t)] = p^s + f[h^s - h(t)], \tag{5.1}$$

where f is some function of $e(t)$ and p^s is the actuating signal at steady state when the
deviation is $h' = 0$; it is commonly referred to as the **controller bias signal**. Our task is to
determine plausible choices of the controller functions – what are also called control laws.
The classical controller functions are explained in the following subsections, but if you need
a more physical feel of how a control system is put together now, you may jump ahead and
read Section 5.2 first.

5.1.1. Proportional Control

The simplest idea is that the compensation signal (actual controller output) is proportional
to the error $e(t)$:

$$p(t) = p^s + K_c e(t) = p^s + K_c[h^s - h(t)], \tag{5.2}$$

where K_c is the proportional gain of the controller. It is obvious that the value of K_c
determines the controller "sensitivity" – how much compensation to enact for a given change
in error.

Figure 5.2. Information around
the summing point in a negative-
feedback system.

For all commercial devices, the proportional gain is a *positive* quantity. Because we use negative feedback (see Fig. 5.2), the controller output moves in the *reverse* direction of the controlled variable.[3] In the liquid-level control example, if the inlet flow is disturbed such that h rises above h^s, then $e < 0$, and that leads to $p < p^s$, i.e., the controller output is decreased. In this case, we of course will have to select or purchase a valve such that a lowered signal means opening the valve (decreasing flow resistance). Mathematically, this valve has a negative steady-state gain $(-K_v)$.[4]

Now what if our valve has a positive gain? (An increased signal means opening the valve.) In this case, we need a *negative* proportional gain. Commercial devices provide such a "switch" on the controller box to invert the signal. Mathematically, we have changed the sign of the compensation term to $p = p^s - K_c e$.[5]

By the definition of a control problem, there should be no error at $t = 0$, i.e., $e^s = 0$, and the deviation variable of the error is simply the error itself:

$$e'(t) = e(t) - e^s = e(t).$$

Hence Eq. (5.2) is a relation between the deviation variables of the error and the controller output:

$$p(t) - p^s = K_c[e(t) - e^s],$$

or

$$p'(t) = K_c e'(t),$$

and the transfer function of a proportional controller is simply

$$G_c(s) = \frac{P(s)}{E(s)} = K_c. \tag{5.3}$$

Generally, the proportional gain is dimensionless [i.e., $p(t)$ and $e(t)$ have the same units]. Many controller manufacturers use the percent **proportional band** (PB), which is defined as

$$PB = \frac{100}{K_c}. \tag{5.4}$$

---

[3] You may come across the terms *reverse* and *direct acting*; they are extremely confusing. Some authors consider the action between the controller output and the controlled variable, and thus a negative-feedback loop with positive K_c is considered reverse acting. However, most commercial vendors consider the action between the error (controller input) and the controller output, and now a controller with a positive K_c is direct acting, exactly the opposite terminology. We will avoid using these confusing terms. The important point is to select the proper signs for all the steady-state gains, and we will get back to this issue in Section 5.4.

[4] Take note that, from the mass balance of the tank, the process gain associated with the outlet flow rate is also negative. A simple-minded check is that in a negative-feedback system there can only be one *net* negative sign – at the feedback summing point. If one unit in the system has a negative steady-state gain, we know something else must have a negative steady-state gain too.

[5] More confusion may have been introduced here than in texts that ignore this tiny detail. To reduce confusion, we keep K_c a positive number. For problems in which the proportional gain is negative, we use the notation $-K_c$. We can imagine that the minus sign is likened to having flipped the action switch on the controller.

A high proportional gain is equivalent to a narrow PB and a low gain is wide PB. We can interpret a PB as the range over which the error must change to drive the controller output over its full range.[6]

Before doing any formal analysis, we state a few qualitative features of each type of controller. This is one advantage of classical control. We can make a fairly easy physical interpretation of the control algorithm. The analyses in Section 5.3 will confirm these qualitative observations.

General Qualitative Features of Proportional Control

- We expect that a proportional controller will improve or accelerate the response of a process. The larger K_c is, the faster and more sensitive is change in compensation with respect to a given error. However, if K_c is too large, we expect the control compensation to overreact, leading to oscillatory response. In the worst case, the system may become unstable.
- There are physical limits to a control mechanism. A controller (like an amplifier) can deliver only so much voltage or current; a valve can deliver only so much fluid when fully opened. At these limits, the control system is **saturated**.[7]
- We expect a system with only a proportional controller to have a **steady-state error** (or an **offset**). A formal analysis is introduced in Section 5.3. This is one simplistic way to see why. Let's say we change the system to a new set point. The proportional controller output, $p = p^s + K_c e$, is required to shift away from the previous bias p^s and move the system to a new steady state. For p to be different from p^s, the error must have a finite nonzero value.[8]
- To tackle a problem, consider a simple proportional controller first. This may be all we need (lucky break!) if the offset is small enough (for us to bear with) and the response is adequately fast. Even if this is not the case, the analysis should help us plan the next step.

5.1.2. Proportional–Integral Control

To eliminate offset, we can introduce integral action into the controller. In other words, we use a compensation that is related to the history of the error:

$$p'(t) = \frac{1}{\tau_I} \int_0^t e'(t)\,dt, \qquad \frac{P(s)}{E(s)} = \frac{1}{s\,\tau_I},$$

where τ_I is the **integral time** constant (**reset time**, or minutes per repeat[9]). Commercial devices may also use $1/\tau_I$, which is called the reset rate (repeats per minute).

---

[6] In some commercial devices, the proportional gain is defined as the ratio of the percentage of controller output to the percentage of controlled variable change [%/%]. In terms of the control system block diagram that we will go through in the next subsection, we just have to add "gains" to do the unit conversion.

[7] Typical ranges of device outputs are 0–10 V, 0–1 V, 4–20 mA, and 3–15 psi.

[8] The exception is when a process contains integrating action, i.e., $1/s$ in the transfer functions – a point that is illustrated in Example 5.5.

[9] Roughly, the reset time is the time that it takes the controller to repeat the proportional action. This is easy to see if we take the error to be a constant in the integral.

The integral action is such that we accumulate the error from $t = 0$ to the present. Thus the integral is not necessarily zero even if the current error is zero. Moreover, the value of the integral will not decrease unless the integrand $e'(t)$ changes its sign. As a result, integral action forces the system to overcompensate and leads to oscillatory behavior, i.e., the closed-loop system will exhibit an underdamped response. If there is too much integral action, the system may become unstable.

In practice, integral action is never used by itself. The norm is a **proportional–integral (PI) controller**. The time-domain equation and the transfer function are

$$p'(t) = K_c \left[e'(t) + \frac{1}{\tau_I} \int_0^t e'(t)\, dt \right], \quad G_c(s) = K_c \left[1 + \frac{1}{\tau_I s} \right]. \tag{5.5}$$

If the error cannot be eliminated within a reasonable period, the integral term can become so large that the controller is saturated – a situation referred to as integral or **reset windup**. This may happen during start-up or large set-point changes. It may also happen if the proportional gain is too small. Many industrial controllers have "antiwindup" that temporarily halts the integral action whenever the controller output becomes saturated.[10]

On the plus side, the integration of the error allows us to detect and eliminate very small errors. To obtain a simple explanation of why integral control can eliminate offsets, refer back to the intuitive explanation of offset with only a proportional controller. If we desire $e = 0$ at steady state and to shift controller output p away from the previous bias p^s, we must have a nonzero term. Here, it is provided by the integral in the first equation of Eqs. (5.5). That is, as time progresses, the integral term takes on a final nonzero value, thus permitting the steady-state error to stay at zero.

General Qualitative Features of PI control

- PI control can eliminate offset. We must use a PI controller in our design if the offset is unacceptably large.
- The elimination of the offset is usually at the expense of a more underdamped system response. The oscillatory response may have a short rise time, but is penalized by excessive overshoot or exceedingly long settling time.[11]
- Because of the inherent underdamped behavior, we must be careful with the choice of the proportional gain. In fact, we usually lower the proportional gain (or detune the controller) when we add integral control.

[10] Another strategy is to implement the PI algorithm in the so-called **reset-feedback** configuration. The basis of internal reset feedback is to rearrange and implement the PI transfer function as

$$1 + \frac{1}{\tau_I s} = \frac{\tau_I s + 1}{\tau_I s} = \frac{1}{\tau_I s / (\tau_I s + 1)} = \frac{1}{1 - 1/(\tau_I s + 1)}.$$

Now, the internal state of the controller, whether it be electronics or a computer algorithm for integration, will have an upper limit. External reset feedback, on the other hand, makes use of measurements of the manipulated variable. You may find such implementation details in more applied control books.

[11] Some texts use the term "sluggish" here without further qualification. The sluggishness in this case refers to the long settling time, not the initial response.

5.1.3. Proportional–Derivative Control

We certainly want to respond very differently if the temperature of a chemical reactor is changing at a rate of 100 °C/s as opposed to 1 °C/s. In a way, we want to "project" the error and make corrections accordingly. In contrast, proportional and integral controls are based on the present and the past. Derivative controller action is based on how fast the error is changing with time (rate action control). We can write

$$p'(t) = \tau_D \frac{de'}{dt}, \qquad \frac{P(s)}{E(s)} = \tau_D s,$$

where τ_D is the **derivative time** constant (sometimes just rate time).

Here, the controller output is zero as long as the error stays constant, that is, even if the error is not zero. Because of the proportionality to the rate of change, the controller response is very sensitive to noise. If there is a sudden change in error, especially when we are just changing the set point, the controller response can be unreasonably large – leading to what is called a **derivative kick**.

Derivative action is never used by itself. The simplest implementation is a **proportional–derivative (PD) controller**. The time-domain equation and the transfer function of an "ideal" PD controller are

$$p'(t) = K_c \left[e'(t) + \tau_D \frac{de'}{dt} \right], \qquad G_c(s) = K_c [1 + \tau_D s]. \tag{5.6}$$

In practice, we cannot build a pneumatic device or a passive circuit that provides ideal derivative action. Commercial (real!) PD controllers are designed on the basis of a **lead–lag element**:

$$G_c(s) = K_c \left[\frac{\tau_D s + 1}{\alpha \tau_D s + 1} \right], \tag{5.7}$$

where α is a small number, typically $0.05 \leq \alpha \leq 0.2$.

In effect, we are adding a very large real pole to the derivative transfer function. Later, after learning root-locus and frequency-response analysis, we can make more rational explanations, including why the function is called a lead–lag element. We will see that this is a nice strategy that is preferable to using the ideal PD controller.

To reduce derivative kick (the sudden jolt in response to set-point changes), the derivative action can be based on the rate of change of the measured (controlled) variable instead of the rate of change of the error. One possible implementation of this idea is in Fig. 5.3. This

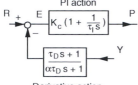

Figure 5.3. Implementation of derivative control on the measured variable.

way, the derivative control action ignores changes in the reference and just tries to keep the measured variable constant.[12]

General Qualitative Features of Derivative Control

- PD control is not useful for systems with large dead time or noisy signals.
- The sign of the rate of change in the error could be opposite that of the proportional or the integral terms. Thus adding derivative action to PI control may counteract the overcompensation of the integrating action. PD control may improve system response while reducing oscillations and overshoot. (Formal analysis in Chap. 7 will show that the problem is more complex than is implied by this simple statement.)
- If simple proportional control works fine (in the sense of acceptable offset), we may try PD control. Similarly, we may try PID on top of PI control. The additional stabilizing action allows us to use a larger proportional gain and obtain a faster system response.

5.1.4. Proportional–Integral–Derivative Control

Finally, we can put all the components together to make a **PID** (or three-mode) **controller**. The time-domain equation and the transfer function of an "ideal" PID controller are

$$p'(t) = K_c \left[e'(t) + \frac{1}{\tau_I} \int_0^t e'(t)\, dt + \tau_D \frac{de'}{dt} \right], \tag{5.8a}$$

$$G_c(s) = K_c \left[1 + \frac{1}{\tau_I s} + \tau_D s \right] = K_c \frac{\tau_I \tau_D s^2 + \tau_I s + 1}{\tau_I s}. \tag{5.8b}$$

We also find it rather usual that the proportional gain is multiplied into the bracket to give the integral and the derivative gains:

$$G_c(s) = K_c + \frac{K_I}{s} + K_D s, \tag{5.8c}$$

where $K_I = K_c/\tau_I$ and $K_D = K_c \tau_D$. With a higher-order polynomial in the numerator, the ideal PID controller is not considered physically realizable. We nevertheless use this ideal controller in analyses because of the cleaner algebra, and more importantly because we can gain valuable insight with it. We can say the same with the use of the ideal PD controller too.

In real life, different manufacturers implement the "real" PID controller slightly differently.[13] One possibility is to modify the derivative action as

$$G_c(s) = K_c \left[1 + \frac{1}{\tau_I s} + \frac{\tau_D s}{\alpha \tau_D s + 1} \right] = K_c \left[\frac{(\alpha + 1)\tau_D s + 1}{\alpha \tau_D s + 1} + \frac{1}{\tau_I s} \right] \tag{5.9a}$$

---

[12] For review after Chap. 7 on root locus: With the strategy in Fig. 5.3, the closed-loop characteristic polynomial and thus the poles remain the same, but not the zeros. You may also wonder how to write the function $G_c(s)$, but it is much easier and more logical just to treat the PI action and derivation action as two function blocks when analyzing a problem.

[13] Not only that, but most implementations are based on some form of the digitized control law. An illustration of the positional digital algorithm along with concepts such as bumpless transfer, external-rate feedback, and bias tracking is in the LabView liquid-level simulation module on the *Web Support*.

Another implementation of the actual PID control is to introduce the derivative control in series with PI control:

$$G_c(s) = K_c \left[\frac{\tau_I s + 1}{\tau_I s} \right] \left[\frac{\tau_D s + 1}{\alpha \tau_D s + 1} \right]. \qquad (5.9b)$$

This configuration is also referred to as interacting PID, series PID, or rate-before-reset. To eliminate derivative kick, the derivative lead–lag element is implemented on the measured (controlled) variable in the feedback loop.

5.2. Closed-Loop Transfer Functions

We first establish the closed-loop transfer functions of a fairly general SISO system. After that, we walk through the diagram block by block to gather the thoughts that we must have in synthesizing and designing a control system. An important detail is the units of the physical properties.

5.2.1. Closed-Loop Transfer Functions and Characteristic Polynomials

Consider the stirred-tank heater again, this time in a closed loop (Fig. 5.4). The tank temperature can be affected by variables such as the inlet and the jacket temperatures and the inlet flow rate. Back in Chap. 2, we derived the transfer functions for the inlet and the jacket temperatures. In Laplace transform, the change in temperature is given in Eq. (2.49b) as

$$T(s) = G_L(s)T_i(s) + G_p(s)T_H(s). \qquad (5.10)$$

This is our process model. Be careful with the context when using the word "input." The inlet and the jacket temperatures are the inputs to the process, but they are not necessarily the inputs to the system. One of them will become the manipulated variable of the system.

In a SISO system, we manipulate only one variable, so we must make a decision. Because our goal is to control the tank temperature, it would be much more sensible to manipulate the steam temperature T_H instead of the inlet temperature. We can arrive at this decision with physical intuition, or we can base it on the fact that, from Chap. 2, the steam temperature has a higher process gain. Hence with respect to the control system, we choose T_H as the **manipulated variable** (M), which is governed by the actuator function G_a and the controller signal P. The tank temperature T is the system output (also the controlled variable C).

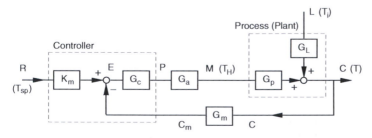

Figure 5.4. Block diagram of a simple SISO closed-loop system.

The system input is the set point T_{sp} (or reference R) – our desired steady-state tank temperature.[14]

There are other system inputs that can affect our closed-loop response, and we consider them load (or disturbance) variables. In this case, the load variable is the inlet temperature, T_i. Now you may understand why we denote the two transfer functions as G_p and G_L. The important point is that input means different things for the process and for the closed-loop system.

For the rest of the control loop, G_c is obviously the controller transfer function. The measuring device (or transducer) function is G_m. Although it is not shown in the block diagram, the steady-state gain of G_m is K_m. The key is that the summing point can compare only quantities with the same units. Hence we need to introduce K_m on the reference signal, which should have the same units as C. The use of K_m, in a way, performs unit conversion between what we "dial in" and what the controller actually uses in comparative tests.[15]

The next order of business is to derive the closed-loop transfer functions. For better readability, we write the Laplace transforms without the s dependence explicitly. Around the summing point, we observe that

$$E = K_m R - G_m C.$$

From the process we have

$$C = G_p(G_a G_c E) + G_L L.$$

Substituting for E, we find that the equation becomes

$$C = G_p G_a G_c(K_m R - G_m C) + G_L L.$$

This step can be rearranged to give

$$C = \left(\frac{K_m G_c G_a G_p}{1 + G_m G_c G_a G_p} \right) R + \left(\frac{G_L}{1 + G_m G_c G_a G_p} \right) L = G_{sp}R + G_{load}L, \tag{5.11}$$

which provides us with the closed-loop transfer functions G_{sp} and G_{load}. Based on Eq. (5.11), the inputs to the system are the reference R and the load variable L; the controlled variable is the system output. The first transfer function G_{sp} accounts for the effect of a set-point change and is also called the command tracking function. The second function G_{load} accounts for the effect of changes in disturbance.

The important point is that the dynamics and the stability of the system are governed by the **closed-loop characteristic polynomial:**

$$1 + G_m G_c G_a G_p = 0, \tag{5.12}$$

which is the *same* whether we are looking at set-point or disturbance changes. As an abbreviation, many authors write $G_{OL} = G_m G_c G_a G_p$ and refer to it as the open-loop transfer

---

[14] There is no standard notation. We could have used Y in place of C for system output or replaced G_a with G_v for valve (G_f is also used), G_L and L with G_d and D for disturbance, and G_m with G_T for transducer. Here P was selected to denote controller output, more or less for pneumatic.

[15] Many texts, especially those in electrical engineering, ignore the need for K_m, and the final result is slightly different. They do not have to check the units because all they deal with are electrical signals.

function as if the loop were disconnected.[16] We may also refer to $G_c G_a G_p$ as the forward-loop transfer function.

Our analyses of SISO systems seldom take into account simultaneous changes in set point and load.[17] We denote the two distinct possibilities as two different problems:

(1) **Servo** problems: Consider changes in set point with no disturbance $(L = 0)$; $C = G_{sp} R$. Ideally (meaning unlikely to be encountered in reality), we would like to achieve perfect tracking of set-point changes: $C = R$. *Reminder*: We are working with deviation variables.

(2) **Regulator** problems: Consider changes in disturbance with a fixed set point $(R = 0)$; $C = G_{load} L$. The goal is to reject disturbances, i.e., keep the system output at its desired value in spite of load changes. Ideally, we would like to have $C = 0$, i.e., perfect disturbance rejection.

5.2.2. How Do We Choose the Controlled and the Manipulated Variables?

In homework problems, by and large, the variables are stated. Things will be different when we are on the job. Here are some simple ideas on how we may make the decision.

Choice of controlled variables:

- those that are dictated by the problem (for instance, temperature of a refrigerator)
- those that are not self-regulating
- those that may exceed equipment or process constraints
- those that may interact with other variables (for example, reactor temperature may affect product yield).

Choice of manipulated variables:

- those that have a direct effect on the process, especially the output variable
- those that have a large steady-state gain (good sensitivity)
- those that have no dead time
- those that have no interaction with other control loops.

After we have chosen the controlled and the manipulated variables, the remaining ones are taken as load variables in a SISO system.

5.2.3. Synthesis of a Single-Loop Feedback System

We now walk through the stirred-tank heater system once again. This time, we will take a closer look at the transfer functions and the units (Fig. 5.5).

---

[16] Can an open loop be still a loop? You may wonder, what is an open loop? Often, we loosely refer to elements or properties of part of a system as open loop, as opposed to a complete closed-loop system. You will see more of this language in Chap. 7.

[17] In real life, we expect probable simultaneous reference and disturbance inputs. As far as analysis goes, the mathematics is much simpler if we consider one case at a time. In addition, either case *shares the same closed-loop characteristic polynomial*. Hence they should also share the same stability and dynamic response characteristics. Later when we talk about integral error criteria in controller design, there are minor differences, but not sufficient to justify analyzing a problem with simultaneous reference and load inputs.

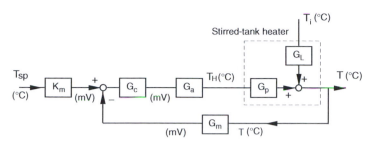

Figure 5.5. Block diagram of a simple SISO closed-loop system with physical units.

Process Model

The first item on the agenda is "process identification." We either derive the transfer functions of the process based on scientific or engineering principles, or we simply do a step-input experiment and fit the data to a model. Either way, we need to decide what the controlled variable is. We then need to decide which should be the manipulated variable. All remaining variables are delegated to become disturbances.

With the stirred-tank heater, we know quite well by now that we want to manipulate the heating-coil temperature to control the tank temperature. The process function G_p is defined based on this decision. In this simple illustration, the inlet temperature is the only disturbance, and the load function is defined accordingly. From Subsection 2.8.2 and Eq. (2.49b), we have the first-order process model:

$$T = G_L T_i + G_p T_H = \left(\frac{K_L}{\tau_p s + 1}\right) T_i + \left(\frac{K_p}{\tau_p s + 1}\right) T_H. \tag{5.13}$$

From Subsection 2.8.2, we know that the steady-state gain and the time constant are dependent on the values of flow rate, liquid density, heat capacity, heat transfer coefficient, and so on. For the sake of illustration, the heat transfer analysis is skipped. Let's presume that we have done our homework, substituted in numerical values, and have found $K_p = 0.85\ °C/°C$ and $\tau_p = 20$ min.

Signal Transmitter

Once we know what to control, we need to find a way to measure the quantity. If the transducer (sensor and transmitter packaged together) is placed far downstream or is too well insulated and the response is slow, the measurement transfer function may appear as

$$\frac{T_m}{T} = G_m = \frac{K_m e^{-t_d s}}{\tau_m s + 1}, \tag{5.14}$$

where K_m is the measurement gain of the transducer, τ_m is the time constant of the device, and t_d accounts for transport lag. In the worst case, the sensor may be nonlinear, meaning that the measurement gain would change with the range of operating conditions.

With temperature, we can use a thermocouple, which typically has a resolution of the order of 0.05 mV/°C. [We could always use a resistance temperature detector (RTD) for better resolution and response time.] That is too small a change in output for most 12-bit analog-digital converters, so we must have an amplifier to boost the signal. This is something we do in a laboratory, but commercially, we should find off-the-shelf transducers with the sensor and amplifier packaged together. Many of them have a scaled output of, for example, 0–1 V or 4–20 mA.

For the sake of illustration, let's presume that the temperature transmitter has a built-in amplifier that allows us to have a measurement gain of $K_m = 5$ mV/°C. Let's also presume that there is no transport lag and that the thermocouple response is rapid. The measurement transfer function in this case is simply

$$G_m = K_m = 5 \text{ mV/°C}.$$

This so-called measurement gain is really the slope of a calibration curve – an idea that we are familiar with. We do a least-squares fit if this curve is linear and find the tangent at the operating point if the curve is nonlinear.

Controller

The amplified signal from the transmitter is sent to the controller, which can be a computer or a little black box. Not much can be said about the controller function now, except that it is likely a PID controller or a software application with a similar interface.

A reminder is that a controller has a front panel with physical units such as degrees celsius. (Some also have relative scales of 0–100%.) Therefore, when we dial a change in the set point, the controller needs to convert the change into electrical signals. That's why K_m is part of the controller in the block diagram (Fig. 5.5).

Actuator Control Valve

Last but not least, designing a proper actuator can create the most headaches. We have to find an actuator that can drive the range of the manipulated variable. We also want the device to have a faster response than the process. After that, we have to find a way to interface the controller to the actuator. A lot of work is masked by the innocent-looking notation G_a.

For the stirred-tank heater example, several comments should be made here. We need to consider **safety**. If the system fails, we want to make sure that no more heat is added to the tank. Thus we want a fail-close valve – meaning that the valve requires energy (or a positive signal change) to open it. In other words, the valve gain is positive. We can check the thinking as follows: If the tank temperature *drops* below the set point, the error increases. With a positive proportional gain, the controller output will *increase*, hence opening up the valve. If the process plant has a power outage, the valve closes and shuts off the steam. But how can the valve shut itself off without power?

This leads to the second comment. One may argue for emergency power or a spring-loaded valve, but to reduce fire hazards, the nominal industrial practice is to use pneumatic (compressed-air-driven) valves that are regulated by a signal of 3–15 psi. The electrical signal to and from the controller is commonly 4–20 mA. A current signal is less susceptible to noise than a voltage signal is over longer transmission distances. Hence, in a more applied setting, we expect to find a **current-to-pressure transducer** (I/P) situated between the controller output and the valve actuator.

Finally, we have been sloppy in associating the flow rate of steam with the heating-coil temperature. The proper analysis that includes a heat balance of the heating medium is in the Review Problems. To sidestep the actual calculations, we have to make a few more assumptions for the valve gain to illustrate what we need to do in reality:

(1) Assume that we have the proper amplifier or transducer to interface the controller output with the valve, i.e., converting electrical information into flow rate.
(2) We use a valve with linear characteristics such that the flow rate varies linearly with the opening.[18]
(3) The change in steam flow rate can be "translated" to changes in heating-coil temperature.

When the steady-state gains of all three assumptions are lumped together, we may arrive at a valve gain K_v with units of degrees celsius per millivolt. For this illustration, let's say the valve gain is 0.6 °C/mV and the time constant is 0.2 min. The actuator controller function would appear as

$$G_v = \frac{K_v}{\tau_v s + 1} = \frac{0.6 \,°\text{C/mV}}{0.2\,s + 1}.$$

The closed-loop characteristic equation of the stirred-tank heater system is hence

$$1 + G_c G_v G_p G_m = 1 + G_c \frac{(0.6)(0.85)(5)}{(0.2s + 1)(20s + 1)} = 0.$$

We will not write the entire closed-loop function C/R or, in this case, T/T_{sp}. The main reason is that our design and analysis will be based on only the characteristic equation. The closed-loop function is handy to do only time-domain simulation, which we can easily compute by using MATLAB. That being said, we need to analyze the closed-loop transfer function for several simple cases so we have a better theoretical understanding.

5.3. Closed-Loop System Response

In this section, we will derive the closed-loop transfer functions for several examples. The scope is limited by how much sense we can make out of the algebra. Nevertheless, the steps that we go through are necessary to learn how to set up problems properly. The analysis also helps us to better understand why a system may have a faster response, why a system may become underdamped, and when there is an offset. When the algebra is clean enough, we can also make observations as to how controller settings may affect the closed-loop system response. The results generally reaffirm the qualitative statements that we have made concerning the characteristics of different controllers.

The actual work is rather cookbook-like:

(1) With a given problem statement, draw the control loop and derive the closed-loop transfer functions.
(2) Pick either the servo or the regulator problem. Reminder: the characteristic polynomial is the same in either case.

---

[18] In reality, the valve characteristic curve is likely nonlinear and we need to look up the technical specification in the manufacturer's catalog. After that, the valve gain can be calculated from the slope of the characteristic curve at the operating point. See Homework Problem I.33 and the *Web Support*.

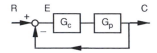

Figure 5.6. Simple unity feedback system.

(3) With the given choices of G_c (P, PI, PD, or PID), G_p, G_a, and G_m, plug their transfer functions into the closed-loop equation. The characteristic polynomial should fall out nicely.

(4) Rearrange the expressions such that we can redefine the parameters as time constants and steady-state gains for the closed-loop system.

All analyses follow the same general outline. What we must accept is that there are no handy-dandy formulas to plug and chug. We must be competent in deriving the closed-loop transfer function, steady-state gain, and other relevant quantities for each specific problem.

In our examples, we will take $G_m = G_a = 1$ and use a servo system with $L = 0$ to highlight the basic ideas. The algebra tends to be more tractable in this simplified unity feedback system with only G_c and G_p (Fig. 5.6), and the closed-loop transfer function is

$$\frac{C}{R} = \frac{G_c G_p}{1 + G_c G_p},\tag{5.15}$$

which has the closed-loop characteristic equation $1 + G_c G_p = 0$.

Example 5.1: Derive the closed-loop transfer function of a system with proportional control and a first-order process. What is the value of the controlled variable at steady state after a unit-step change in set point?

In this case, we consider $G_c = K_c$ and $G_p = [K_p/(\tau_p s + 1)]$, and substitution into Eq. (5.15) leads to[19]

$$\frac{C}{R} = \frac{K_c K_p}{\tau_p s + 1 + K_c K_p}.\tag{E5.1}$$

---

[19] You may wonder how transfer functions are related to differential equations. This is a simple illustration. We use y to denote the controlled variable. The first-order process function G_p arises from Eq. (3.6):

$$\tau_p \frac{dy}{dt} + y = K_p x.$$

In the unity feedback loop with $G_c = K_c$, we have $x = K_c(r - y)$. Substitution for x in the ODE leads to

$$\tau_p \frac{dy}{dt} + y = K_c K_p (r - y)$$

or

$$\tau_p \frac{dy}{dt} + (1 + K_c K_p)y = K_c K_p r.$$

It is obvious that Eq. (E5.1) is the Laplace transform of this equation. This same idea can be applied to all other systems, but, of course, nobody does that. We all work within the Laplace transform domain.

We now divide both the numerator and denominator with $(1 + K_c K_p)$ to obtain

$$\frac{C}{R} = \frac{K_c K_p / (1 + K_c K_p)}{[\tau_p / (1 + K_c K_p)]s + 1} = \frac{K}{\tau s + 1}, \tag{E5.2}$$

where

$$K = \frac{K_c K_p}{1 + K_c K_p}, \qquad \tau = \frac{\tau_p}{1 + K_c K_p}$$

are the closed-loop steady-state gain and time constant.

Recall Eq. (5.11); the closed-loop characteristic equation is the denominator of the closed-loop transfer function, and the probable locations of the *closed-loop pole* are given by

$$s = -(1 + K_c K_p)/\tau_p.$$

There are two key observations. First, $K < 1$, meaning that the controlled variable will change in magnitude less than a given change in set point, the source of offset. The second is that $\tau < \tau_p$, meaning that the system has a faster response than the open-loop process. The system time constant becomes smaller as we increase the proportional gain. This is consistent with the position of the closed-loop pole, which should "move away" from the origin as K_c increases.

We now take a formal look at the steady-state error (offset). Let's consider a more general step change in set point, $R = M/s$. The eventual change in the controlled variable, by means of the final-value theorem, is

$$c'(\infty) = \lim_{s \to 0} s \frac{K}{\tau s + 1} \frac{M}{s} = MK.$$

The offset is the relative error between the set point and the controlled variable at steady state, i.e., $(r - c_\infty)/r$:

$$e_{ss} = \frac{M - MK}{M} = 1 - K = 1 - \frac{K_c K_p}{1 + K_c K_p} = \frac{1}{1 + K_c K_p}. \tag{E5.3}$$

We can reduce the offset if we increase the proportional gain.

Let's take another look at the algebra for evaluating the steady-state error. The error that we have derived in the example is really the difference between the change in controlled variable and the change in set point in the block diagram (Fig. 5.6). Thus we can write

$$E = R - C = R\left(1 - \frac{G_c G_p}{1 + G_c G_p}\right) = R\left(\frac{1}{1 + G_c G_p}\right).$$

Now if we have a unit-step change $R = 1/s$, the steady-state error by means of the final-value theorem is (recall that $e = e'$)

$$e_{ss} = \lim_{s \to 0} s \frac{1}{1 + G_c G_p} \frac{1}{s} = \frac{1}{1 + \lim_{s \to 0} G_c G_p} = \frac{1}{1 + K_{err}}, \tag{5.16}$$

where $K_{err} = \lim_{s \to 0} G_c G_p$. We call K_{err} the **position error constant**.[20] For the error to

---

[20] In many control texts, we also find the derivation of the velocity error constant (by using $R = s^{-2}$) and acceleration error constant (by using $R = s^{-3}$) and a subscript p is used on what we call K_{err} here.

approach zero, K_{err} must approach infinity. In Example 5.1, the error constant and the steady-state error are

$$K_{err} = \lim_{s \to 0} G_c G_p = \frac{K_c K_p}{\tau_p s + 1} = K_c K_p, \quad \text{and again } e_{ss} = \frac{1}{1 + K_c K_p}. \tag{5.17}$$

Example 5.2: Derive the closed-loop transfer function of a system with proportional control and a second-order overdamped process. If the second-order process has time constants 2 and 4 min and process gain 1.0 (units), what proportional gain would provide us with a system with damping ratio of 0.7?

In this case, we consider $G_c = K_c$, and $G_p = \{K_p/[(\tau_1 s + 1)(\tau_2 s + 1)]\}$, and substitution into Eq. (5.15) leads to

$$\frac{C}{R} = \frac{K_c K_p}{(\tau_1 s + 1)(\tau_2 s + 1) + K_c K_p} = \frac{K_c K_p/(1 + K_c K_p)}{\left(\frac{\tau_1 \tau_2}{1 + K_c K_p}\right) s^2 + \left(\frac{\tau_1 + \tau_2}{1 + K_c K_p}\right) s + 1}. \tag{E5.4}$$

The key is to recognize that the system may exhibit underdamped behavior even though the open-loop process is overdamped. The closed-loop characteristic polynomial can have either real or complex roots, depending on our choice of K_c. (This is much easier to see when we work with a root locus in Chap. 7.) For now, we rewrite the closed-loop function as

$$\frac{C}{R} = \frac{K}{\tau^2 s^2 + 2\zeta \tau s + 1}, \tag{E5.4a}$$

where the closed-loop steady-state gain is $K = [(K_c K_p)/(1 + K_c K_p)]$, and the system natural time period and damping ratio are

$$\tau = \sqrt{\frac{\tau_1 \tau_2}{1 + K_c K_p}}, \quad \zeta = \frac{1}{2} \frac{(\tau_1 + \tau_2)}{\sqrt{\tau_1 \tau_2 (1 + K_c K_p)}}. \tag{E5.5}$$

If we substitute $\zeta = 0.7$, $K_p = 1$, $\tau_1 = 2$, and $\tau_2 = 4$ into the second expression, we should find the proportional gain K_c to be 1.29.

Last, we should see immediately that the system steady-state gain in this case is the same as that in Example 5.1, meaning that this second-order system will have the same steady-state error.

In terms of controller design, we can take an entirely analytical approach when the system is simple enough. Of course, such circumstances are not common in real life. Furthermore, we often have to compromise between conflicting criteria. For example, we cannot require a system to have both a very fast rise time and a very short settling time. If we want to provide a smooth response to a set-point change without excessive overshoot, we cannot also expect a fast and snappy initial response. As engineers, it is our job to decide.

In terms of design specification, it is not uncommon to use the decay ratio (DR) as the design criterion. Repeating Eq. (3.29), we know that the DR [or the overshoot (OS)] is a function of the damping ratio:

$$\text{DR} = (\text{OS})^2 = \exp\left(\frac{-2\pi \zeta}{\sqrt{1 - \zeta^2}}\right). \tag{5.18}$$

From this equation we can derive

$$\zeta^2 = \frac{(\ln \mathrm{DR})^2}{4\pi^2 + (\ln \mathrm{DR})^2}. \tag{5.19}$$

If we have a second-order system, we can derive an analytical relation for the controller. If we have a proportional controller with a second-order process, as in Example 5.2, the solution is unique. However, if we have, for example, a PI controller (two parameters) and a first-order process, there are no unique answers as we only have one design equation. We must specify one more design constraint in order to have a well-posed problem.

Example 5.3: Derive the closed-loop transfer function of a system with PI control and a first-order process. What is the offset in this system?

We substitute $G_c = K_c[(\tau_I s + 1)/\tau_I s]$ and $G_p = [K_p/(\tau_p s + 1)]$ into Eq. (5.15), and we find that the closed-loop servo transfer function is

$$\frac{C}{R} = \frac{K_c K_p (\tau_I s + 1)}{\tau_I s(\tau_p s + 1) + K_c K_p (\tau_I s + 1)} = \frac{(\tau_I s + 1)}{\left(\frac{\tau_I \tau_p}{K_c K_p}\right)s^2 + \frac{\tau_I(1 + K_c K_p)}{K_c K_p}s + 1}. \tag{E5.6}$$

There are two noteworthy items. First, the closed-loop system is now second order. The integral action adds another order. Second, the system steady-state gain is unity and will not have an offset. This is a general property of using PI control. [If this is not immediately obvious, try taking $R = 1/s$ and apply the final-value theorem. We should find the eventual change in the controlled variable to be $c'(\infty) = 1$.]

With the expectation that the second-order system may exhibit underdamped behavior, we rewrite the closed-loop function as

$$\frac{C}{R} = \frac{(\tau_I s + 1)}{\tau^2 s^2 + 2\zeta \tau s + 1}, \tag{E5.6a}$$

where the natural time period and DR of the system are

$$\tau = \sqrt{\frac{\tau_I \tau_p}{K_c K_p}}, \qquad \zeta = \frac{1}{2}(1 + K_c K_p)\sqrt{\frac{\tau_I}{K_c K_p \tau_p}}. \tag{E5.7}$$

Although we have the analytical results, it is not obvious how choices of the integral time constant and the proportional gain may affect the closed-loop poles or the system DR. (We may get a partial picture if we consider circumstances under which $K_c K_p \gg 1$.) Again, the analysis is deferred until we cover root locus; we should find that to be a wonderful tool in assessing how controller design may affect system response.

Example 5.4: Derive the closed-loop transfer function of a system with PD control and a first-order process.

Closed-loop transfer function (5.15) with $G_c = K_c(1 + \tau_D s)$ and $G_p = [K_p/(\tau_p s + 1)]$ is

$$\frac{C}{R} = \frac{K_c K_p (\tau_D s + 1)}{(\tau_p s + 1) + K_c K_p (\tau_D s + 1)} = \frac{K_c K_p (\tau_D s + 1)}{(\tau_p + K_c K_p \tau_D)s + 1 + K_c K_p}. \tag{E5.8}$$

The closed-loop system remains first order and the function is that of a lead–lag element. We can rewrite the closed-loop transfer function as

$$\frac{C}{R} = \frac{K(\tau_D s + 1)}{\tau s + 1},$$ (E5.8a)

where the system steady-state gain and time constant are

$$K = \frac{K_c K_p}{1 + K_c K_p}, \qquad \tau = \frac{\tau_p + K_c K_p \tau_D}{1 + K_c K_p}.$$

The system steady-state gain is the same as that with proportional control in Example 5.1. We, of course, expect the same offset with PD control too. The system time constant depends on various parameters. Again, this analysis is deferred until we discuss root locus.

Example 5.5: Derive the closed-loop transfer function of a system with proportional control and an integrating process. What is the offset in this system?

Let's consider $G_c = K_c$ and $G_p = 1/As$; substitution into Eq. (5.15) leads to

$$\frac{C}{R} = \frac{K_c}{As + K_c} = \frac{1}{(A/K_c)s + 1}.$$ (E5.9)

We can see quickly that the system has unity gain and there should be no offset. The point is that integral action can be introduced by the process and we do not need PI control under such circumstances. We come across processes with integral action in the control of rotating bodies and liquid levels in tanks connected to pumps (e.g., Example 3.1).

Example 5.6: Provide illustrative closed-loop time-response simulations. Most texts have schematic plots to illustrate the general properties of a feedback system. This is something that we can do ourselves by using MATLAB. Simulate the observations that we have made in previous examples. Use a unity feedback system.

We consider Example 5.3 again; let's pick τ_p to be 5 min and K_p be 0.8 (unit). Instead of using the equation that we derived in Example 5.3, we can use the following statements in MATLAB to generate a simulation for the case of a unit-step change in the set point. This approach is much faster than using Simulink.

```
kc=1;                        % The two tuning parameters to
taui=10;                     % be varied

% The following statements are best saved in an M-file

Gc=tf(kc*[taui 1],[taui 0]); % The PI controller function
Gp=tf(0.8,[5 1]);            % The process function
Gcl=feedback(Gc*Gp,1)        % Unity closed-loop function
                             % GcGp/(1 + GcGp)

step(Gcl);                   % Personalize your own plotting
                             % and put a hold for additional
                             % curves
```

In these statements, we have used `feedback()` to generate the closed-loop function C/R. The unity feedback loop is indicated by the "1" in the function argument. Try first with $K_c = 1$ and τ_I with values of 10, 1, and 0.1. Next, select $\tau_I = 0.1$ and repeat with $K_c = 0.1$, 1, 5, and 10. In both cases, the results should follow the qualitative trends that we anticipate. If we repeat the calculation with a larger integral time $\tau_I = 1$ and use $K_c = 0.1$, 1, 5, 10, and 50, we may find the results to be rather unexpected. However, we do not have enough theory to explain them now. Keep the results in mind; it is hoped that this will be motivation to explore the later chapters.

We could also modify the M-file by changing the PI controller to a PD or a PID controller to observe the effects of changing the derivative time constant. (Help is in MATLAB Session 5.) We will understand the features of these dynamic simulations better in later chapters. For now, the simulations should give us a qualitative feel of the characteristics of a PID controller and (we hope) also the feeling that we need a better way to select controller settings.

Example 5.7: We have to design a servo controller for a mixing system (see Fig. E5.7A). A blue dye for making denim is injected into a stream of water. The injected dye is blended into the pipe flow with the aid of in situ static mixers. A photodetector downstream is used to monitor the dye concentration. The analog output of the detector is transmitted to a controller, which in turn sends a signal to the dye-injection regulating valve. In designing this mixing system, we have to be careful with the location of the photodetector. It has to be downstream enough to ensure a good reading of the mixed stream. However, if the photodetector is too far downstream, the larger transport lag can destabilize the system.

The water flow rate in the pipe is 2 L/s and the pipe has a cross-sectional area of 5 cm². The regulating valve is especially designed so that the dye dosage, in milliliters per second, varies linearly with the valve position. The regulating valve is thus first order with a time constant of 0.2 s and a steady-state gain of 0.6 mL s^{-1} mV^{-1}. The mixing process itself can also be modeled as first order with a steady-state gain of 0.8 ppm s mL^{-1} (where ppm indicates parts per million). A previous experiment indicated that a step change in the regulating valve resulted in a response in a dye concentration that is 99.3% complete in 20 s. The magic photodetector is extremely fast, and the response is linear over a large concentration range. The manufacturer provided the calibration as

$$v = 0.3 + 2.6 \text{ (dye)},$$

where the voltage output is in millivolts and the concentration of the dye is in parts per million.

This is a problem that we have to revisit many times in later chapters. For now, let's draw the block diagram of the dye control system and provide the necessary transfer functions. We identify units in the diagram and any possible disturbances to the system. In addition,

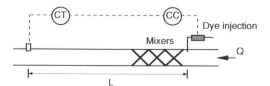

Figure E5.7A.

Analysis of Single-Loop Control Systems

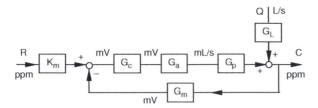

Figure E5.7B.

we do not know where to put the photodetector at this point. Let's just presume that the photodetector is placed 290 cm downstream.

The block diagram is shown in Fig. E5.7B, where the dye concentration is denoted by C and the set point by R. The flow rate is one probable source of disturbance.

Based on the information given, the transfer functions are

$$G_p = \frac{K_p}{\tau_p s + 1}, \quad G_a = \frac{K_v}{\tau_v s + 1}, \quad G_m = K_m e^{-t_d s},$$

and we do not know the steady-state gain of the load function G_L. The values of various parameters are $K_p = 0.8$ ppm s mL^{-1}, $\tau_p \approx 20/5 = 4$ s, $K_a = 0.6$ mL s^{-1} mV^{-1}, $\tau_a = 0.2$ s, and $K_m = 2.6$ mV/ppm. The average fluid velocity is $2000/5 = 400$ cm/s. The transport lag is hence $t_d = 290/400 = 0.725$ s. We presumably will use a PID transfer function for the controller G_c; we will continue with this problem in Example 5.7A.

5.4. Selection and Action of Controllers

A few words need to be added on the action of controllers. The point to make is that we have to do a bit of physical reasoning when we design a real system. We also have to factor in safety and determine what the controller and actuator may do if there is a system failure.

A standard textbook system has a controller with a positive proportional gain. All the other blocks such as the process and actuator have positive steady-state gains as well. However, this is not always the case. Here liquid-level control is used to illustrate the idea. Keep Fig. 5.7 in mind in the discussion below.

Say we want to control the liquid level in a tank by manipulating the *inlet* flow rate (Fig. 5.8). If the liquid level *drops* below the set point, the controller will *increase* its output signal to open up the inlet valve and increase liquid flow. The changes in controlled variable and controller output are in opposite directions. This is a consequence of how the error is defined in a negative-feedback system.

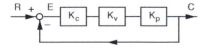

Figure 5.7. Simple system used in the discussion of controller actions. Depending on the situation, K_c, K_v, and K_p can be either positive or negative.

102

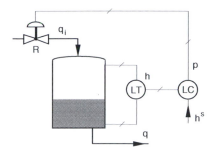

Figure 5.8. Manipulating the liquid level with an inlet valve.

In this particular case, we use an **air-to-open valve**, meaning that we need to increase the signal to open up the valve. That is, the valve has a *positive* steady-state gain ($+K_v$). A pneumatic air-to-open valve also means that energy is required for keeping it open. Under a system failure in which power is lost, the valve closes and prevents flooding the tank. We refer to the valve here as a **fail-close valve**, which is the preferred safety design in Fig. 5.8.

The opposite case is to use an **air-to-close valve**, which has a *negative* steady-state gain ($-K_v$); an increase in signal will close the valve.[21] Hence this is a **fail-open valve**, which, for safety reasons, is not a wise choice here. Nevertheless, if we insist on installing this air-to-close valve, we will need a controller with a negative gain ($-K_c$). Now, if the liquid level *drops*, the controller output signal will also *decrease*, opening up the air-to-close valve.

Let's take a look at the logic when we control the liquid level by manipulating the *outlet* valve (Fig. 5.9). In this case the process gain K_p associated with the outlet flow is negative. If the liquid level drops below the set point, we now want to reduce the outlet flow rate by closing up the valve. Again, there are two possible cases.

If we use a controller with positive gain ($+K_c$), the controller output increases as the liquid level drops. We can reduce the flow only if we use an air-to-close valve ($-K_v$). In case of power outage, the valve will stay open. This fail-open valve can drain the entire tank, an event that we may not want to happen.

On the other hand, we can choose an air-to-open valve ($+K_v$) at the outlet location. Now the only way to reduce the flow rate as the liquid level drops is to "switch" to a controller

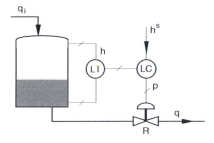

Figure 5.9. Manipulating the liquid level with an outlet valve. The process gain is negative in this case.

---

[21] This point can easily get lost in the long explanation: An air-to-open valve has a positive gain and is failed close. An air-to-close valve has a negative gain ($-K_v$) and is failed open.

with a negative gain $(-K_c)$. With $-K_c$ as the proportional gain, a drop in liquid level will lead to a decrease in controller output. In this case, we have a fail-close valve, which is desirable if we do not want perturbations to propagate downstream.

There is no question that the terminology is confusing. Do not let it confuse you. The best strategy is to "walk" through the sign (action) of every single steady-state gain of the block diagram, including the process and the sensor, and determine the probable and logical signs (actions) of the controller and the actuator. As a consistency check, we should note that within the feedback loop there should only be one net negative sign. There is no getting away from doing some thinking of your own.

5.4.1. A Few Comments on the Choice of Controllers

In process engineering, the most common types of controlled variables are liquid level, flow rate, temperature, pressure, and sometimes concentration. Here are some very general ideas. To fully appreciate these tidbits, we also need to know something about the actual hardware – actuators or control elements – that we find in handbooks or equipment manuals.

Flow Control

PI controllers are the most common. They eliminate offsets and have acceptable speeds of response in most industrial settings. We usually pick a low to intermediate gain (wide proportional band, PB ≈ 150) to reduce the effect of noisy signals (from flow turbulence; also why we do not use D control). We also use a low reset time (≈ 0.1 min/repeat; i.e., relatively large I action) to get fast set-point tracking.

Level Control

We usually need to keep the liquid level within a certain range only around the desired set point. Speed is not a great concern. Recall that, depending on how we implement the outlet pump, we can have a process with integrating action itself. Also, if the outlet flow rate is used as the manipulated variable, the controller setting must be conservative to avoid sudden surges in the exit flow rate. Thus a simple P controller is usually adequate, presuming that we have no complications such as boiling or vaporization. Make sure you check whether the valve that you are using (or buying) is air-to-open or air-to-close.

Pressure Control

The control of pressure depends on the situation and cannot be generalized. We can see this from a couple of examples:

For the vapor pressure in a flash drum (and thus also vapor flow rate), we need a fast and tight response loop. We need at least a PI controller as in flow control.

For the top of a distillation column, we usually control the pressure indirectly by means of the condensation of vapor in the condenser, which in turn is controlled by the amount of cooling water. Heat transfer through a heat exchanger has very slow dynamics. Thus we cannot use PI control. We either use P control or, when response time is important, PID.

Temperature Control

Heat transfer lags can be significant, and the nature of the problem can be quite different in various processes. If there is a sensor lag, it is mostly due to heat transfer between the sensor and the fluid medium. (Thermocouples, depending on how we make them, can have very

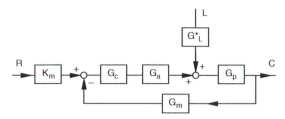

Figure R5.1.

fast response times.) The overall response is sluggish, and PI control will make it more so. It is unlikely we can live with any offsets. PID control is the appropriate choice.

Concentration Control

The strategy depends on the situation and how we measure the concentration. If we can rely on pH or absorbance (UV, visible, or IR spectrometer), the sensor response time can be reasonably fast, and we can make our decision based on the actual process dynamics. Most likely we would be thinking along the lines of PI or PID controllers. If we can use only gas chromatography (GC) or other slow analytical methods to measure concentration, we must consider discrete data-sampling control. Indeed, prevalent time delay makes chemical process control unique and, in a sense, more difficult than many mechanical or electrical systems.

In terms of the situation, if we use a PI controller on a slow multicapacity process, the resulting system response will be even more sluggish. We should use PID control to increase the speed of the closed-loop response (being able to use a higher proportional gain) while maintaining stability and robustness. This comment applies to other cases such as temperature control as well.

Review Problems

(1) An alternative way of drawing a feedback system block diagram is shown in Fig. R5.1. How is G_L^* related to G_L as in Fig. 5.4?

(2) Try to obtain the closed-loop transfer functions in Eq. (5.11) by means of observation, i.e., without using the algebraic steps in the text.

(3) Consider the liquid flow-rate controller in Fig. R5.3. We want to keep the flow rate q constant no matter how the upstream pressure fluctuates. Consider what happens if the upstream flow Q drops below the steady-state value. How would you choose the regulating valve when you have (a) a positive and (b) a negative proportional gain?

(4) What is the OS and DR if we pick $\zeta = 0.707$? If the DR is 1/4, what is the damping ratio?

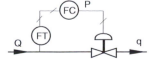

Figure R5.3.

(5) Refer back to Example 5.1. What is the offset if we consider the regulating problem ($R = 0$, $L = 1/s$)?

(6) When we developed the model for the stirred-tank heater, we ignored the dynamics of the heating coil. Provide a slightly more realistic model that takes into consideration the flow rate of condensing steam.

(7) Do the time-domain simulations with MATLAB in Example 5.6. Try also with a PD or a PID controller.

Hints:

(1) $G_L^* = G_L/G_p$.

(2) The G_{sp} function is obvious. To obtain the load function G_{load}, set $R = 0$, and try to visualize the block diagram such that it is unity in the forward path and all the functions are in the feedback loop.

(4) OS = 4.32% and DR = 1.87×10^{-3}. When DR = 0.25, $\zeta = 0.215$.

(5) Now,

$$C = \frac{G_L}{1 + G_c G_p} L,$$

$R = 0$, and thus

$$E = R - C = 0 - \frac{G_L}{1 + G_c G_p} L.$$

With $L = 1/s$ and the final-value theorem,

$$e(\infty) = - \lim_{s \to 0} \frac{G_L}{1 + G_c G_p}.$$

Substitution of the first-order functions and a proportional controller gives

$$e(\infty) = - \frac{K_L}{1 + K_c K_p},$$

which becomes smaller if we increase K_c.

(6) How we model the stirred-tank heater is subject to the actual situation. At a slightly more realistic level, we may assume that heat is provided by condensing steam and that the coil metal is at the same temperature as the condensing steam. The heat balance and the Laplace transform of the tank remains identical to those given in Chap. 2:

$$\rho C_p V \frac{dT}{dt} = \rho C_p Q (T_i - T) + U A (T_H - T),$$

$$T = \left(\frac{K_d}{\tau_p s + 1} \right) T_i + \left(\frac{K_p}{\tau_p s + 1} \right) T_H.$$

We also need a heat balance for the heating coil, which can be written as

$$M_H C_H \frac{dT_H}{dt} = m_s \lambda - U A (T_H - T),$$

where T_H is the temperature and M_H and C_H are the mass and the heat capacity, respectively, of the heating coil. The steam mass flow rate is m_s, and λ is the heat of condensation. We should obtain the Laplace transform of the form

$$T_H = \left(\frac{1}{\tau_H s + 1}\right) T + \left(\frac{K_s}{\tau_H s + 1}\right) M_s.$$

You should be able to fill in the gaps and finish the rest of the work in deriving the transfer functions. In this case, you may want to use the steam mass flow rate as the manipulated variable. The transfer function relating its effect on T will be second order, and the characteristic polynomial does not have the clean form in our simpler examples.

(7) The basic statements are provided already in the example. For more details, see the *Web Support* for the MATLAB statements and plots.

6

Design and Tuning of Single-Loop Control Systems

W
e will go through a whole bundle of tuning methods. We only need to "pick" three numbers for a PID controller, but this is one of the most confusing parts of learning control. Different tuning techniques give similar but not identical results. There are no "best" or "absolutely correct" answers. The methods all have pros and cons and, working together, they complement each other. We need to make proper selection and sound judgment – very true to the act (and art) of design.

What Are We Up to?

- Tuning a controller with empirical relations
- Tuning a controller with internal model control relations

6.1. Tuning Controllers with Empirical Relations

Let's presume that we have selected the valves and the transducers and even installed a controller. We now need to determine the controller settings – a practice that is called tuning a controller. Trial-and-error tuning can be extremely time consuming (and dumb!), to the extent that it may not be done. A large distillation column can take hours to reach steady state. A chemical reactor may not reach steady state at all if you have reactor "runaway." Some systems are unstable at high and low feedback gains; they are stable only in some intermediate range. These are reasons why we have to go through all the theories to learn how to design and tune a controller with well-educated (or so we hope) guesses.

Empirical tuning roughly involves doing either an open-loop or a closed-loop experiment and fitting the response to a model. The controller gains are calculated on the basis of this fitted function and some empirical relations. When we use empirical tuning relations, we cannot dictate system dynamic response specifications. The controller settings are seldom optimal and most often require field tuning after installation to meet more precise dynamic response specifications. Empirical tuning may not be appealing from a theoretical viewpoint, but it gives us a quick-and-dirty starting point. Two remarks before we begin.

- Most empirical tuning relations that we use here are based on open-loop data fitted to a first-order with dead-time transfer function. This feature is unique to process engineering, in which most units are self-regulating. The dead time is either an approximation of multistage processes or a result of transport lag in the measurement. With large uncertainties and the need for field tuning, models more elaborate than the first-order with dead-time function are usually not warranted with empirical tuning.
- Some empirical tuning relations, such as that of Cohen and Coon, are developed to achieve a one-quarter DR response in handling disturbances. When we apply the settings of these relations to a servo problem, it tends to be very oscillatory and is not what is considered as slightly underdamped.[1] The controller design depends on the specific problem at hand. We certainly need to know how to tune a controller after using empirical tuning relations to select the initial settings.[2]

6.1.1. Controller Settings Based on a Process-Reaction Curve

To make use of empirical tuning relations, one approach is to obtain the so-called **process-reaction curve**. We disable the controller and introduce a step change to the actuator. We then measure the **open-loop step response**. This practice can simply be called an open-loop step test. Although we disconnect the controller in the schematic diagram (Fig. 6.1), we usually only need to turn the controller to the "manual" mode in reality. As shown in the block diagram, what we measure is a lumped response, representing the dynamics of the blocks G_a, G_p, and G_m. We denote the lumped function as G_{PRC}, the process-reaction-curve function:

$$G_{\text{PRC}} = \frac{C_m}{P} = G_a G_p G_m. \tag{6.1}$$

From the perspective of doing the experiment, we need the actuator to effect a change in the manipulated variable and the sensor to measure the response.

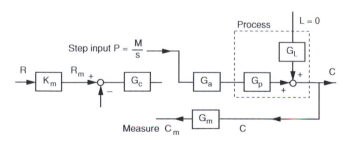

Figure 6.1. Block-diagram illustration of an open-loop step test.

---

[1] If we assume that an oscillatory system response can be fitted to a second-order underdamped function. With Eq. (3.29), we can calculate that with a DR of 0.25, the damping ratio ζ is 0.215, and the maximum percent OS is 50%, which is *not* insignificant. (These values came from Revew Problem (4) back in Chap. 5.)

[2] By and large, a quarter DR response is acceptable for disturbances but not desirable for set-point changes. Theoretically, we can pick any DR of our liking. Recall from Section 2.7 that the position of the closed-loop pole lies on a line governed by $\theta = \cos^{-1} \zeta$. In the next chapter, we will locate the pole position on a root-locus plot based on a given damping ratio.

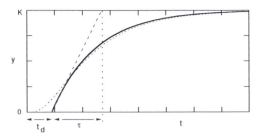

Figure 6.2. Illustration of fitting Eq. (6.2) (solid curve) to open-loop step-test data representative of self-regulating and multicapacity processes (dotted curve). The time-constant estimation shown here is based on the initial slope and a visual estimation of dead time. The Ziegler–Nichols tuning relation (see Table 6.2) also uses the slope through the inflection point of the data (not shown). Alternative estimation methods are provided on the *Web Support*.

The measurement of G_{PRC} is how we may design a system if we know little about our process and are incapable of constructing a model (what an excuse!). Even if we know what the functions G_a and G_m should be, we do not need them because the controller empirical tuning relations were developed for the lumped function G_{PRC}. On the other hand, if we know precisely what the functions G_a, G_p, and G_m are, we may use them to derive G_{PRC} as a reduced-order approximation of the product of $G_a G_p G_m$.

The real-time data (the process-reaction curve) in most processing unit operations take the form of a sigmoidal curve, which is fitted to a first-order with dead-time function (Fig. 6.2):

$$G_{PRC} = \frac{C_m}{P} \approx \frac{K e^{-t_d s}}{\tau s + 1}. \tag{6.2}$$

One reason why this approximation works is that process unit operations are generally open-loop stable, and many are multicapacity in nature. *Reminder*: An underdamped response of the system is due to the controller, which is taken out in the open-loop step test.

Using the first-order with dead-time function, we can go ahead and determine the controller settings with empirical tuning relations. The most common ones are the **Ziegler–Nichols** relations. In process unit operation applications, we can also use the **Cohen–Coon** or the **Ciancone–Marlin** relations. These relations are listed in Table 6.1.

6.1.2. Minimum Error-Integral Criteria

The open-loop test response fitted to a first-order with dead-time function G_{PRC} can be applied to other tuning relations. One such possibility is a set of relations derived from the minimization of error integrals. Here, only the basic idea behind the use of error integrals is provided.

To derive the tuning equations, we would use the theoretical time-domain closed-loop system response as opposed to a single quantity, such as the DR. The time-domain solution is dependent on the type of controller and the nature of the input (set-point or disturbance changes) and, in our case, a "process" function that is first order with dead-time. We can also calculate the error – the difference between the set point and the controlled variable. We then find controller settings that may minimize the error over time (the error integral), using for instance, Lagrange multipliers as in introductory calculus. Of course, we are not doing the work; the actual analysis is better left for a course in optimal control.

There are different ways to define the error function to be minimized. A few possibilities are as follows:

(1) Integral of the square error (ISE):

$$\text{ISE} = \int_0^\infty [e'(t)]^2 \, dt. \tag{6.3}$$

The ISE magnifies large errors – squaring a small number (< 1) makes it even smaller. Thus minimization of this integral should help to suppress large initial errors. The resulting controller setting tends to have a high proportional gain and the system is very underdamped.

(2) Integral of the absolute error (IAE):

$$\text{IAE} = \int_0^\infty |e'(t)| \, dt. \tag{6.4}$$

The IAE simply integrates the absolute value and puts equal weight to large and small errors.

(3) Integral of the time-weighted absolute error (ITAE):

$$\text{ITAE} = \int_0^\infty t|e'(t)| \, dt. \tag{6.5}$$

The time-weighting function puts a heavy penalty on errors that persist for long periods of time. This weighting function also helps to derive controller settings that allow for low settling times.

Before we move on, a few comments and reminders:

- As far as we are concerned, using the error-integral criteria is just another empirical method. We are simply using the results of minimization obtained by other people, not to mention that the first-order with dead-time function is from an open-loop test.
- The controller setting is different depending on which error integral we minimize. Set-point and disturbance inputs have different differential equations, and because the optimization calculation depends on the time-domain solution, the result will depend on the type of input. The closed-loop poles are the same, but the zeros, which affect the time-independent coefficients, are not.
- The time integral is from $t = 0$ to $t = \infty$, and we can minimize it only if it is bounded. In other words, we cannot minimize the integral if there is a steady-state error. Only PI and PID controllers are applicable to this design method.[3]
- Theoretically, we can minimize the integral by using other criteria. On the whole, the controller settings based on minimizing the ITAE provide the most conservative

---

[3] If you come across a proportional controller here, it is possible only if the derivation has ignored the steady-state error or shifted the reference such that the so-called offset is zero.

controller design and are highly recommended. This is the only set of tuning relations included in Table 6.1.

6.1.3. Ziegler–Nichols Ultimate-Cycle Method

This empirical method is based on closed-loop testing (also called on-line tuning) of processes that are inherently stable, but in which the system may become unstable. We use only proportional control in the experiment. If it is not possible to disable the integral and derivative control modes, we set the integral time to its maximum value and the derivative time to its minimum. The proportional gain is slowly increased until the system begins to exhibit sustained oscillations with a given small-step set-point or load change. The proportional gain and the period of oscillation at this point are the **ultimate gain** K_{cu} and the **ultimate period** T_u. These two quantities are used in a set of empirical tuning relations developed by Ziegler and Nichols – again listed in Table 6.1.

Two more comments:

- A preview: We can derive the ultimate gain and the ultimate period (or frequency) with stability analyses. In Chap. 7, we use the substitution $s = j\omega$ in the closed-loop characteristic equation. In Chap. 8, we make use of what is called the Nyquist stability criterion and Bode plots.
- One may question the meaning of "sustained oscillations." We may gather that the ultimate gain and ultimate period are associated with marginal stability – the instance when the system is just about to become unstable. Of course, we never want to push that far in real life. With large uncertainties involved in empirical tuning and field tuning, it is not necessary to have accurate measurements of K_{cu} and T_u. When we do an experiment, we just increase the proportional gain until we achieve a fairly underdamped response.

Example 5.7A: What would be the PID controller settings for the dye-mixing problem in Example 5.7?

Based on what we have obtained in Example 5.7, if we did an open-loop experiment as suggested in Eq. (6.1), our step response would fit well to the function:

$$G_{\text{PRC}} = G_a G_p G_m = \frac{(0.8)(0.6)(2.6)e^{-0.725s}}{(4s+1)(0.2s+1)}.$$

However, to use the empirical tuning relations, we need to fit the data to a first-order function with dead-time. Thus at this stage, we probably would have obtained the following approximation:

$$G_{\text{PRC}} \approx \frac{1.25e^{-0.9s}}{4s+1}.$$

Here we assume that the data fitting allows us to recover the time constant of the dominant pole reasonably well, and the dead-time is roughly $0.9s$. We are not adding exactly 0.2 to 0.725 as a way to emphasize that, in reality, we would be doing data fitting and the result will vary. How good an approximation is depends very much on the relative differences in the time constants. (Try with MATLAB simulation to see how good the approximation is. For the numbers chosen in this example, it is easy to obtain a good approximation.)

Now, with Table 6.1,[4] we can calculate the following PID controller settings:

	K_c	τ_I	τ_D
Cohen–Coon	4.9	2.0	0.31
Ziegler–Nichols	4.3	1.8	0.45
ITAE (set point)	2.8	3.1	0.31
Ciancone–Marlin (set point)	1.2	4.4	0.07

All tuning relations provide different results. Generally, the Cohen–Coon relation has the largest proportional gain and the dynamic response tends to be the most underdamped. The Ciancone–Marlin relation provides the most conservative setting, and it uses a very small derivative time constant and a relatively large integral time constant. In a way, their correlation reflects a common industrial preference for PI controllers.

We'll see how they compare in time response simulations when we come back to this problem in Example 5.7C. A point to be made is that empirical tuning is a very imprecise science. There is no reason to worry about the third significant figure in your tuning parameters. The calculation only serves to provide us with an initial setting with which we begin to do field or computational tuning.

Although the calculations in the preceding example may appear as simple plug-and-chug, we should take a closer look at the tuning relations. The Cohen–Coon equations for the proportional gain taken from Table 6.1 are

$$\text{P:} \qquad K_c K = \left(\frac{\tau}{t_d} + \frac{1}{3} \right), \tag{6.6}$$

$$\text{PI:} \qquad K_c K = \left(0.9 \frac{\tau}{t_d} + \frac{1}{12} \right), \tag{6.7a}$$

$$\text{PID:} \quad K_c K = \left(\frac{4}{3} \frac{\tau}{t_d} + \frac{1}{4} \right). \tag{6.8a}$$

The choice of the proportional gain is affected by two quantities: the product $K_c K$ and the ratio of dead time to time constant t_d/τ. It may not be obvious why the product $K_c K$ is important now, but we shall see in Section 6.2 how it arises from direct synthesis and appreciate in Chap. 8 how it helps determine system stability.

Under circumstances in which the dead time is relatively small, only the first term on the RHS is important in the three tuning equations. When dead time becomes larger (or τ/t_d smaller), we need to decrease the proportional gain, and this is how the tuning relations are constructed. When we add integral control, we need to decrease K_c. Indeed, in Eq. (6.7a), the τ/t_d term is decreased by 10% and the constant term is reduced to 1/12. With the implementation of PID control, we can afford to have a larger K_c. This is reflected in Eq. (6.8a). We can make similar observations with the Ziegler–Nichols relations in Table 6.1. Furthermore, we may also see in Table 6.1 that if the dead-time increases, we should raise the integral time constant.

[4] Really calculated with our M-file recipe.m, which can be found on our *Web Support*.

Design and Tuning of Single-Loop Control Systems

Table 6.1. Table of tuning relations

A. Tuning relations based on open-loop testing and response fitted to a first-order with dead-time function:

$$G_{PRC} = \frac{K e^{-t_d s}}{\tau s + 1}$$

Controller	Cohen–Coon		Ziegler–Nichols	
P	$K_c K = \left(\dfrac{\tau}{t_d} + \dfrac{1}{3}\right)$	(6.6)	$K_c K = \dfrac{\tau}{t_d}$	(6.9)
PI	$K_c K = \left(0.9\dfrac{\tau}{t_d} + \dfrac{1}{12}\right)$	(6.7a)	$K_c K = 0.9\dfrac{\tau}{t_d}$	(6.10a)
	$\tau_I = t_d\dfrac{30 + 3(t_d/\tau)}{9 + 20(t_d/\tau)}$	(6.7b)	$\tau_I = 3.3 t_d$	(6.10b)
PID	$K_c K = \left(\dfrac{4}{3}\dfrac{\tau}{t_d} + \dfrac{1}{4}\right)$	(6.8a)	$K_c K = 1.2\dfrac{\tau}{t_d}$	(6.11a)
	$\tau_I = t_d\dfrac{32 + 6(t_d/\tau)}{13 + 8(t_d/\tau)}$	(6.8b)	$\tau_I = 2t_d$	(6.11b)
	$\tau_D = t_d\dfrac{4}{11 + 2(t_d/\tau)}$	(6.8c)	$\tau_D = 0.5 t_d$	(6.11c)

Minimum ITAE criterion

For load change:

$$K_c = \frac{a_1}{K}\left(\frac{\tau}{t_d}\right)^{b_1}, \quad \tau_I = \frac{\tau}{a_2}\left(\frac{t_d}{\tau}\right)^{b_2}, \quad \tau_D = a_3\tau\left(\frac{t_d}{\tau}\right)^{b_3}. \tag{6.12}$$

Controller	a_1	b_1	a_2	b_2	a_3	b_3
PI	0.859	0.977	0.674	0.680	—	—
PID	1.357	0.947	0.842	0.738	0.381	0.995

For set-point change:

$$K_c = \frac{a_1}{K}\left(\frac{\tau}{t_d}\right)^{b_1}, \quad \tau_I = \frac{\tau}{a_2 - b_2(t_d/\tau)}, \quad \tau_D = a_3\tau\left(\frac{t_d}{\tau}\right)^{b_3}. \tag{6.13}$$

Controller	a_1	b_1	a_2	b_2	a_3	b_3
PI	0.586	0.916	1.03	0.165	—	—
PID	0.965	0.855	0.796	0.147	0.308	0.929

B. Tuning relations based on closed-loop testing and the Ziegler–Nichols ultimate-gain (cycle) method with given ultimate proportional gain K_{cu} and ultimate period T_u.

Ziegler–Nichols ultimate-gain method
Controller

P	$K_c = 0.5K_{cu}$	(6.14)					
PI	$K_c = 0.455K_{cu}$	(6.15a)					
	$\tau_I = 0.833T_u$	(6.15b)					

PID	Quarter decay		Just a bit of overshoot		No overshoot	
	$K_c = 0.6K_{cu}$	(6.16a)	$K_c = 0.33K_{cu}$	(6.17a)	$K_c = 0.2K_{cu}$	(6.18a)
	$\tau_I = 0.5T_u$	(6.16b)	$\tau_I = 0.5T_u$	(6.17b)	$\tau_I = 0.5T_u$	(6.18b)
	$\tau_D = 0.125\ T_u$	(6.16c)	$\tau_D = 0.333T_u$	(6.17c)	$\tau_D = 0.333T_u$	(6.18c)

Note: All formulas in Table 6.1, and the PID settings in Table 6.2 later, are implemented in the M-file recipe.m, available from our *Web Support*. The Ciancone and Marlin tuning relations are graphical, and we have omitted them from the tables. The correlation plots, explanations, and the interpolation calculations are provided by our M-file ciancone.m, which is also used by recipe.m.

6.2. Direct Synthesis and Internal Model Control

We now apply a different philosophy to controller design. Up until now, we have had a preconceived idea of what a controller should be, and we tune it until we have the desired system response. On the other hand, we can be more proactive: We define what our desired closed-loop response should be and design the controller accordingly. The resulting controller is not necessarily a PID controller. This is acceptable with computer-based controllers because we are not restricted to off-the-shelf hardware.

In this chapter, however, our objective is more restricted. We purposely choose simple cases and make simplifying assumptions such that the results are PID controllers. We will see how the method helps us select controller gains based on process parameters (i.e., the process model). The method provides us with a more rational controller design than do the empirical tuning relations. Because the result depends on the process model, this method is what we consider a **model-based** design.

6.2.1. Direct Synthesis

We consider a servo problem (i.e., $L = 0$) and set $G_m = G_a = 1$. The closed-loop function is the familiar

$$\frac{C}{R} = \frac{G_c G_p}{1 + G_c G_p},$$

(6.19)

which we now rearrange as

$$G_c = \frac{1}{G_p}\left(\frac{C/R}{1 - C/R}\right).$$

(6.20)

The implication is that if we define our desired system response C/R, we can derive the appropriate controller function for a specific process function G_p.

A couple of quick observations: First, G_c is the reciprocal of G_p. The poles of G_p are related to the zeros of G_c and vice versa – this is the basis of the so-called *pole-zero cancellation*.[5] Second, the choice of C/R is not entirely arbitrary; it must satisfy the closed-loop characteristic equation:

$$1 + G_c G_p = 1 + \left(\frac{C/R}{1 - C/R} \right) = 0. \tag{6.21}$$

From Eq. (6.20), it is immediately clear that we cannot have an ideal servo response where $C/R = 1$, which would require an infinite controller gain. Now Eq. (6.21) indicates that C/R cannot be some constant either. To satisfy Eq. (6.21), the closed-loop response C/R must be some function of s, meaning that the system cannot respond instantaneously and must have some finite response time.

Let's select a more realistic system response, say, a simple first-order function with unity steady-state gain:

$$\frac{C}{R} = \frac{1}{\tau_c s + 1}, \tag{6.22}$$

where τ_c is the system time constant, a design parameter that we specify. The unity gain means that we should eliminate offset in the system. Substitution of Eq. (6.22) into Eq. (6.20) leads to the controller function:

$$G_c = \frac{1}{G_p} \left(\frac{1}{\tau_c s} \right). \tag{6.23}$$

The closed-loop characteristic equation, corresponding to Eq. (6.21), is

$$1 + \frac{1}{\tau_c s} = 0, \tag{6.24}$$

which really is $1 + \tau_c s = 0$, as dictated by Eq. (6.22). The closed-loop pole is at $s = -1/\tau_c$. This result is true no matter what G_p is – as long as we can physically build or program the controller on a computer. Because the system time constant τ_c is our design parameter, it appears that direct synthesis magically allows us to select whatever response time we want. Of course this cannot be the case in reality. There are physical limitations, such as saturation.

Example 6.1: Derive the controller function for a system with a first-order process and a system response dictated by Eq. (6.22).

The process transfer function is $G_p = [K_p/(\tau_p s + 1)]$, and the controller function according to Eq. (6.23) is

$$G_c = \frac{(\tau_p s + 1)}{K_p} \frac{1}{\tau_c s} = \frac{\tau_p}{K_p \tau_c} \left(1 + \frac{1}{\tau_p s} \right), \tag{E6.1}$$

---

[5] The controller function will take on a positive pole if the process function has a positive zero. It is not desirable to have an inherently unstable element in our control loop. This is an issue that internal model control will address.

Table 6.2. Summary of PID controller settings based on internal model control or direct synthesis

Process model	Controller	K_c	τ_I	τ_D
$\dfrac{K_p}{\tau_p s + 1}$	PI	$\dfrac{\tau_p}{K_p \tau_c}$	τ_p	—
$\dfrac{K_p}{(\tau_1 s + 1)(\tau_2 s + 1)}$	PID	$\dfrac{\tau_1 + \tau_2}{K_p \tau_c}$	$\tau_1 + \tau_2$	$\dfrac{\tau_1 \tau_2}{\tau_1 + \tau_2}$
	PID with $\tau_1 > \tau_2$	$\dfrac{\tau_1}{K_p \tau_c}$	τ_1	τ_2
	PI (underdamped)	$\dfrac{\tau_1}{4 K_p \zeta^2 \tau_2}$	τ_1	—
$\dfrac{K_p}{\tau^2 s^2 + 2\zeta \tau s + 1}$	PID	$\dfrac{2\zeta \tau}{K_p \tau_c}$	$2\zeta \tau$	$\dfrac{\tau}{2\zeta}$
$\dfrac{K_p}{s(\tau_p s + 1)}$	PD	$\dfrac{1}{K_p \tau_c}$	—	τ_p
$\dfrac{K_p e^{-t_d s}}{\tau_p s + 1}$	PI	$\dfrac{\tau_p}{K_p (\tau_c + t_d)}$	τ_p	—
	PID	$\dfrac{1}{K_p} \dfrac{2\tau_p/t_d + 1}{2\tau_c/t_d + 1}$	$\tau_p + t_d/2$	$\dfrac{\tau_p}{2\tau_p/t_d + 1}$
$\dfrac{K_p}{s}$	P	$\dfrac{1}{K_p \tau_c}$	—	—

which is obviously a PI controller with $K_c = \tau_p/K_p \tau_c$ and $\tau_I = \tau_p$. Note that the proportional gain is inversely proportional to the process gain. Specification of a small system time constant τ_c also leads to a large proportional gain.

Reminder: The controller settings K_c and τ_I are governed by the process parameters, and the system response, which we choose. The *one and only tuning parameter* is the system response time constant τ_c.

Example 6.2: Derive the controller function for a system with a second-order overdamped process and system response as dictated by Eq. (6.22).

The process transfer function is $G_p = \{K_p/[(\tau_1 s + 1)(\tau_2 s + 1)]\}$, and the controller function according to Eq. (6.23) is

$$G_c = \frac{(\tau_1 s + 1)(\tau_2 s + 1)}{K_p} \frac{1}{\tau_c s}.$$

We may see that this is a PID controller. Nevertheless, there are two ways to manipulate the function. One is to expand the terms in the numerator and factor out $(\tau_1 + \tau_2)$ to obtain

$$G_c = \frac{(\tau_1 + \tau_2)}{K_p \tau_c} \left[1 + \frac{1}{(\tau_1 + \tau_2)} \frac{1}{s} + \left(\frac{\tau_1 \tau_2}{\tau_1 + \tau_2} \right) s \right]. \tag{E6.2}$$

The proportional gain, integral time, and derivative time constants are provided by the respective terms in the transfer function. If you have trouble spotting them, they are summarized in Table 6.2.

The second approach is to consider the controller function as a series PID such that we write

$$G_c = \frac{\tau_1}{K_p \tau_c}\left(1 + \frac{1}{\tau_1 s}\right)(\tau_2 s + 1), \tag{E6.3}$$

with $\tau_1 > \tau_2$. We can modify the derivative term to be the "real" derivative action as written in Eqs. (5.9a) and (5.9b).

From our experience that the derivative time constant should be smaller than the integral time constant, we should pick the larger time constant as the integral time constant. Thus we select τ_1 to be the integral time constant and τ_2 to be the derivative time constant. In the limit $\tau_1 \gg \tau_2$, both arrangements [Eqs. (E6.2) and (E6.3)] of the controller function are the same.

When dead-time is inherent in a process, it is difficult to avoid dead-time in the system. Thus we define the system response as

$$\frac{C}{R} = \frac{e^{\theta s}}{\tau_c s + 1}, \tag{6.25}$$

where θ is the dead-time in the system. The controller function, by means of Eq. (6.20), is hence

$$G_c = \frac{1}{G_p}\left[\frac{e^{-\theta s}}{(\tau_c s + 1) - e^{-\theta s}}\right] \approx \frac{1}{G_p}\left[\frac{e^{-\theta s}}{(\tau_c + \theta)s}\right]. \tag{6.26}$$

To arrive at the last term, we use a simple Taylor expansion ($e^{-\theta s} \approx 1 - \theta s$) of the exponential term. This is purposely done to simplify the algebra, as shown in Example 6.3. [We could have used the Padé approximation in Eq. (6.26), but the result will not be the simple PI controller.]

Example 6.3: Derive the controller function for a system with a first-order process with dead-time and system response as dictated by Eq. (6.25).

The process transfer function is $G_p = [(K_p e^{-t_d s})/(\tau_p s + 1)]$. To apply Eq. (6.26), we make an assumption about the dead time, that $\theta = t_d$. The result is a PI controller:

$$G_c = \frac{\tau_p}{K_p(\tau_c + \theta)}\left(1 + \frac{1}{\tau_p s}\right). \tag{E6.4}$$

Even though this result is based on what we say is a process function, we could apply Eq. (E6.4) as if the derivation were for the first-order with dead-time function G_{PRC} obtained from an open-loop step test.

This is a question that invariably arises: What is a reasonable choice of the system time constant τ_c? Various sources in the literature have different recommendations. For example, one guideline suggests that we need to ensure that $\tau_c > 1.7\theta$ for a PI controller and $\tau_c > 0.25\theta$ for a PID controller. A reasonably conservative choice has been programmed into the M-file recipe.m available from the *Web Support*. The important reminder is that we should have a habit of checking the τ_c setting with time-response simulation and tuning analysis.

In contrast to Eq. (6.22), we can dictate a second-order underdamped system response:

$$\frac{C}{R} = \frac{1}{\tau^2 s^2 + 2\zeta\tau s + 1},\tag{6.27}$$

where τ and ζ are the system natural period and damping ratio yet to be determined. Substitution of Eq. (6.27) into Eq. (6.20) leads to

$$G_c = \frac{1}{G_p}\left(\frac{1}{\tau^2 s^2 + 2\zeta\tau s}\right),\tag{6.28}$$

which is a slightly more complicated form than that of Eq. (6.23). Again, with simplified cases, we can arrive at PID-type controllers.

Example 6.4: Derive the controller function for a system with a second-order overdamped process but an underdamped system response as dictated by Eq. (6.27).

The process transfer function is

$$G_p = \frac{K_p}{(\tau_1 s + 1)(\tau_2 s + 1)},$$

and the controller function according to Eq. (6.28) is

$$G_c = \frac{(\tau_1 s + 1)(\tau_2 s + 1)}{K_p \tau s(\tau s + 2\zeta)}.$$

We now define $\tau_f = \tau/2\zeta$, and G_c becomes

$$G_c = \frac{(\tau_1 s + 1)(\tau_2 s + 1)}{2\zeta K_p \tau s(\tau_f s + 1)}.$$

Suppose that τ_2 is associated with the slower pole ($\tau_2 > \tau_1$); we now require $\tau_f = \tau_2$ such that the pole and the zero cancel each other. The result is a PI controller:

$$G_c = \frac{1}{2\zeta K_p \tau} \frac{(\tau_1 s + 1)}{s}.$$

With our definition of τ_f and the requirement that $\tau_f = \tau_2$, we can write $\tau = 2\zeta\tau_2$, and the final form of the controller is

$$G_c = \frac{\tau_1}{4K_p \zeta^2 \tau_2}\left(1 + \frac{1}{\tau_1 s}\right) = K_c\left(1 + \frac{1}{\tau_I s}\right).\tag{E6.5}$$

The integral time constant is $\tau_I = \tau_1$, and the group multiplying the term in parentheses is the proportional gain K_c. In this problem, the system damping ratio ζ is the only tuning parameter.

6.2.2. Pole-Zero Cancellation

We used the term pole-zero cancellation at the beginning of this section. A few more words should be said as we can better appreciate the idea behind direct synthesis. Pole-zero cancellation is also referred to as **cancellation compensation** or dominant-pole design. Of course, it is unlikely to have perfect pole-zero cancellation in real life, and this discussion is aimed more toward helping our theoretical understanding.

The idea is that we may cancel the (undesirable open-loop) poles of our process and replace them with a desirable closed-loop pole. Recall in Eq. (6.20) that G_c is sort of the reciprocal of G_p. The zeros of G_c are by choice the poles of G_p. The product of G_cG_p cancels everything out – hence the term pole-zero cancellation. To be redundant, we can rewrite the general design equation as

$$G_cG_p = \left(\frac{C/R}{1 - C/R}\right). \tag{6.20a}$$

That is, no matter what G_p is, we define G_c such that their product is dictated entirely by a function (the RHS) in terms of our desired system response (C/R). For the specific closed-loop response as dictated by Eq. (6.22), we can also rewrite Eq. (6.23) as

$$G_cG_p = \left(\frac{1}{\tau_c s}\right). \tag{6.23a}$$

Because the system characteristic equation is $1 + G_cG_p = 0$, our closed-loop poles are dependent on only our design parameter τ_c. A closed-loop system designed on the basis of pole-zero cancellation has drastically different behavior than a system without such cancellation.

Let's try to illustrate with a PI controller and a first-order process function, and the simplification that $G_m = G_a = 1$. The closed-loop characteristic equation is

$$1 + G_cG_p = 1 + K_c\left(\frac{\tau_I s + 1}{\tau_I s}\right)\frac{K_p}{\tau_p s + 1} = 0. \tag{6.29}$$

Under normal circumstances, we would pick a τ_I that we deem appropriate. Now if we pick τ_I to be identical to τ_p, the zero of the controller function cancels the pole of the process function. We are left with only one open-loop pole at the origin. Equation (6.29), when $\tau_I = \tau_p$, is reduced to

$$1 + \frac{K_cK_p}{\tau_p s} = 0, \quad \text{or } s = -\frac{K_cK_p}{\tau_p}.$$

There is now only one real and negative closed-loop pole (presuming that $K_c > 0$). This situation is exactly what direct synthesis leads us to.

Recall from Example 6.1 that, based on the chosen C/R in Eq. (6.22), the PI controller function is

$$G_c = K_c\left(\frac{\tau_I s + 1}{\tau_I s}\right) = \frac{\tau_p}{K_p\tau_c}\left(\frac{\tau_p s + 1}{\tau_p s}\right),$$

where $\tau_I = \tau_p$ and $K_c = \tau_p/K_p\tau_c$. Substitution of K_c one step back into the characteristic equation shows that the closed-loop pole is indeed at $s = -1/\tau_c$. The product G_cG_p is also consistent with Eq. (6.23a) and τ_c.

6.2.3. Internal Model Control

A more elegant approach than direct synthesis is **internal model control (IMC)**. The premise of IMC is that, in reality, we have only an approximation of the actual process. Even if we

6.2. Direct Synthesis and Internal Model Control

(a)

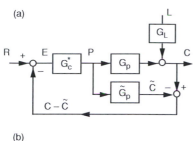

(b)

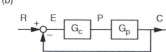

Figure 6.3. A system with (a) IMC compared with (b) a conventional system (b).

have the correct model, we may not have accurate measurements of the process parameters. Thus the imperfect model should be factored as part of the controller design.

In the block diagram that implements IMC [Fig. 6.3(a)], our conventional controller G_c consists of the (theoretical) model controller G_c^* and the approximate function $\tilde{G}_p$. Again, our objective is limited. We use the analysis in very restrictive and simplified cases to arrive at results in Example 6.5 to help us tune PID controllers as in Fig. 6.3(b).

We first need to derive the closed-loop functions for the system. Based on the block diagram, the error is

$$E = R - (C - \tilde{C})$$

and the model controller output is

$$P = G_c^* E = G_c^*(R - C + \tilde{C}).$$

If we substitute $\tilde{C} = \tilde{G}_p P$, we have

$$P = G_c^*(R - C + \tilde{G}_p P), \tag{6.30}$$

from which we can rearrange to obtain

$$P = \frac{G_c^*}{1 - G_c^* \tilde{G}_p}(R - C). \tag{6.28a}$$

The gist of this step is to show the relationship between the conventional controller function G_c and the other functions:

$$G_c = \frac{G_c^*}{1 - G_c^* \tilde{G}_p}. \tag{6.31}$$

This is an equation that we will use to retrieve the corresponding PID controller gains. For now, we substitute Eq. (6.28a) in an equation around the process:

$$C = G_L L + G_p P = G_L L + \frac{G_p G_c^*}{1 - G_c^* \tilde{G}_p}(R - C).$$

From this step, we derive the closed-loop equation:

$$C = \left[\frac{(1 - G_c^*\tilde{G}_p)G_L}{1 + G_c^*(G_p - \tilde{G}_p)}\right]L + \left[\frac{G_p G_c^*}{1 + G_c^*(G_p - \tilde{G}_p)}\right]R. \tag{6.32}$$

The terms in the brackets are the two closed-loop transfer functions. As always, they have the same denominator – the closed-loop characteristic polynomial.

There is still one piece of unfinished business. We do not know how to choose G_c^* yet. Before we make this decision, we may recall that, in direct synthesis, the poles of G_c are "inherited" from the zeros of G_p. If G_p has positive zeros, it will lead to a G_c function with positive poles. To avoid that, we "split" the approximate function as a product of two parts:

$$\tilde{G}_p = \tilde{G}_{p+}\tilde{G}_{p-}, \tag{6.33}$$

with $\tilde{G}_{p+}$ containing all the positive zeros, if present. The controller will be designed on the basis of $\tilde{G}_{p-}$ only. We now define the model controller function in a way similar to direct synthesis[6]:

$$G_c^* = \frac{1}{\tilde{G}_{p-}}\left[\frac{1}{\tau_c s + 1}\right]^r, \tag{6.34}$$

where $r = 1, 2$, etc. Like direct synthesis, τ_c is the closed-loop time constant and our *only* tuning parameter. The first-order function raised to an integer power of r is used to ensure that the controller is physically realizable.[7] Again, we would violate this intention in our simple example just so that we can obtain results that resemble an ideal PID controller.

Example 6.5: Repeat the derivation of a controller function for a system with a first-order process with dead time using an IMC.

Say we model our process (read: fitting the open-loop step-test data) as a first-order function with time delay; expecting experimental errors or uncertainties, we find that our measured or approximate model function $\tilde{G}_p$ is

$$\tilde{G}_p = \frac{K_p e^{-t_d s}}{\tau_p s + 1}.$$

We use the first-order Padé approximation for the dead-time and isolate the positive zero term as in Eq. (6.33):

$$\tilde{G}_p \approx \frac{K_p}{(\tau_p s + 1)\left(\frac{t_d}{2}s + 1\right)}\left(-\frac{t_d}{2}s + 1\right) = \tilde{G}_{p-}\tilde{G}_{p+}, \tag{E6.6}$$

---

[6] If the model is perfect, $G_p = \tilde{G}_p$ and Eq. (6.32) becomes simply $C = G_p G_c^* R$ if we also set $L = 0$. We choose C/R to be a first-order response with unity gain, and we arrive at a choice of G_c^* very similar to the definition in Eq. (6.34).

[7] The literature refers the term as a first-order filter. It makes sense only if you recall your linear circuit analysis or if you wait until the chapter on frequency-response analysis.

where

$$\tilde{G}_{p+} = \left(-\frac{t_d}{2}s + 1 \right).$$

If we choose $r = 1$, Eq. (6.34) gives

$$G_c^* = \frac{(\tau_p s + 1)\left(\frac{t_d}{2}s + 1\right)}{K_p} \frac{1}{(\tau_c s + 1)}. \tag{E6.7}$$

Substitution of Eq. (E6.7) and approximation (E6.6) into Eq. (6.31), after some algebraic work, will lead to the tuning parameters of an ideal PID controller:

$$K_c = \frac{1}{K_p} \frac{2\frac{\tau_p}{t_d} + 1}{2\frac{\tau_c}{t_d} + 1}; \quad \tau_I = \tau_p + \frac{t_d}{2}; \quad \tau_D = \frac{\tau_p}{2\frac{\tau_p}{t_d} + 1}. \tag{E6.8}$$

Example 5.7B: What would be the PID controller settings for the dye-mixing problem if we use IMC-based tuning relations?

With the same first-order with dead-time approximation as that in Example 5.7A, and the choice of τ_c being two thirds the value of dead time, the IMC relations in Eqs. (E6.8) provide the following PID settings (as computed with our M-file recipe.m):

	K_c	τ_I	τ_D
IMC	3.4	4.5	0.4

Compare this result by using other tuning relations in Example 5.7A. The IMC proportional gain falls in between the Cohen–Coon and ITAE settings, but the integral time constant is relatively high. With less integrating action, we expect this IMC tuning to be less oscillatory. Indeed, we shall see that when we do Example 5.7C (or you can cheat and read the plotting result from the *Web Support*).

Example 5.7C: How do the different controller settings affect the system time response in the dye-mixing problem?

We can use the following MATLAB statements to do time-response simulations (explanations are in MATLAB Session 5). Better yet, save them in an M-file. The plotting can be handled differently to suit your personal taste. (Of course, you can use Simulink instead.)

```
alfa=0.1;                       % Real PID
Gc=tf(kc*[taui*taud (taui+taud) 1],[alfa*taui*taud taui 0]);
td=0.725;
Gm=tf([-td/2 1],[td/2 1]);      % Padé approximation for dead time
Km=2.6;                         % Move Km into the forward path
Gp=tf(0.8,[4 1]);
Ga=tf(0.6,[0.2 1]);
Gcl=feedback(Km*Gc*Ga*Gp,Gm);   % The closed-loop function

step(Gcl)                       % Plotting...
```

Table 6.3. Summary of methods to select controller gains

Method	What to do?	What is evaluated?	Comments
Transient response criteria • Analytical derivation	Derive closed-loop damping ratio from a second-order system characteristic polynomial. Relate the damping ratio to the proportional gain of the system.	Usually the proportional gain	• Limited to second-order systems. No unique answer other than a P controller • Theoretically can use other transient response criteria. • 1/4 DR provides a 50% OS.
Empirical tuning with open-loop step test	Measure open-loop step response, the so-called process-reaction curve. Fit data to first-order with dead-time function.		
• Cohen–Coon • Ziegler–Nichols • Ciacone–Marlin	Apply empirical design relations.	Proportional gain, integral, and derivative time constants to PI and PID controllers	• Cohen–Coon was designed to handle disturbances by preventing a large initial deviation from the set point. The one-quarter DR response is generally too underdamped for set point changes.
• Time integral performance criteria (ISE, IAE, ITAE)	Apply design relations derived from minimization of an error integral of the theoretical time-domain response.	Proportional gain, integral, and derivative time constants to PI and PID controllers	• Different settings for load and set-point changes • Different settings for different definitions of the error integral • The minimum ITAE criterion provides the least oscillatory response
Ziegler–Nichols continuous cycling (empirical tuning with closed-loop test)	Increase proportional gain of only a proportional controller until system sustains oscillation. Measure ultimate gain and ultimate period. Apply empirical design relations.	Proportional gain, integral, and derivative time constants of PID controllers	• Experimental analog of the $s = j\omega$ substitution calculation • Not necessarily feasible with chemical systems in practice • Tuning relations allow for choices from one-quarter DR to little oscillations.

Stability analysis methods

Method	Procedure	Comments
• Routh–Hurwitz criterion	Apply the Routh test on the closed-loop characteristic polynomial to find if there are closed-loop poles on the RHP. Establish limits on the controller gain.	• Usually applies to relatively simple systems with the focus on the proportional gain • Need be careful on interpretation when the lower limit on proportional gain is negative • Result on ultimate gain is consistent with the Routh array analysis • Limited to relatively simple systems
• Direct substitution	Substitute $s = j\omega$ into characteristic polynomial and solve for closed-loop poles on the Im axis. The Im and the Re parts of the equation allow the ultimate gain and ultimate frequency to be solved. Ultimate gain and ultimate period $(P_u = 2\pi/\omega_u)$ that can be used in the Ziegler–Nichols continuous-cycling relations	
• Root-locus	With each chosen value of proportional gain, plot the closed-loop poles. Generate the loci with either hand-sketching or computer. The loci of closed-loop poles reveal the effect of controller gain on the probable closed-loop dynamic response. Together with specifications of damping ratio and time constant, the loci can be a basis of selecting proportional gain.	• Rarely used in the final controller design because of difficulty in handling dead time • Method is instructive and great pedagogical tool
(Model-based) direct synthesis	For a given system, synthesize the controller function according to a specified closed-loop response. The system time constant, τ_c, is the only tuning parameter. Proportional gain, integral, and derivative time constants where appropriate	• The design is not necessarily PID, but where the structure of a PID controller results, this method provides insight into the selection of the controller mode (PI, PD, PID) and settings. • Theory is based on pole-zero cancellation, which is rare in practice. • Especially useful with system that has no dead time

(cont.)

Table 6.3. (cont.)

Method	What to do?	What is evaluated?	Comments
• IMC	Extension of direct synthesis. Controller design includes an internal approximation process function.	For a first-order function with dead time, the proportional gain, integral, and derivative time constants of an ideal PID controller	• Can handle dead time easily and rigorously • The Nyquist criterion allows the use of open-loop functions in Nyquist or Bode plots to analyze the closed-loop problem • The stability criteria have no use for simple first- and second-order systems with no positive open-loop zeros.
Frequency-domain methods			
• Nyquist plot • Bode plot	Nyquist plot is a frequency parametric plot of the magnitude and the argument of the open-loop transfer function in polar coordinates. Bode plot is magnitude vs. frequency and phase angle vs. frequency plotted individually.	Calculate proportional gain needed to satisfy the gain or phase margin.	• These plots address the stability problem but need other methods to reveal the probable dynamic response.
• Nichols chart	Nichols chart is a frequency parametric plot of open-loop function magnitude vs. phase angle. The closed-loop magnitude and phase angle are overlaid as contours.	With gain or phase margin, calculate proportional gain. Can also estimate the peak amplitude ratio, and assess the degree of oscillation.	• Nichols chart is usually constructed for unity feedback loops only.
• Maximum closed-loop log modulus	A plot of the magnitude vs. frequency of the closed-loop transfer function	The peak amplitude ratio for a chosen proportional gain	

We reset the three controller parameters each time we execute the M-file. For example, to use the Cohen–Coon results, we would take from Example 5.7A

```
kc=4.9; taui=2; taud=0.31;
```

MATLAB calculation details and plots can be found on the *Web Support*. You should observe that Cohen–Coon and Ziegler–Nichols tuning relations lead to roughly 74% and 64% OS, respectively, which are more significant than what we expect with a quarter DR criterion. The ITAE, with ~14% OS, is more conservative. Ciancone–Marlin tuning relations are ultraconservative; the system is slow and overdamped.

With the IMC tuning setting in Example 5.7B, the resulting time-response plot is (very nicely) slightly underdamped, even though the derivation in Example 6.4 is predicated on a system response without oscillations. Part of the reason lies in the approximation of the dead-time function and part of the reason is due to how the system time constant was chosen. Generally, it is important to double check IMC settings with simulations.

At this point, you may be sufficiently confused with respect to all the different controller tuning methods. Use Table 6.3 as a guide to review and compare different techniques from this chapter and also from Chaps. 7 and 8.

Review Problems

(1) Repeat Example 6.1 when we have

$$G_p = \frac{K_p}{s(\tau_p s + 1)}.$$

What is the offset of the system?

(2) What are the limitations to an IMC? (Especially with respect to the choice of τ_c?)

(3) What control action increases the order of the system?

(4) Refer back to Example 6.4. If we have a third-order process,

$$G_p = \frac{K_p}{(\tau_1 s + 1)(\tau_2 s + 1)(\tau_3 s + 1)},$$

what is the controller function if we follow the same strategy as that in the example?

(5) Complete the time-response simulations in Example 5.7C using settings in Example 5.7A.

(6) (*Optional*) How would you implement the PID algorithm in a computer program?

Hints:

(1) The result is an ideal PD controller with the choice of $\tau_D = \tau_p$. See that you can obtain the same result with an IMC too. Here, take the process function as the approximate model; it has no parts that we need to consider as having positive zeros. There is no offset; the integrating action is provided by G_p.

(2) Too small a value of τ_c means too large a K_c and therefore saturation. System response is subject to imperfect pole-zero cancellation.

(3) Integration is $1/s$.

(4) The intermediate step is

$$G_c = \frac{(\tau_1 s + 1)(\tau_2 s + 1)(\tau_3 s + 1)}{2\zeta K_p \tau s(\tau_f s + 1)},$$

where $\tau_f = \tau/2\zeta$, and now we require that $\tau_f = \tau_3$, presuming it is the largest time constant. The final result, after some of the ideas also taken from Example 6.2, is an ideal PID controller with the form

$$G_c = \frac{(\tau_1 + \tau_2)}{4 K_p \zeta^2 \tau_3}\left(1 + \frac{1}{\tau_1 + \tau_2}\frac{1}{s} + \frac{\tau_1 \tau_2}{\tau_1 + \tau_2}s\right).$$

The necessary choices of K_c, τ_I, and τ_D are obvious. Again, ζ is the only tuning parameter.

(5) See the *Web Support* for the simulations.

(6) Use finite difference. The ideal PID in Eq. (5.8a) can be discretized as

$$p_n = p^s + K_c\left[e_n + \frac{\Delta t}{\tau_I}\sum_{k=1}^{n} e_k + \frac{\tau_D}{\Delta t}(e_n - e_{n-1})\right],$$

where p_n is the controller output at the nth sampling period, p^s is the controller bias, Δt is the sampling period, and e_n is the error. This is referred to as the *position-form* algorithm. The alternative approach is to compute the change in the controller output based on the difference between two samplings:

$$\Delta p_n = p_n - p_{n-1} = K_c\left[(e_n - e_{n-1}) + \frac{\Delta t}{\tau_I}e_n + \frac{\tau_D}{\Delta t}(e_n - 2e_{n-1} + e_{n-2})\right].$$

This is the *velocity-form* algorithm, which is considered to be more attractive than the position form. The summation of error is not computed explicitly and thus the velocity form is not as susceptible to reset windup.

7

Stability of Closed-Loop Systems

W hen we design a closed-loop system, the specifications may dictate features in dynamic response. However, we cannot do that unless the system is stable. Thus the foremost concern in a control system design is to keep the system stable, which in itself can be used as a design tool.

What Are We Up to?

Analyzing stability of a closed-loop system with three techniques:

- Routh–Hurwitz criterion for the stability region
- Substitution of $s = j\omega$ to find roots at marginal stability
- Root-locus plots of the closed-loop poles

7.1.   Definition of Stability

Our objective is simple. We want to make sure that the controller settings will not lead to an unstable system. Consider the closed-loop system response that we derived in Section 5.2:

$$C = \left(\frac{K_m G_c G_a G_p}{1 + G_m G_c G_a G_p} \right) R + \left(\frac{G_L}{1 + G_m G_c G_a G_p} \right) L \tag{7.1}$$

with the characteristic equation

$$1 + G_m G_c G_a G_p = 0. \tag{7.2}$$

The closed-loop system is stable if all the roots of the characteristic polynomial have negative real parts. Or we can say that all the poles of the closed-loop transfer function lie in the left-hand plane (LHP). When we make this statement, the stability of the system is defined entirely on the inherent dynamics of the system and not on the input functions. In other words, the results apply to both servo and regulating problems.

We also see another common definition – **bounded-input bounded-output (BIBO) stability**: A system is BIBO stable if the output response is bounded for any bounded input.

One illustration of this definition is to consider a hypothetical situation with a closed-loop pole at the origin. In such a case, we know that if we apply an impulse input or a rectangular pulse input, the response remains bounded. However, if we apply a step input, which is bounded, the response is a ramp, which has no upper bound. For this reason, we cannot accept any control system that has closed-loop poles lying on the imaginary axis. They must be in the LHP.[1]

The addition of a feedback control loop can stabilize or destabilize a process. We will see plenty examples of the latter. For now, we use the classic example of trying to stabilize an open-loop unstable process.

Example 7.1: Consider the unstable process function $G_p = [K/(s-a)]$, which may arise from a linearized model of an exothermic chemical reactor with an improper cooling design. The question is whether we can make a stable control system by using simply a proportional controller. For illustration, we consider a unity feedback loop with $G_m = 1$. We also take the actuator transfer function to be unity, $G_a = 1$.

With a proportional controller, $G_c = K_c$, the transfer function of this closed-loop servo system is

$$\frac{Y}{R} = \frac{G_c G_p}{1 + G_c G_p} = \frac{K_c K}{s - a + K_c K}.$$

The characteristic equation is

$$s - a + K_c K = 0,$$

which means that if we want a stable system, the closed-loop poles must satisfy

$$s = a - K_c K < 0.$$

In other words, the closed-loop system is stable if $K_c > a/K$.

For a more complex problem, the characteristic polynomial will not be as simple, and we need tools to help us. The two techniques that we will learn are the Routh–Hurwitz criterion and root locus. Root locus is, by far, the more important and useful method, especially when we can use a computer. Where circumstances allow (i.e., the algebra is not too ferocious), we can also find the roots on the imaginary axis – the case of marginal stability. In the simple example above, this is where $K_c = a/K$. Of course, we have to be smart enough to pick $K_c > a/K$, and not $K_c < a/K$.

7.2. The Routh–Hurwitz Criterion

The time-honored (i.e., ancient!) Routh–Hurwitz criterion is introduced for stability testing. It is not proved here – hardly any text does anymore. Nonetheless, two general polynomials

---

[1] Do not be confused by the integral control function; its pole at the origin is an open-loop pole. This point should be cleared up when we get to the root-locus section. Furthermore, conjugate poles on the imaginary axis are BIBO stable – a step input leads to a sustained oscillation that is bounded in time. However, we do not consider this oscillatory steady state as stable, and hence we exclude the entire imaginary axis. In an advanced class, you should find more mathematical definitions of stability.

are used to illustrate some simple properties. First, consider a second-order polynomial with the leading coefficient $a_2 = 1$. If the polynomial has two real poles p_1 and p_2, it can be factored as

$$P(s) = s^2 + a_1 s + a_0 = (s - p_1)(s - p_2). \tag{7.3}$$

We may observe that if a_1 is zero, both roots, $\pm j\sqrt{a_0}$, are on the imaginary axis. If a_0 is zero, one of the two roots is at the origin. We can expand the pole form to give

$$P(s) = s^2 - (p_1 + p_2)s + p_1 p_2 \tag{7.4}$$

and compare the coefficients with the original polynomial in Eq. (7.3). If both p_1 and p_2 are negative, the coefficients a_1 and a_0 must be positive definite, which is the mathematicians' way of saying that $a_1 > 0$ and $a_0 > 0$.

Next, consider a third-order polynomial with leading coefficient $a_3 = 1$ and the form in terms of the poles:

$$P(s) = s^3 + a_2 s^2 + a_1 s + a_0 = (s - p_1)(s - p_2)(s - p_3). \tag{7.5}$$

We expand the pole form to

$$P(s) = s^3 - (p_1 + p_2 + p_3)s^2 + (p_1 p_2 + p_1 p_3 + p_2 p_3)s - p_1 p_2 p_3. \tag{7.6}$$

Once again, if all three poles are negative, the coefficients a_2, a_1, and a_0 must be positive definite.

The idea is that the signs of the pole are related to the coefficients $a_n, a_{n-1}, \ldots, a_0$ of an nth-order characteristic polynomial. If we require all the poles to have negative-real parts, there must be some way that we can tell from the coefficients without actually having to solve for the roots. The idea is that all of the coefficients in the characteristic polynomial must be positive definite. We could develop a comprehensive theory, which Routh did. The attractiveness of the Routh criterion is that *without solving* for the closed-loop poles, we can derive inequalities that would provide a bound for stable controller design.

The complete Routh array analysis allows us to find, for example, the number of poles on the imaginary axis. Because we require that all poles lie in the LHP, we will not bother with these details (which are still in many control texts). Consider the fact that we can easily calculate the exact roots of a polynomial with MATLAB; we use the Routh criterion to the extent that it serves its purpose.[2] That would be to derive inequality criteria for proper selection of controller gains of relatively simple systems. The technique loses its attractiveness when the algebra becomes too messy. Now the simplified Routh–Hurwitz recipe without proof follows.

(1) **Hurwitz test for the polynomial coefficients.** For a given nth-order polynomial

$$P(s) = a_n s^n + a_{n-1} s^{n-1} + \cdots + a_2 s^2 + a_1 s + a_0, \tag{7.7}$$

all the roots are in the LHP if and only if all the coefficients $a_0, \ldots, a_n$ are positive definite.

---

[2] MATLAB does not even bother with a Routh function. Such an M-file is provided on the *Web Support* for demonstration purposes.

If any one of the coefficients is negative, at least one root has a positive-real part [i.e., in the right hand plane (RHP)]. If any of the coefficients is zero, not all of the roots are in the LHP: it is likely that some of them are on the imaginary axis. Either way, stop. This test is a necessary condition for BIBO stability. There is no point in doing more other than to redesign the controller.

(2) **Routh array construction.** If the characteristic polynomial passes the coefficient test, we then construct the Routh array to find the necessary and sufficient conditions for stability. This is one of the few classical techniques that is not emphasized and the general formula is omitted. The array construction up to a fourth-order polynomial is used to illustrate the concept.

Generally, for an nth-order polynomial, we need $(n + 1)$ rows. The first two rows are filled in with the coefficients of the polynomial in a column-wise order. The computation of the array entries is very much like the negative of a normalized determinant anchored by the first column. Even without the general formula, you may pick out the pattern as you read the following three examples.

The **Routh criterion** states that, in order to have a stable system, all the coefficients in the first column of the array must be positive definite. If any of the coefficients in the first column is negative, there is at least one root with a positive-real part. The number of sign changes is the number of positive poles.

Here is the array for a second-order polynomial, $p(s) = a_2 s^2 + a_1 s + a_0$:

1: $\qquad\qquad a_2 \qquad\qquad\qquad a_0$

2: $\qquad\qquad a_1 \qquad\qquad\qquad 0$

3: $\quad b_1 = \dfrac{a_1 a_0 - (0)a_2}{a_1} = a_0$

In the case of a second-order system, the first column of the Routh array reduces to simply the coefficients of the polynomial. The coefficient test is sufficient in this case. Or we can say that both the coefficient test and the Routh array provide the same result.

The array for a third-order polynomial, $p(s) = a_3 s^3 + a_2 s^2 + a_1 s + a_0$, is

1: $\qquad\qquad a_3 \qquad\qquad\qquad\qquad a_1 \qquad\qquad\qquad 0$

2: $\qquad\qquad a_2 \qquad\qquad\qquad\qquad a_0 \qquad\qquad\qquad 0$

3: $\quad b_1 = \dfrac{a_2 a_1 - a_3 a_0}{a_2} \qquad b_2 = \dfrac{(a_2)(0) - (a_3)0}{a_2} = 0$

4: $\quad c_1 = \dfrac{b_1 a_0 - b_2 a_2}{b_1} = a_0 \qquad\qquad\qquad 0$

In this case, we have added one column of zeros; they are needed to show how b_2 is computed. Because $b_2 = 0$ and $c_1 = a_0$, the Routh criterion adds one additional constraint in the case of a third-order polynomial:

$$b_1 = \frac{a_2 a_1 - a_3 a_0}{a_2} > 0. \qquad\qquad (7.8)$$

We follow with the array for a fourth-order polynomial, $p(s) = a_4 s^4 + a_3 s^3 + a_2 s^2 + a_1 s + a_0$:

1:          a_4 a_2 a_0

2:          a_3 a_1 0

3:      $b_1 = \dfrac{a_3 a_2 - a_1 a_4}{a_3}$ $b_2 = \dfrac{a_3 a_0 - (0) a_4}{a_3} = a_0$ 0

4:      $c_1 = \dfrac{b_1 a_1 - b_2 a_3}{b_1}$ $c_2 = \dfrac{b_1(0) - (0) a_3}{b_1} = 0$

5:     $d_1 = \dfrac{c_1 b_2 - (0) b_1}{c_1} = b_2 = a_0$ 0

The two additional constraints from the Routh array are hence

$$b_1 = \frac{a_3 a_2 - a_1 a_4}{a_3} > 0, \tag{7.9}$$

$$c_1 = \frac{b_1 a_1 - b_2 a_3}{b_1} = \frac{b_1 a_1 - a_0 a_3}{b_1} > 0. \tag{7.10}$$

Example 7.2: If we have only a proportional controller (i.e., one design parameter) and negative-real open-loop poles, the Routh–Hurwitz criterion can be applied to a fairly high-order system with ease. For example, for the following closed-loop system characteristic equation,

$$1 + K_c \frac{1}{(s+3)(s+2)(s+1)} = 0,$$

find the stability criteria.

We expand and rearrange the equation to the polynomial form:

$$s^3 + 6s^2 + 11s + (6 + K_c) = 0.$$

The Hurwitz test requires that $K_c > -6$ or simply $K_c > 0$ for positive controller gains.
The Routh array is

1:     1   11

2:     6 $6 + K_c$

3:    b_1 0

4: $6 + K_c$ 0

The Routh criterion requires, as in Eq. (7.8), that

$$b_1 = \frac{(6)(11) - (6 + K_c)}{6} > 0 \quad \text{or} \quad 60 > K_c.$$

The range of proportional gain to maintain system stability is hence $0 < K_c < 60$.

Example 7.3: Consider a second-order process function $G_p = [1/(s^2 + 2s + 1)]$, which is critically damped. If we synthesize a control system with a PI controller, what are the stability constraints?

For illustration, we take $G_m = G_a = 1$ and the closed-loop transfer function for a servo problem is simply

$$\frac{C}{R} = \frac{G_c G_p}{1 + G_c G_p}.$$

In this problem, the closed-loop characteristic equation is

$$1 + G_c G_p = 1 + K_c \left(1 + \frac{1}{\tau_I s}\right) \frac{1}{s^2 + 2s + 1} = 0$$

or

$$\tau_I s^3 + 2\tau_I s^2 + \tau_I (1 + K_c)s + K_c = 0.$$

With the Routh–Hurwitz criterion, we need immediately $\tau_I > 0$ and $K_c > 0$. (The s term requires that $K_c > -1$, which is overridden by the last constant coefficient.) The Routh array for this third-order polynomial is

1:    τ_I $\tau_I (1 + K_c)$

2:    $2\tau_I$ K_c

3:    b_1 0

4:    K_c

With the use of Eq. (7.8), we require that

$$b_1 = \frac{2\tau_I^2(1 + K_c) - \tau_I K_c}{2\tau_I} = \tau_I(1 + K_c) - \frac{K_c}{2} > 0.$$

The inequality is rearranged to

$$\tau_I > \frac{K_c}{2(1 + K_c)} \quad \text{or} \quad \frac{2\tau_I}{1 - 2\tau_I} > K_c,$$

which can be interpreted in two ways. For a given K_c, there is a minimum integral time constant. If the proportional gain is sufficiently large such that $K_c \gg 1$, the rough estimate for the integral time constant is $\tau_I > 1/2$. Or if the value of τ_I is less than 0.5, there is an upper limit on how large K_c could be.

If the given value of τ_I is larger than 0.5, the inequality simply infers that K_c must be larger than some negative number. To be more specific, if we pick $\tau_I = 1$,[3] the Routh criterion becomes

$$2 > \frac{K_c}{(1 + K_c)},$$

which of course can be satisfied for all $K_c > 0$. No new stability requirement is imposed in this case. Let us try another choice of $\tau_I = 0.1$. In this case, the requirement for the proportional gain is

$$0.2(1 + K_c) > K_c \quad \text{or} \quad K_c < 0.25.$$

---

[3] Note that, with this very specific case, if $\tau_I = 1$ is chosen, the open-loop zero introduced by the PI controller cancels one of the open-loop poles of the process function at -1. If we do a root-locus plot, we will see how the root loci change to that of a purely second-order system. With respect to this example, the value is not important as long as $\tau_I > 1/2$.

The entire range of stability for $\tau_I = 0.1$ is $0 < K_c < 0.25$. We will revisit this problem when we cover root-locus plots; we can make much better sense without doing any algebraic work!

7.3. Direct-Substitution Analysis

The closed-loop poles may lie on the imaginary axis at the moment a system becomes unstable. We can substitute $s = j\omega$ into the closed-loop characteristic equation to find the proportional gain that corresponds to this stability limit (which may be called marginal unstable). The value of this specific proportional gain is called the **critical** or **ultimate gain**. The corresponding frequency is called the **crossover** or **ultimate frequency**.

Example 7.2A: Apply direct substitution to the characteristic equation in Example 7.2:

$$s^3 + 6s^2 + 11s + (6 + K_c) = 0.$$

Substitution of $s = j\omega$ leads to

$$-j\omega^3 - 6\omega^2 + 11\omega j + (6 + K_c) = 0.$$

The real- and imaginary-part equations are

Re: $\quad -6\omega^2 + (6 + K_c) = 0 \quad$ or $\quad K_c = 6(\omega^2 - 1)$,

Im: $\quad -\omega^3 + 11\omega = 0 \quad$ or $\quad \omega(11 - \omega^2) = 0.$

From the imaginary-part equation, the ultimate frequency is $\omega_u = \sqrt{11}$. Substituting this value into the real-part equation leads to the ultimate gain $K_{c,u} = 60$, which is consistent with the result of the Routh criterion.

If we have chosen the other possibility, $\omega_u = 0$, meaning that the closed-loop poles are on the real axis, the ultimate gain is $K_{c,u} = -6$, which is consistent with the other limit obtained with the Routh criterion.

Example 7.3A: Repeat Example 7.3 to find the condition for the ultimate gain.
If we make the $s = j\omega$ substitution into

$$\tau_I s^3 + 2\tau_I s^2 + \tau_I(1 + K_c)s + K_c = 0,$$

it becomes

$$-\tau_I \omega^3 j - 2\tau_I \omega^2 + \tau_I(1 + K_c)\omega j + K_c = 0.$$

We have two equations after collecting all the real and the imaginary parts and requiring both to be zero:

Re: $\quad K_c - 2\tau_I \omega^2 = 0,$

Im: $\quad \tau_I \omega[-\omega^2 + (1 + K_c)] = 0.$

Thus we have either $\omega = 0$ or $-\omega^2 + (1 + K_c) = 0$. Substitution of the real-part equation into the nontrivial imaginary-part equation leads to

$$-\omega^2 + 1 + 2\tau_I \omega^2 = 0 \quad \text{or} \quad \omega_u^2 = \frac{1}{1 - 2\tau_I},$$

where in the second form, we have added a subscript to denote the ultimate frequency, ω_u. Substitution of the ultimate frequency back into the real-part equation gives the relation for the ultimate proportional gain:

$$K_{c,u} = \frac{2\tau_I}{1 - 2\tau_I}.$$

Note that if we have chosen the other possibility of $\omega_u = 0$, meaning where the closed-loop poles are on the real axis, the ultimate gain is $K_{c,u} = 0$, which is consistent with the other limit obtained with the Routh criterion. The result of direct substitution confirms the inequality derived from the Routh criterion, which should not be a surprise.

We may question whether direct substitution is a better method. There is no clear-cut winner here. By and large, we are less prone to making algebraic errors when we apply the Routh–Hurwitz recipe, and the interpretation of the results is more straightforward. With direct substitution, we do not have to remember any formulas, and we can find the ultimate frequency, which, however, can be obtained with a root-locus plot or frequency-response analysis – techniques that are covered in Section 7.4 and Chap. 8.

When the system has dead-time, we must make an approximation, such as the Padé approximation, on the exponential dead-time function before we can apply the Routh–Hurwitz criterion. The result is hence only an estimate. Direct substitution allows us to solve for the ultimate gain and ultimate frequency exactly. The next example illustrates this point.

Example 7.4: Consider a system with a proportional controller and a first-order process but with dead-time. The closed-loop characteristic equation is given as

$$1 + K_c \frac{0.8e^{-2s}}{5s + 1} = 0.$$

Find the stability criteria of this system.

Let us first use the first-order Padé approximation for the time-delay function and apply the Routh–Hurwitz criterion. The approximate equation becomes

$$1 + K_c \frac{0.8}{5s + 1} \frac{(-s + 1)}{(s + 1)} = 0$$

or

$$5s^2 + (6 - 0.8K_c)s + (1 + 0.8K_c) = 0.$$

The Routh–Hurwitz criterion requires that $6 - 0.8K_c > 0$ or $K_c < 7.5$, and $K_c > -1/0.8$. When K_c is kept positive, the approximate range of the proportional gain for system stability is $0 < K_c < 7.5$.

We now repeat the problem with the $s = j\omega$ substitution into the characteristic equation and rewrite the time-delay function with Euler's identity:

$$(5\omega j + 1) + 0.8K_c(\cos 2\omega - j \sin 2\omega) = 0.$$

Collecting terms of the real and the imaginary parts provides the two equations:

Re: $\quad 1 + 0.8K_c \cos 2\omega = 0 \quad$ or $\quad K_c = -1/(0.8 \cos 2\omega)$,

Im: $\quad 5\omega - 0.8K_c \sin 2\omega = 0.$

Substitution of the real-part equation into the imaginary-part equation gives

$$5\omega + \tan 2\omega = 0.$$

The solution of this equation is the ultimate frequency $\omega_u = 0.895$, and, from the real-part equation, the corresponding ultimate proportional gain is $K_{c,u} = 5.73$. Thus the more accurate range of K_c that provides system stability is $0 < K_c < 5.73$.

Note 1: This result is consistent with the use of frequency-response analysis in Chap. 8.

Note 2: The iterative solution in solving the ultimate frequency is tricky. The equation has poor numerical properties – arising from the fact that $\tan \theta$ "jumps" from infinity at $\theta = (\pi/2)_-$ to negative infinity at $\theta = (\pi/2)_+$. To better see why, use MATLAB to make a plot of the function (LHS of the equation) with $0 < \omega < 1$. With MATLAB, we can solve the equation with the `fzero()` function. Create an M-file named `f.m`, and enter these two statements in it:

```
function y=f(x)
y = 5*x + tan(2*x);
```

After you have saved the file, enter, at the MATLAB prompt;

```
fzero('f',0.9)
```

where 0.9 is the initial guess. MATLAB should return 0.8953. If you shift the initial guess just a bit, say by using 0.8, you may get a "solution" of 0.7854. Note that (2)(0.7854) is $\pi/2$. If you blindly accept this incorrect value, $K_{c,u}$ will be infinity according to the real-part equation. MATLAB is handy, but it is not foolproof!

7.4. Root-Locus Analysis

The idea of a root-locus plot is simple – if we have a computer. We pick *one* design parameter, say, the proportional gain K_c, and write a small program to calculate the roots of the characteristic polynomial for each chosen value of K_c as in 0, 1, 2, 3, ..., 100, . . . , etc. The results (the values of the roots) can be tabulated or, better yet, plotted on the complex plane. Even though the idea of plotting a root locus sounds so simple, it is one of the most powerful techniques in controller design and analysis *when there is no time delay*.

Root locus is *a graphical representation* of the roots of the closed-loop characteristic polynomial (i.e., the closed-loop poles) as a chosen parameter is varied. Only the roots are plotted. The values of the parameter are not shown explicitly. The analysis most commonly uses the proportional gain as the parameter. The value of the proportional gain is varied from 0 to infinity, or in practice, just "large enough." Now a simple example is needed to get this idea across.

Example 7.5: Construct the root-locus plots of a first- and a second-order system with a proportional controller. See how the loci approach infinity.

(a) Consider the characteristic equation of a simple system with a first-order process and a proportional controller:

$$1 + K_c \frac{K_p}{\tau_p s + 1} = 0.$$

(a)

(b)

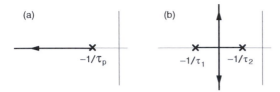

$-1/\tau_p$

$-1/\tau_1$

$-1/\tau_2$

Figure E7.5.

The solution, meaning the closed-loop poles of the system, is

$$s = \frac{-(1 + K_c K_p)}{\tau_p}.$$

The root-locus plot (Fig. E7.5a) is simply a line on the real axis starting at $s = -1/\tau_p$ when $K_c = 0$ and extends to negative infinity as K_c approaches infinity. As we increase the proportional gain, the system response becomes faster. Would there be an upper limit in reality? (Yes, saturation.)

(b) We repeat the exercise with a second-order overdamped process function. The closed-loop characteristic equation of the closed-loop system is

$$1 + K_c \frac{K_p}{(\tau_1 s + 1)(\tau_2 s + 1)} = 0 \quad \text{or} \quad \tau_1 \tau_2 s^2 + (\tau_1 + \tau_2)s + (1 + K_c K_p) = 0.$$

The two closed-loop poles are

$$s = \frac{-(\tau_1 + \tau_2) \pm \sqrt{(\tau_1 + \tau_2)^2 - 4\tau_1 \tau_2 (1 + K_c K_p)}}{2\tau_1 \tau_2}.$$

In the mathematical limit of $K_c = 0$, we should find two negative real poles at

$$s = \frac{-(\tau_1 + \tau_2) \pm \sqrt{(\tau_1 - \tau_2)^2}}{2\tau_1 \tau_2} = -\frac{1}{\tau_1} \quad \text{or} \quad -\frac{1}{\tau_2},$$

which are the locations of the two open-loop poles. (This result is easy to spot if we use the first step without expanding the terms.) As K_c becomes larger, we should come to a point where we have two repeated roots at

$$s = \frac{-(\tau_1 + \tau_2)}{2\tau_1 \tau_2}.$$

If we increase further the value of K_c, the closed-loop poles will branch off (or break away) from the real axis and become two complex conjugates (Fig. E7.5b). No matter how large K_c becomes, these two complex conjugates always have the same real part as given by the repeated root. Thus what we find are two vertical loci extending toward positive and negative infinity. In this analysis, we also see how, as we increase K_c, the *system* changes from overdamped to become underdamped, but it is always stable.

This is the idea behind the plotting of the closed-loop poles – in other words, construction of root-locus plots. Of course, we need mathematical or computational tools when we have more complex systems. An important observation from Example 7.5 is that with simple first- and second-order systems with *no open-loop zeros* in the RHP, the closed-loop system is *always stable*.

We can now state the problem in more general terms. Let us consider a closed-loop characteristic equation $1 + K_c G_0 = 0$, where $K_c G_0$ is referred to as the "open-loop" transfer function, G_{OL}. The proportional gain is K_c and G_0 is "everything else." If we have only a proportional controller, then $G_0 = G_m G_a G_p$. If we have other controllers, then G_0 would contain parts of the controller function. We further denote G_0 as a ratio of two polynomials, $Q(s)/P(s)$, and rewrite the polynomials in the pole-zero form[4]:

$$1 + K_c G_0 = 1 + K_c \frac{Q(s)}{P(s)} = 1 + K_c \frac{(s - z_1)(s - z_2) \cdots (s - z_n)}{(s - p_1)(s - p_2) \cdots (s - p_m)} = 0 \qquad (7.11)$$

or as

$$[(s - p_1)(s - p_2) \cdots (s - p_m)] + K_c[(s - z_1)(s - z_2) \cdots (s - z_n)] = 0. \qquad (7.11a)$$

The roots of the mth-order $P(s) = 0$, $p_1, p_2, \ldots, p_m$, are the open-loop poles. The roots of the nth-order $Q(s) = 0$, $z_1, z_2, \ldots, z_n$, are the open-loop zeros. The roots of the entire characteristic equation (7.11) are the closed-loop poles that constitute the root loci.

There will be m root loci, matching the order of the characteristic polynomial. We can easily see that when $K_c = 0$ the poles of the closed-loop system characteristic polynomial $(1 + K_c G_0)$ are essentially the same as the poles of the open loop. When K_c approaches infinity, the poles of the closed-loop system are the zeros of the open-loop. These are important mathematical features.

In other words, on a root-locus plot, we expect the "trace" of the root loci to begin at the open-loop poles and terminate at the open-loop zeros (if there is one). For real systems, $m > n$ and $n \geq 0$. In these cases, the $(m - n)$ root loci will originate from an open-loop pole and extend toward infinity somehow, depending on the specific problem.

Before reading further, it is very important that you go through at least the first half of MATLAB Session 6 and do computations with sample numbers while reading these root-locus examples.

Example 7.2B: Do the root-locus plot and find the ultimate gain of Example 7.2. The closed-loop equation from that example is

$$1 + K_c \frac{1}{(s + 3)(s + 2)(s + 1)} = 0.$$

We can easily use MATLAB to find that the ultimate gain is roughly 60. The statements to use are

```
G=zpk([],[-1 -2 -3],1);
k=0:1:100;    % We have to use our own gain vector in this example
rlocus(G,k)   % because the MATLAB default plot does not cross the
              % Im axis
rlocfind(G)
```

---

[4] If you cannot follow the fancy generalization, think of a simple problem such as a unity feedback loop with a PD controller and a first-order process. The closed-loop characteristic equation is

$$1 + K_c \frac{K_p(\tau_D s + 1)}{(\tau_p s + 1)} = 0.$$

The closed-loop pole "runs" from the point $s = -1/\tau_p$ at the mathematical limit of $K_c = 0$ to the point $s = -1/\tau_D$ as K_c approaches infinity.

After we enter the `rlocfind()` command, MATLAB will prompt us to click a point on the root-locus plot. In this problem, we select the intersection between the root locus and the imaginary axis for the ultimate gain.

Example 7.3B: Repeat Example 7.3 with a root-locus analysis.
   The closed-loop characteristic equation from Example 7.3 is

$$1 + K_c \frac{(\tau_I s + 1)}{\tau_I s (s^2 + 2s + 1)} = 0.$$

Select various values of τ_I and use MATLAB to construct the root-locus plots. Sample statements are

```
taui=0.2;    % Open-loop zero at -5
G=tf([taui 1],conv([taui 0],[1 2 1]));
rlocus(G)
```

We should find that for values of $\tau_I > 0.5$ the system stays stable. For $\tau_I = 0.5$, the system may become unstable, but only at infinitely large K_c. The system may become unstable for $\tau_I < 0.5$ if K_c is too large. Finally, for the choice of $\tau_I = 0.1$, we should find with the MATLAB function `rlocfind` that the ultimate gain is roughly 0.25, the same answer from Example 7.3. How close you get depends on how accurately you can click the axis crossover point.

Even as we rely on MATLAB to generate root-locus plots, it is important to appreciate how they are generated. To say the least, we need to know how to identify the open-loop poles and zeros and the direction of the loci. These hints are given in the *Web Support*. The following example illustrates some of the more common ones that we may encounter in control analysis. There are many other possibilities, but this exercise should be a good starting point. MATLAB Session 6 has more suggestions regarding the plots that are associated with the use of controllers.

Example 7.6: Construct the root-locus plots of some of the more common closed-loop equations with numerical values. Make sure you try them yourself with MATLAB.
   (a) A sample first-order system and the MATLAB statement:

$$1 + K \frac{1}{(s + 2)} = 0 \qquad \qquad \texttt{rlocus(tf(1,[1 2]))}$$

   (b) A second-order system:

$$1 + K \frac{1}{(s + 2)(s + 1)} = 0 \qquad \texttt{rlocus(zpk([],[-1 -2],1))}$$

   (c) Second-order system with repeated open-loop poles:

$$1 + K \frac{1}{(s + 2)^2} = 0 \qquad \qquad \texttt{rlocus(zpk([],[-2 -2],1))}$$

   (d) Second-order system with different open-loop zeros:

$$1 + K \frac{(s + 0.5)}{(s + 2)(s + 1)} = 0 \qquad \texttt{rlocus(zpk(-0.5,[-1 -2],1))}$$

$$1 + K \frac{(s + 1.5)}{(s + 2)(s + 1)} = 0 \qquad \text{rlocus(zpk(-1.5,[-1 -2],1))}$$

$$1 + K \frac{(s + 4)}{(s + 2)(s + 1)} = 0 \qquad \text{rlocus(zpk(-4,[-1 -2],1))}$$

(e) Third-order system:

$$1 + K \frac{1}{(s + 3)(s + 2)(s + 1)} = 0 \qquad \text{rlocus(zpk([-1 -2 -3],1))}$$

(f) Third-order system with an open-loop zero:

$$1 + K \frac{s + 1.5}{(s + 3)(s + 2)(s + 1)} = 0 \qquad \text{rlocus(zpk(-1.5,[-1 -2 -3],1))}$$

See also what the plot is like if the open-loop zero is at -0.5, -2.5, and -3.5.

Rough sketches of what you should obtain with MATLAB are in Fig. E7.6. The root locus of the system in (a) is a line on the real axis extending to negative infinity (Fig. E7.6A). The root loci in (b) approach each other (arrows not shown) on the real axis and then branch off toward infinity at 90°. The repeated roots in (c) simply branch off toward infinity.

With only open-loop poles, examples (a) – (c) can represent only systems with a proportional controller. In case (a), the system contains a first-order process and in (b) and (c) are overdamped and critically damped second-order processes.

The sketches for (d) illustrate how an open-loop zero, say, in an ideal PD controller, may affect the root-locus plot and dynamics of a system containing an overdamped second-order process. Underdamped system behavior is expected only when the open-loop zero is large enough (i.e., τ_D sufficiently small). On the left-hand panel of Fig. E7.6B, one locus goes from the -1 open-loop pole to the open-loop zero at -0.5 (arrow not shown). The second locus goes from -2 to negative infinity on the real axis. In the middle panel, one locus goes from -1 to the open-loop zero at -1.5 (arrow not shown). On the right, where the open-loop zero is at -4, two root loci move toward each from -1 and -2 (arrows not shown), then branch off. The branches break in later onto the real axis; one locus approaches the zero at -4, the other toward negative infinity.

Systems (e) and (f) would contain a third-order process. Of course, we can have a proportional control only in case (e), and (f) represents one probable example of using an ideal PD controller.

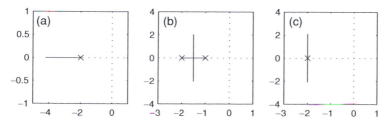

Figure E7.6A.

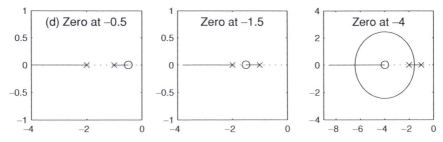

Figure E7.6B.

The system in (e) can become unstable, whereas a proper addition of an open-loop zero, as in (f), can help stabilize the system (Fig. E7.6C). In (e), the two loci from −1 and −2 approach each other (arrows not shown). They then break away, and the closed-loop poles become unstable. The two loci approach positive infinity at ±60°. In (f), the system is always stable. The dominant closed-loop pole is the locus moving from −1 to −1.5 (arrow not shown). This system is faster than if we had put the open-loop zero at, say, −0.5.

In most classic control texts, we find **plotting guidelines** to help hand-sketching of root-locus plots. After going over Example 7.6, you should find that some of them are quite intuitive. These are the simple guidelines:

(1) The root-locus plot is symmetric about the real axis.
(2) The number of loci equals the number of open-loop poles (or the order of the system).
(3) A locus (closed-loop root path) starts at an open-loop pole and either terminates at an open-loop zero or extends to infinity.
(4) On the real axis, a root locus exists only to the *left* of an odd number of real poles and zeros. (The explanation of this point is on the *Web Support*.)
(5) The points at which the loci cross the imaginary axis can be found by the Routh–Hurwitz criterion or by substituting $s = j\omega$ in the characteristic equation. (Of course, we can also use MATLAB to do that.)

To determine the shape of a root-locus plot, we need other rules to determine the locations of the so-called breakaway and break-in points, the corresponding angles of departure and arrival, and the angle of the asymptotes if the loci approach infinity. They all arise from the analysis of the characteristic equation. These features, including item (4) above, are explained on the *Web Support*. With MATLAB, our need for them is minimal.

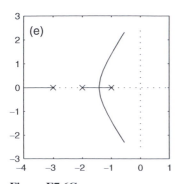
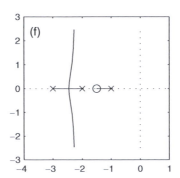

Figure E7.6C.

Note: There are two important steps that you must follow. First, make sure you go through the MATLAB tutorial (Session 6) carefully to acquire a feel on the probable shapes of root-locus plots. Secondly, test guidelines (3) and (4) listed previously for every plot that you make in the tutorial. These guidelines can become your most handy tool to deduce, without doing any algebra, whether a system will exhibit underdamped behavior, or, in other words, whether a system will have complex closed-loop poles.

7.5.   Root-Locus Design

In terms of controller design, the closed-loop poles (or now the root loci) also tell us about the system dynamics. We can extract much more information from a root-locus plot than from a Routh criterion analysis or an $s = j\omega$ substitution. In fact, it is common to impose, say, a time constant or a damping ratio specification on the system when we use root-locus plots as a design tool.

Example 7.5A: Consider the second-order system in Example 7.5: What should the proportional gain be if we specify the controlled system to have a damping ratio of 0.7?

The second-order closed-loop characteristic equation in Example 7.5 can be rearranged as

$$\frac{\tau_1 \tau_2}{(1 + K_c K_p)} s^2 + \frac{(\tau_1 + \tau_2)}{(1 + K_c K_p)} s + 1 = 0.$$

We can compare this equation with the general form $\tau^2 s^2 + 2\zeta\tau s + 1 = 0$, where now τ is the system time period and ζ is the system damping ratio. From Example 5.2, we have derived that

$$\tau = \sqrt{\frac{\tau_1 \tau_2}{(1 + K_c K_p)}}, \qquad \zeta = \frac{1}{2}\frac{(\tau_1 + \tau_2)}{\sqrt{\tau_1 \tau_2(1 + K_c K_p)}}.$$

Thus we can solve for K_c with a given choice of ζ in a specific problem.

However, MATLAB allows us to get the answer with very little work – something that is very useful when we deal with more complex systems. Consider a numerical problem with values of the process gain $K_p = 1$ and process time constants $\tau_1 = 2$ and $\tau_2 = 4$ such that the closed-loop equation is

$$1 + K_c \frac{1}{(2s + 1)(4s + 1)} = 0.$$

We enter the following statements in MATLAB[5]:

```
G=tf(1,conv([2 1],[4 1]));
rlocus(G)
sgrid(0.7,1)                    % plot the 0.7 damping ratio lines
[kc,cpole]=rlocfind(G)
```

---

[5] The technique of using the damping ratio line $\theta = \cos^{-1}\zeta$ in Eq. (2.34) is applied to higher-order systems. When we do so, we are implicitly making the *assumption* that we have chosen the dominant closed-loop pole of a system and that this system can be approximated as a second-order underdamped function at sufficiently large times. For this reason, root locus is also referred to as **dominant-pole design**.

Where the root locus intersects the 0.7 damping ratio line, we should find, from the result returned by `rlocfind()`, the proportional gain to be 1.29 (1.2944 to be exact) and the closed-loop poles at $-0.375 \pm 0.382j$. The real and imaginary parts are not identical because $\cos^{-1} 0.7$ is not exactly $45°$. We can confirm the answer by substituting the values into our analytical equations. We should find that the real part of the closed-loop pole agrees with what we derived in Example 7.5 and the value of the proportional gain agrees with the expression that we derived in this example.

Example 7.7: Consider installing a PI controller in a system with a first-order process such that we have no offset. The process function has a steady-state gain of 0.5 and a time constant of 2 min. Take $G_a = G_m = 1$. The system has the simple closed-loop characteristic equation:

$$1 + K_c \frac{0.5(\tau_I s + 1)}{\tau_I s(2s + 1)} = 0.$$

We also want to have a slightly underdamped system with a reasonably fast response and a damping ratio of 0.7. How should we design the controller? To restrict the example for illustration, we consider (a) $\tau_I = 3$ min and (b) $\tau_I = 2/3$ min.

To begin with, this is a second-order system with no positive zeros, and so stability is not an issue. Theoretically speaking, we could have derived and proved all results with the simple second-order characteristic equation, but we take the easy way out with root-locus plots.

(a) The open-loop poles are at -0.5 and at the origin. If the integral time constant is $\tau_I = 3$ min, the open-loop zero is at $-1/3$, and all we have are negative and real closed-loop poles [Fig. E7.7(a)]. The system will not become underdamped. The "speed" of the response, as we increase the proportional gain, is limited by the zero position at $-1/3$. This is where the dominant closed-loop pole will approach.

(b) The situation is more interesting with a smaller integral time constant, $\tau_I = 2/3$ min, now with the open-loop zero at -1.5. The root loci first approach one another (arrows not shown) before breaking away to form the circle [Fig. E7.7(b)]. As K_c increases further, the loci break into the real axis. The dominant closed-loop pole will approach the zero at -1.5.

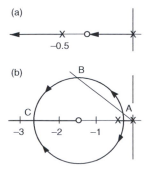

(a)

−0.5

(b)

B

C

−3   −2   −1

A

Figure E7.7.

We use the following MATLAB statements to draw the 0.7 damping ratio line, and it intersects the root loci at two points (Fig. E7.7): at point A, $-0.312 + 0.323j$, when $K_c = 0.55$, and at point B, $-1.15 + 1.17j$, when $K_c = 7.17$:

```
kc=1; taui=2/3;
Gc=tf(kc*[taui 1], [taui 0]);
Gp=tf(0.5, [2 1]);
rlocus(Gc*Gp)
sgrid(0.7,1)
[kc,cpole]=rlocfind(Gc*Gp)
```

If saturation is not a problem, the proportional gain $K_c = 7.17$ (point B) is preferred. The corresponding closed-loop pole has a faster time constant. (The calculation of the time period or frequency and confirmation of the damping ratio are left as homework.)

Note 1: Theoretically speaking, point C on the root-locus plot is ideal – the fastest possible response without any oscillation. We rarely can do that in practice; the proportional gain would be so large that the controller would be saturated.

Note 2: As we reduce the integral time constant from $\tau_I = 3$ min to exactly 2 min, we have the situation of pole-zero cancellation. The terms in the closed-loop characteristic equation cancel out, and the response in this example is reduced to that of a first-order system. If τ_I is only slightly less than 2 min, we have a very slightly underdamped system; the "circle" in Fig. E7.7(b) is very small (try this with MATLAB). In reality, it is very difficult to have perfect pole-zero cancellation, and if we design the controller with τ_I too close to τ_p, the system response may "bounce back and forth."

7.6. Final Remark on Root-Locus Plots

Instructive as root-locus plots appear to be, this technique does have its limitations. The most important one is that it cannot handle dead-time easily. When we have a system with dead-time, we must make an approximation with the Padé formulas. This is the same limitation that applies to the Routh–Hurwitz criterion.

In this computer age, you may question why nobody would write a program that can solve for the roots with dead-time accurately. Someone did. There are even refined hand-sketching techniques to account for the lag that is due to dead time. However, these tools are not as easy to apply and are rarely used. Few people use them because frequency-response analysis in Chap. 8 can handle dead-time accurately and extremely easily.

A second point on root-locus plots is that they can give us only the so-called absolute stability, but not relative stability, meaning that there is no easy way to define a good, general "safety margin." We may argue that we can define a certain distance that a closed-loop pole must stay away from the imaginary axis, but such an approach is very problem specific. Recall that the proportional gain is an implicit parameter along a locus, and is very difficult to tell what effects we may have with slight changes in the proportional gain (the sensitivity problem). Frequency-response analysis once again does better and allows us to define *general* relative stability criteria. Furthermore, frequency-response analysis can help us to understand why a certain choice of, for example, an integral time constant may destabilize a system. (Jumping ahead, it has to do with when we bring in phase lead. We shall see that in Chap. 8 and Example 10.1.)

On the other hand, frequency-response analysis cannot reveal information on dynamic response easily – something root locus does very well. Hence controller design is always an iterative procedure. There is no one-stop shopping. There is never a unique answer.

Finally, you may wonder if you can use the integral or the derivative time constant as the parameter. Theoretically, you can. You may rearrange the characteristic equation in such a way that can take advantage of prepackaged programs that use the proportional gain as the parameter. In practice, nobody does that. One main reason is that the resulting loci plot will not have this nice interpretation that we have by varying the proportional gain.

Review Problems

(1) Repeat Examples 7.2 and 7.4 with the general closed-loop characteristic polynomial:

$$a_3 s^3 + a_2 s^2 + a_1 s + a_0 + K_c = 0.$$

Derive the ultimate frequency and the ultimate gain.

(2) Revisit Example 5.4. What choice of τ_D should we make?

(3) No additional reviews are needed as long as you go through each example carefully with MATLAB.

Hint:

(2) We can use Example 7.6(d) as a hint. To have a faster system response, we want to choose $\tau_D < \tau_p$ such that the open-loop zero $-1/\tau_D$ lies to the left of the open-loop pole $-1/\tau_p$ on the real-axis.

8

Frequency-Response Analysis

Т he response of a stable system *at large times* is characterized by its amplitude and phase shift when the input is a sinusoidal wave. These two quantities can be obtained from the transfer function, of course, without inverse transform. The analysis of this **frequency response** can be based on a simple substitution (mapping) $s = j\omega$, and the information is given by the magnitude (modulus) and the phase angle (argument) of the transfer function. Because the analysis begins with a Laplace transform, we are still limited to linear or linearized models.

What Are We Up to?

- Theoretically, we are making the presumption that we can study and understand the dynamic behavior of a process or system by imposing a sinusoidal input and measuring the frequency response. With chemical systems that cannot be subject to frequency-response experiments easily, it is very difficult for a beginner to appreciate what we will go through. So until then, take frequency response as a math problem.
- Both the magnitude and the argument are functions of the frequency. The so-called Bode and Nyquist plots are nothing but graphical representations of this functional dependence.
- Frequency-response analysis allows us to derive a general relative stability criterion that can easily handle systems with time delay. This property is used in controller design.

8.1. Magnitude and Phase Lag

Our analysis is based on the mathematical property that, given a *stable* process (or system) and a sinusoidal input, the response will eventually become a purely sinusoidal function. This output will have the same frequency as the input, but with different amplitude and phase angle. The two latter quantities can be derived from the transfer function.

The idea of frequency response is first illustrated with an inverse Laplace transform. Consider our good old familiar first-order model equation,[1]

$$\tau_p \frac{dy}{dt} + y = K_p f(t) \tag{8.1}$$

with its transfer function

$$G(s) = \frac{Y(s)}{F(s)} = \frac{K_p}{\tau_p s + 1}. \tag{8.2}$$

If the input is a sinusoidal function such that $f(t) = A \sin \omega t$, the output $Y(s)$ is

$$Y(s) = \frac{K_p}{\tau_p s + 1} \frac{A\omega}{s^2 + \omega^2}. \tag{8.3}$$

If we do the partial-fraction expansion and inverse transform, we should find, after some hard work, the time-domain solution:

$$y(t) = \frac{A K_p \tau_p \omega}{\tau_p^2 \omega^2 + 1} e^{-t/\tau_p} + \left(\frac{A K_p}{\sqrt{\tau_p^2 \omega^2 + 1}} \right) \sin(\omega t + \phi), \tag{8.4}$$

where

$$\phi = \tan^{-1}(-\omega \tau_p) \tag{8.5}$$

is the phase lag.[2] The algebraic details of deriving Eq. (8.4) are not important. The important aspects will be derived by means of an alternative route just a few steps ahead. For now, the crucial point is to observe that if time is sufficiently large (as relative to τ_p), the exponential term in Eq. (8.4) will decay away and the time response becomes a purely sinusoidal function (Fig. 8.1).

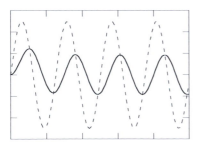

Figure 8.1. Schematic response (solid curve) of a first-order function to a sinusoidal input (dashed curve). The response has a smaller amplitude, a phase lag, and its exponential term decays away quickly to become a pure sinusoidal response.

[1] Although we retain our usual notation of using a subscript p for a first-order function in illustrations, the analysis is general; it applies to an open-loop process or a closed-loop system. In other words, we can apply a sinusoidal change to the set point of a system if we need to.

[2] In these calculations, the units of frequency are radians per time. If the period T is given in seconds, then frequency $\omega = 1/T$ Hz (hertz or cycles per second) or $\omega = 2\pi/T$ (rad/s).

The time response at sufficiently large times can be normalized with respect to the amplitude of the input sine wave:

$$\frac{y_\infty(t)}{A} = \left(\frac{K_p}{\sqrt{\tau_p^2 \omega^2 + 1}} \right) \sin(\omega t + \phi), \tag{8.6}$$

where now the amplitude of this normalized large-time response is called the **amplitude ratio (AR)**. The response has a different amplitude and phase lag in relation to the input wave. If we further normalize the equation such that the amplitude of the sine wave is bounded by one, we obtain

$$\frac{y_\infty(t)}{AK_p} = \left(\frac{1}{\sqrt{\tau_p^2 \omega^2 + 1}} \right) \sin(\omega t + \phi). \tag{8.7}$$

The amplitude of this normalized response, y_∞/AK_p, is called the *magnitude ratio*.[3]

Next, let's substitute $s = j\omega$ into the transfer function, i.e., $G_p(s) = G_p(j\omega)$, which makes G_p a complex number[4]:

$$G_p(j\omega) = \frac{K_p}{j\omega\tau_p + 1} \left(\frac{-j\omega\tau_p + 1}{-j\omega\tau_p + 1} \right) = \left(\frac{K_p}{\tau_p^2 \omega^2 + 1} \right) - j \left(\frac{K_p \omega \tau_p}{\tau_p^2 \omega^2 + 1} \right). \tag{8.8}$$

If we put Eq. (8.8) in polar coordinates, $G_p(j\omega) = |G_p(j\omega)|e^{j\phi}$, the **magnitude** and the **phase angle** are

$$|G_p(j\omega)| = \frac{K_p}{\sqrt{\tau_p^2 \omega^2 + 1}}, \qquad \phi = \angle G_p(j\omega) = \tan^{-1}(-\omega\tau_p). \tag{8.9}$$

A comparison of Eqs. (8.9) with Eq. (8.6) shows that the magnitude and the phase angle of $G_p(j\omega)$ are exactly the same as the amplitude and the phase lag of the normalized "large-time" time-domain solution.

We should note that the magnitude and the phase angle of $G_p(j\omega)$ are functions of the input frequency. The larger the frequency, the lower the magnitude and the larger the phase lag. We can make this observation by writing $\tau_p \omega = \omega/\omega_p$. When the imposed frequency is large with respect to the process "natural" frequency, ω_p, the process cannot respond fast enough, resulting in low magnitude and large phase lag. When the imposed frequency is relatively small, the magnitude approaches the steady-state gain and the phase lag approaches zero.

8.1.1. The General Analysis

We now generalize our simple illustration. Consider a general transfer function of a *stable* model $G(s)$, which we also denote as the ratio of two polynomials, $G(s) = Q(s)/P(s)$. We

---

[3] This is a major source of confusion in many texts. The magnitude ratio is *not* the magnitude of a transfer function. It is the AR that is the same as the magnitude of $G(s)$. To avoid confusion, we stick strictly with the mathematical property, i.e., the magnitude of $G(s)$. We will use neither AR nor magnitude ratio in our terminology. It is much more sensible to consider the magnitude of a transfer function.

[4] If you need that, a brief summary of complex variable definitions is on the *Web Support*.

impose a sinusoidal input $f(t) = A \sin \omega t$ such that the output is

$$Y(s) = G(s)\left(\frac{A\omega}{s^2 + \omega^2}\right) = \frac{Q(s)}{P(s)}\frac{A\omega}{(s + j\omega)(s - j\omega)}.$$

Because the model is stable, all the roots of $P(s)$, whether they be real or complex, have negative-real parts and their corresponding time-domain terms will decay away exponentially. Thus if we are interested in only the time-domain response at sufficiently large times, we need to consider only the partial-fraction expansion of the two purely sinusoidal terms associated with the input:

$$Y_\infty(s) = \frac{a}{(s + j\omega)} + \frac{a^*}{(s - j\omega)}. \tag{8.10}$$

We can find their inverse transform easily. Apply the Heaviside expansion,

$$a = (s + j\omega)Y(s)|_{s=-j\omega} = G(-j\omega)\frac{A\omega}{-2j\omega} = \frac{AG(-j\omega)}{-2j},$$

and its conjugate redundantly just in case you do not believe the result is correct:

$$a^* = (s - j\omega)Y(s)|_{s=+j\omega} = G(j\omega)\frac{A\omega}{2j\omega} = \frac{AG(j\omega)}{2j}.$$

The time-domain solution (at large times) is hence

$$y_\infty(t) = \left[\frac{AG(-j\omega)}{-2j}\right]e^{-j\omega t} + \left[\frac{AG(j\omega)}{2j}\right]e^{j\omega t}.$$

Note that $G(j\omega) = |G(j\omega)|e^{j\phi}$ and $G(-j\omega) = |G(j\omega)|e^{-j\phi}$, and we can write

$$\frac{y_\infty(t)}{A} = |G(j\omega)|\left(\frac{e^{-j\phi}e^{-j\omega t}}{-2j} + \frac{e^{j\phi}e^{j\omega t}}{2j}\right).$$

Apply Euler's identity and the final result for the normalized response is

$$\left[\frac{y_\infty(t)}{A}\right] = |G(j\omega)| \sin(\omega t + \phi), \tag{8.11}$$

where $\phi = \angle G(j\omega)$. This is a crucial result. It constitutes the basis of frequency-response analysis, where in general, all we need are the magnitude and the argument of the transfer function $G(s)$ after the substitution $s = j\omega$.

8.1.2. Some Important Properties

We need to appreciate some basic properties of transfer functions when they are viewed as complex variables. They are important in performing frequency-response analysis. Consider that any given transfer function can be "broken up" into a *product* of simpler ones:

$$G(s) = G_1(s)G_2(s)\cdots G_n(s). \tag{8.12}$$

We do not need to expand the entire function into partial fractions. The functions G_1, G_2, etc., are better viewed as simply first-order and, at the most, second-order functions. In frequency-response analysis, we make the $s = j\omega$ substitution and further write the function in terms of magnitude and phase angle as

$$G(j\omega) = G_1(j\omega)G_2(j\omega) \cdots G_n(j\omega) = |G_1(j\omega)|e^{j\phi_1}|G_2(j\omega)|e^{j\phi_2} \cdots |G_n(j\omega)|e^{j\phi_n}$$

or

$$G(j\omega) = |G_1(j\omega)||G_2(j\omega)| \cdots |G_n(j\omega)|e^{j(\phi_1+\phi_2+\cdots+\phi_n)}.$$

The magnitude of $G(j\omega)$ is

$$|G(j\omega)| = |G_1(j\omega)||G_2(j\omega)| \cdots |G_n(j\omega)| \tag{8.13}$$

or

$$\log|G(j\omega)| = \log|G_1| + \log|G_2| + \cdots + \log|G_n|.$$

The phase angle is

$$\phi = \angle G(j\omega) = \angle G_1(j\omega) + \angle G_2(j\omega) + \cdots + \angle G_n(j\omega). \tag{8.14}$$

Example 8.1: Derive the magnitude and phase lag of the following transfer function:

$$G(s) = \frac{(\tau_a s + 1)}{(\tau_1 s + 1)(\tau_2 s + 1)}.$$

We can rewrite the function as a product of

$$G(s) = (\tau_a s + 1)\frac{1}{(\tau_1 s + 1)}\frac{1}{(\tau_2 s + 1)}.$$

The magnitude and the phase angle of these terms with the use of Eqs. (8.9) are

$$|G(j\omega)| = \sqrt{1 + \tau_a^2\omega^2}\frac{1}{\sqrt{1 + \tau_1^2\omega^2}}\frac{1}{\sqrt{1 + \tau_2^2\omega^2}},$$

$$\phi = \tan^{-1}(\omega\tau_a) + \tan^{-1}(-\omega\tau_1) + \tan^{-1}(-\omega\tau_2).$$

We have not covered the case of a zero term in the numerator. Here, we are just guessing that its result is the reciprocal of the first-order function result in Eqs. (8.9). The formal derivation comes later in Example 8.4. Note that τ has units of time, ω is in radians per time, and $\tau\omega$ is in radians.

We could have put $G(s)$ in a slightly different form:

$$G(j\omega) = \frac{G_a(j\omega)G_b(j\omega) \cdots G_m(j\omega)}{G_1(j\omega)G_2(j\omega) \cdots G_n(j\omega)} = \frac{|G_a||G_b| \cdots}{|G_1||G_2| \cdots}e^{j(\theta_a+\theta_b+\cdots-\theta_1-\theta_2-\cdots)}. \tag{8.15}$$

In this case, the equivalent form of Eq. (8.13) is

$$\log|G(j\omega)| = (\log|G_a| + \log|G_b| + \cdots + \log|G_m|)$$
$$- (\log|G_1| + \log|G_2| + \cdots + \log|G_n|) \tag{8.16}$$

and the equivalent to Eq. (8.14) is

$$\phi = (\angle G_a + \angle G_b + \cdots + \angle G_m) - (\angle G_1 + \angle G_2 + \cdots + \angle G_n). \tag{8.17}$$

With these results, we are ready to construct plots used in frequency-response analysis. The important message is that we can add up the contributions of individual terms to construct the final curve. The magnitude, of course, would be on the logarithmic scale.

8.2. Graphical Analysis Tools

We know that both $|G(j\omega)|$ and $\angle G(j\omega)$ are functions of frequency ω. We certainly would like to see the relationships graphically. There are three common graphical representations of the frequency dependence. All three methods are described briefly. The introduction relies on the use the so-called Bode plots, and more details will follow with respective examples.

8.2.1. Magnitude and Phase Plots: log $G(j\omega)$ versus log ω and $\angle G(j\omega)$ versus log ω

Here, we simply plot the magnitude (modulus) and phase angle (argument) individually against frequency – the so-called **Bode plots**. From Eq. (8.16), we should use a log scale for the magnitude. We also use a log scale for the frequency to cover a larger range. Thus we use a log–log plot for $|G(j\omega)|$ versus ω and a semilog plot for $\angle G(j\omega)$ versus ω. The unit of phase angle is commonly converted to degrees, and frequency is in radians per unit time.

In most electrical engineering or industrial control books, the magnitude is plotted in units of decibels (dB) as

$$1\text{ dB} = 20\log|G(j\omega)|. \tag{8.18}$$

Even with MATLAB, we should still know the expected shape of the curves and its "telltale" features. This understanding is crucial in developing our problem-solving skills. Thus doing a few simple hand constructions is very instructive. When we sketch the Bode plot, we must identify the corner (break) frequencies, slopes of the magnitude asymptotes, and the contributions of phase lags at small and large frequencies. We will pick up the details in the examples.

Another advantage of frequency-response analysis is that we can "identify" the process transfer function with experimental data. With either a frequency-response experiment or a pulse experiment with the proper Fourier transform, we can construct the Bode plot by using the open-loop transfer functions and use the plot as the basis for controller design.[5]

8.2.2. Polar-Coordinate Plots – $G(j\omega)$ in Polar Coordinates or $Re[G(j\omega)]$ versus $Im[G(j\omega)]$

We can plot the real and the imaginary parts of $G(j\omega)$ on the s plane with ω as the parameter – the so-called **Nyquist plot**. Because a complex number can be put in polar coordinates, the Nyquist plot is also referred to as the polar plot:

$$G(j\omega) = Re[G(j\omega)] + Im[G(j\omega)] = |G(j\omega)|e^{j\phi}.$$

This plotting format contains the same information as the Bode plot. The polar plot is more compact, but the information on the frequency is not shown explicitly. If we do not have

---

[5] The pulse experiment is not crucial for our understanding of frequency-response analysis and is provided on the *Web Support*, but we will do the design calculations in Section 8.3.

a computer, we theoretically could read numbers off a Bode plot to construct the Nyquist plot. The use of Nyquist plots is more common in multiloop or multivariable analyses. A Bode plot, on the other hand, is easier to interpret and is a good learning tool.

There are hand-plotting techniques, but we will rely on the computer. Still, we need to know the qualitative features of the plot resulting from several simple transfer functions.

8.2.3. Magnitude versus Phase Plot – log $G(j\omega)$ versus $\angle G(j\omega)$

In the third rendition of good old $G(j\omega)$, we can plot the logarithmic magnitude against the phase lag – the so-called **Nichols chart**. Generally, this plot is made with the open-loop transfer function of a unity feedback system. The magnitude and the argument contours of the closed-loop transfer function are overlaid on it. The plot allows for a frequency-response design analysis that better relates to probable closed-loop dynamic response.

For now, we take a look at the construction of Bode and Nyquist plots of transfer functions that we have discussed in Chaps. 2 and 3. Keep in mind that these plots contain the same information: $G(j\omega)$. It is important that you run MATLAB with sample numerical values while reading the following examples. Yes, you need to go through MATLAB Session 7 first.

Example 8.2: What are the Bode and Nyquist plots of a first-order transfer function? We use the time-constant form of transfer functions. The magnitude and the phase angle of

$$G(s) = \frac{Y(s)}{F(s)} = \frac{K_p}{\tau_p s + 1}$$

are derived in Eqs. (8.9) as

$$|G_p(j\omega)| = \frac{K_p}{\sqrt{\tau_p^2 \omega^2 + 1}}, \qquad \phi = \angle G_p(j\omega) = \tan^{-1}(-\omega\tau_p).$$

We may try the following MATLAB statements to generate Fig. E8.2:

```
kp=1; % Just arbitrary values. Try different ones yourself.
tau=2;
G=tf(kp,[tau 1]);
figure(1), bode(G);
```

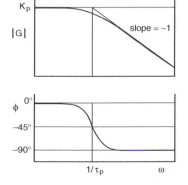
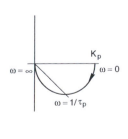

Figure E8.2.

```
figure(2),nyquist(G);% If the Nyquist default plot is
                     % confusing,
                     % follow the instructions in MATLAB Session 7
```

We plot, in theory,

$$\log |G_p(j\omega)| = \log K_p - \frac{1}{2} \log \left(1 + \tau_p^2 \omega^2\right).$$

To make the phase-angle plot, we simply use the definition of $\angle G_p(j\omega)$. As for the polar (Nyquist) plot, we do a frequency parametric calculation of $|G_p(j\omega)|$ and $\angle G_p(j\omega)$, or we can simply plot the real part versus the imaginary part of $G_p(j\omega)$.[6] To check that a computer program is working properly, we need to use only the high- and the low-frequency asymptotes – the same if we had to do the sketch by hand, as in the old days. In the limit of low frequencies,

$$\omega \to 0, \quad |G_p| = K_p, \quad \phi = 0.$$

On the magnitude plot, the low-frequency (also called zero-frequency) asymptote is a horizontal line at K_p. On the phase-angle plot, the low-frequency asymptote is the $0°$ line. On the polar plot, the zero-frequency limit is represented by the point K_p on the real axis. In the limit of high frequencies,

$$\omega \to \infty, \quad |G_p| = \frac{K_p}{\tau_p \omega}, \quad \phi = \tan^{-1}(-\infty) = -90°.$$

With the phase lag, we may see why a first-order function is also called a first-order lag. On the magnitude log–log plot, the high-frequency asymptote has a slope of -1. This asymptote also intersects the horizontal K_p line at $\omega = 1/\tau_p$. On the phase-angle plot, the high-frequency asymptote is the $-90°$ line. On the polar plot, the infinity-frequency limit is represented by the origin as the $G_p(j\omega)$ locus approaches it from the $-90°$ angle.

The frequency at which $\omega = 1/\tau_p$ is called the corner frequency (also the break frequency). At this position,

$$\omega = 1/\tau_p, \quad |G_p| = K_p/\sqrt{2}, \quad \phi = \tan^{-1}(-1) = -45°.$$

We may question the significance of the break frequency, $\omega = 1/\tau$. Let's take the first-order transfer function as an illustration. If the time constant is small, the break frequency is large. In other words, a fast process or system can respond to a large range of input frequencies without a diminished magnitude. On the contrary, a slow process or system has a large time constant and a low break frequency. The response magnitude is attenuated quickly as the input frequency increases. Accordingly, the phase lag also decreases quickly to the theoretical high-frequency asymptote.

A common term used in control engineering is **bandwidth**, which is defined as the frequency at which any given $|G(j\omega)|$ drops to 70.7% of its low-frequency asymptotic value (Fig. 8.2). It is obvious that the 70.7% comes from the $1/\sqrt{2}$ of the first-order function

---

[6] All comments on Nyquist plots are made without the need of formal hand-sketching techniques. Strictly speaking, the polar plot is a mapping of the imaginary axis from $\omega = 0_+$ to ∞. You will see that in texts that provide a more thorough discussion on choosing the so-called Nyquist path.

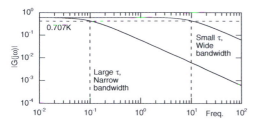

Figure 8.2. Schematic illustration of two systems with wide and narrow bandwidths.

calculation in Example 8.2.[7] A large bandwidth is related to a fast system with a small time constant. On the other hand, slow responses or large time constants are cases with narrow bandwidths. From another perspective, the first-order function is the simplest example of a **low-pass filter**. Signals with frequencies below the bandwidth are not attenuated, and their magnitude diminishes with higher frequencies.

Example 8.3: What are the Bode and Nyquist plots of a second-order transfer function? We make the $s = j\omega$ substitution into the transfer function,

$$G(s) = \frac{K}{\tau^2 s^2 + 2\zeta\tau s + 1},$$

to obtain

$$G(j\omega) = \frac{K}{(1 - \tau^2\omega^2) + j2\zeta\tau\omega} = \frac{K[(1 - \tau^2\omega^2) - j2\zeta\tau\omega]}{(1 - \tau^2\omega^2)^2 + (2\zeta\tau\omega)^2}.$$

After a few algebraic steps, the resulting magnitude and phase-angle of $G(j\omega)$ are

$$|G(j\omega)| = \frac{K}{\sqrt{(1 - \tau^2\omega^2)^2 + (2\zeta\tau\omega)^2}}, \qquad \phi = \angle G(j\omega) = \tan^{-1}\left(\frac{-2\zeta\tau\omega}{1 - \tau^2\omega^2}\right).$$

These are sample MATLAB statements to plot the magnitude and the phase angle as in Fig. E8.3:

```
k=1; % Just arbitrary values. Try different ones yourself.
tau=2;
zeta=0.2;
G=tf(k,[tau*tau 2*zeta*tau 1]);
damp(G)     % confirm the damping ratio
figure(1), bode(G);
figure(2), nyquist(G);
```

In the limit of low frequencies,

$$\omega \to 0, \quad |G| = K, \qquad \phi = 0.$$

In the limit of high frequencies,

$$\omega \to \infty, \quad |G| = \frac{K}{\tau^2\omega^2}, \qquad \phi \approx \tan^{-1}\left(\frac{-2\zeta\tau\omega}{-\tau^2\omega^2}\right) = \tan^{-1}\left(\frac{2\zeta}{\tau\omega}\right) = \tan^{-1}(0) = -180°.$$

---

[7] This is also referred to as the 3-dB bandwidth. The term comes from $20\log(1/\sqrt{2}) = -3.01$ dB.

155

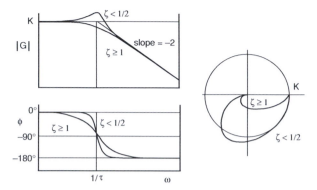

Figure E8.3.

We choose $-180°$ (not $0°$) because we know that there must be a phase lag. On the magnitude log–log plot, the high-frequency asymptote has a slope of -2. This asymptote intersects the horizontal K line at $\omega = 1/\tau$.

At the corner frequency,

$$\omega = 1/\tau, \quad |G(j\omega)| = \frac{K}{2\zeta}, \quad \phi = \tan^{-1}(-\infty) = -90°.$$

For a process or system that is sufficiently underdamped, $\zeta < 1/2$, the magnitude curve will rise above the low-frequency asymptote or the polar plot will extend beyond the K-radius circle.

We can take the derivative of the magnitude equation $|G(j\omega)| = K/\sqrt{(1 - \tau^2\omega^2)^2 + (2\zeta\tau\omega)^2}$ to find the actual maximum and its associated frequency, the so-called **resonant frequency**,[8] ω_r,

$$\omega_r = \frac{\sqrt{1 - 2\zeta^2}}{\tau} = \omega\sqrt{1 - 2\zeta^2}, \tag{8.19}$$

and the maximum magnitude $M_{p\omega}$ is

$$M_{p\omega} = |G(j\omega)|_{\max} = \frac{K}{2\zeta\sqrt{1 - \zeta^2}}. \tag{8.20}$$

From Eq. (8.19), there would be a maximum only if $0 < \zeta < 1/\sqrt{2}$ (or 0.707).

We can design a controller by specifying an upper limit on the value of $M_{p\omega}$. The smaller the system damping ratio ζ, the larger the value of $M_{p\omega}$ and the more OS, or underdamping we expect in the time-domain response. Needless to say, excessive resonance is undesirable.

---

[8] The step just before the result in Eq. (8.19) is $2(1 - \tau^2\omega^2)(-2\omega\tau^2) + 2(2\zeta\tau\omega)(2\zeta\tau) = 0$. In most texts, the derivation is based on unity gain $K = 1$ and so it will not show up in Eq. (8.20). Most texts also plot Eqs. (8.19) and (8.20) as functions of ζ. However, with MATLAB, we can do that ourselves as an exercise.

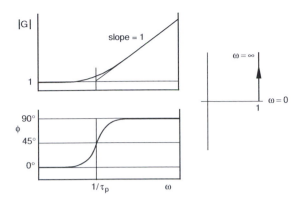

Figure E8.4.

Example 8.4: What are the Bode and Nyquist plots of a **first-order lead** $G(s) = (\tau_p s + 1)$?
After the $s = j\omega$ substitution, we have

$$G(j\omega) = 1 + j\omega\tau_p,$$

where

$$|G(j\omega)| = \sqrt{1 + \omega^2\tau_p^2}, \qquad \angle G(j\omega) = \tan^{-1}(\omega\tau_p).$$

We may try the following MATLAB statements for Fig. E8.4:

```
taud=2; % Just an arbitrary value
G=tf([taud 1],1);
figure(1), bode(G);
figure(2), nyquist(G);
```

The magnitude and the phase-angle plots are sort of "upside-down" versions of first-order
lag, with the phase angle increasing from $0°$ to $90°$ in the high-frequency asymptote. The
polar plot, on the other hand, is entirely different. The real part of $G(j\omega)$ is always 1 and
not dependent on frequency.

Example 8.5: What are the Bode and Nyquist plots of a dead-time function $G(s) = e^{-\theta s}$?
Again, we make the $s = j\omega$ substitution into the transfer function to obtain

$$G(j\omega) = e^{-\theta\omega j}.$$

The magnitude is simply unity, $|G(j\omega)| = 1$, and the phase angle is

$$\angle G(j\omega) = -\omega\theta.$$

When $\omega = \pi/\theta$, $\angle G(j\omega) = -\pi$. On the polar plot, the dead-time function is a unit circle.
The phase-angle plot is not a straight line because frequency is on a log scale.
   We need hints from our MATLAB Session to plot this function in Fig. E8.5. We'll do it
together in the next example.

   The important point is that the phase lag of the dead-time function increases without
bound with respect to frequency. This is what is called a **nonminimum-phase system**,
as opposed to the first and second transfer functions, which are minimum-phase systems.

157

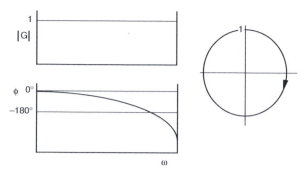

Figure E8.5.

Formally, a **minimum-phase system** is one that has no dead time and has neither poles nor zeros in the RHP. (See the Review Problems.)

From here on, only the important analytical equations or plots of asymptotes are provided in the examples. You should generate plots with sample numerical values by using MATLAB as you read them.

Example 8.6: What are the Bode and Nyquist plots of a first-order lag with dead time?

This example shows us the important reason why and how frequency-response analysis can handle dead time so easily. We substitute $s = j\omega$ into $G(s) = [(K_p e^{-t_d s})/(\tau_p s + 1)]$, and we have

$$G(j\omega) = \frac{K_p e^{-j\omega t_d}}{j\omega \tau_p + 1}.$$

From Example 8.5, we know that the magnitude of the dead time function is 1. Combining also with the results in ·Example 8.2, we find that the magnitude and the phase angle of $G(j\omega)$ are

$$|G(j\omega)| = \frac{K_p}{\sqrt{1 + \omega^2 \tau_p^2}}, \qquad \angle G(j\omega) = \tan^{-1}(\omega \tau_p) - \omega t_d.$$

The results are exact – we do not need to make approximations as we had to with the root-locus or the Routh array. The magnitude plot is the same as the first-order function, but the phase lag increases without bound because of the dead-time contribution in the second term. We will see that this is a major contribution to instability. On the Nyquist plot, the $G(j\omega)$ locus starts at K_p on the real axis and then "spirals" into the origin of the s plane.

This is how we may do the plots with time delay (details in MATLAB Session 7). Half of the work is taken up by the plotting statements.

```
kp=1; % Some arbitrary values
taup=10;
G=tf(kp,[taup 1]);
tdead=2;
freq=logspace(-1,1);                    % Make a frequency vector
[mag,phase]=bode(G,freq);
mag=mag(1,:); phase=phase(1,:);         % MATLAB specific step

phase = phase - ((180/pi)*tdead*freq); % Add dead-time phase lag
```

```
figure(1);
subplot(211), loglog(freq,mag)
               ylabel('Magnitude'),title('Bode Plot'), grid
subplot(212), semilogx(freq,phase)
               ylabel('Phase (degree)'),xlabel('Frequency'), grid

figure(2)              % We have to switch over to the polar plot
phase=phase*pi/180; % function to do this Nyquist plot
polar(phase,mag)
```

Example 8.7: What are the Bode and Nyquist plots of a pure integrating function $G(s) = K/s$?

The $s = j\omega$ substitution into the integrating function leads to a pure imaginary number: $G(j\omega) = K/j\omega = -jK/\omega$. The magnitude and the phase angle are

$$|G(j\omega)| = K/\omega, \qquad \angle G(j\omega) = \tan^{-1}(-\infty) = -90°.$$

Sample MATLAB statements are

```
G=tf(1,[1 0]);
figure(1), bode(G)
figure(2), nyquist(G)
```

The magnitude log–log plot is a line with slope -1. The phase-angle plot is a line at $-90°$. The polar plot is the negative-imaginary axis, approaching from negative infinity with $\omega = 0$ to the origin with $\omega \to \infty$.

Example 8.8: What are the Bode and Nyquist plots of a first-order lag with an integrator? Our first impulse may be an $s = j\omega$ substitution into the transfer function:

$$G(s) = \frac{K_p}{s(\tau_p s + 1)}.$$

However, the result is immediately obvious if we consider the function as the product of a first-order lag and an integrator. Combining the results from Examples 8.2 and 8.7, we find that the magnitude and the phase angle are

$$|G(j\omega)| = \frac{K_p}{\omega\sqrt{1 + \tau_p^2 \omega^2}},$$

$$\angle G(j\omega) = \tan^{-1}(-\infty) + \tan^{-1}(-\tau_p \omega) = -90° + \tan^{-1}(-\tau_p \omega).$$

Sample MATLAB statements are

```
kp=1;
taup=2;
G=tf(kp,[taup 1 0]);
figure(1), bode(G)
figure(2), nyquist(G)
```

Because of the integrator, the magnitude log–log plot does not have a low- or a high-frequency asymptote. The plot is a curve in which magnitude decreases with frequency. The phase-angle plot starts at $-90°$ at the low-frequency asymptote and decreases to $-180°$ at the high-frequency asymptote. The polar-plot curve approaches from negative infinity along the vertical line $-K_p\tau_p$ and approaches the origin as $\omega \to \infty$.

Example 8.9: Sketch the Bode plot of the following transfer function:

$$G(s) = \frac{(5s+1)}{(10s+1)(2s+1)}.$$

The MATLAB statements are

```
G=tf([5 1],conv([10 1],[2 1]));
bode(G);
```

Formally, we would plot

$$|G(j\omega)| = \sqrt{1+5^2\omega^2}\,\frac{1}{\sqrt{1+10^2\omega^2}}\,\frac{1}{\sqrt{1+2^2\omega^2}},$$

$$\angle G(j\omega) = \tan^{-1}(5\omega) + \tan^{-1}(-10\omega) + \tan^{-1}(-2\omega).$$

With MATLAB, what you find is that the actual curves are very smooth; it is quite different from hand-sketching. Nevertheless, understanding the asymptotic features is important in helping us check whether the results are correct. This is particularly easy (and important) with the phase-lag curve.

To help us understand MATLAB results, a sketch of the low- and the high-frequency asymptotes is provided in Fig. E8.9. A key step is to identify the corner frequencies. In this case, the corner frequency of the first-order lead is at 1/5 or 0.2 rad/s, whereas the two first-order lag terms have their corner frequencies at 1/10 and 1/2 rad/s. The final curve is a superimposition of the contributions from each term in the overall transfer function.

In addition, if you want to better see the little phase-lag "hump" that you expect from hand-sketching, change the term in the denominator from $(2s+1)$ to $(s+1)$ so that the phase lag of this term will not kick in too soon.

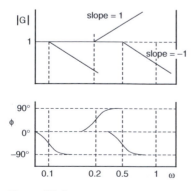

Figure E8.9.

8.3. Stability Analysis

With frequency-response analysis, we can derive a general relative stability criterion. The result is applicable to systems with dead time. The analysis of the closed-loop system can be reduced to using only the open-loop transfer functions in the computation.

8.3.1. Nyquist Stability Criterion

Consider the characteristic equation of a closed-loop system,

$$1 + G_m G_c G_a G_p = 0, \tag{7.2}$$

where often G_{OL} is used to denote the open-loop transfer function: $G_{OL} = G_m G_c G_a G_p$. To "probe" the property on the imaginary axis, we can make a substitution of $s = j\omega$ and rewrite the equation as

$$G_m(j\omega)G_c(j\omega)G_a(j\omega)G_p(j\omega) = -1 \quad \text{or} \quad G_{OL}(j\omega) = -1. \tag{7.2a}$$

This equation, of course, contains information regarding stability and, as it is written, implies that we may match properties on the LHS with the point $(-1, 0)$ on the complex plane. The form in Eq. (7.2a) also implies that in the process of analyzing the closed-loop stability property, the calculation procedures (or computer programs) require only the open-loop transfer functions. For complex problems, this fact eliminates unnecessary algebra. Just the Nyquist stability criterion is given here.[9]

Nyquist stability criterion: Given the closed-loop equation $1 + G_{OL}(j\omega) = 0$, if the function $G_{OL}(j\omega)$ has P open-loop poles and if the polar plot of $G_{OL}(j\omega)$ encircles the $(-1, 0)$ point N times as ω is varied from $-\infty$ to ∞, the number of unstable closed-loop poles in the RHP is $Z = N + P$. [Z is named after the number of zeros to $1 + G_{OL}(j\omega) = 0$.]

Do not panic! Without the explanation in the *Web Support*, this statement makes little sense. On the other hand, we do not really need this full definition because we know that just one unstable closed-loop pole is bad enough. Thus the implementation of the Nyquist stability criterion is much simpler than the theory.

A *simplified statement* of Nyquist stability criterion (Fig. 8.3): Given the closed-loop equation $1 + G_{OL}(j\omega) = 0$, the closed-loop system is stable if the polar plot of $G_{OL}(j\omega)$ does not encircle the $(-1, 0)$ point in the G_{OL}-plane.

In this statement, the term polar plot of G_{OL} has been used to replace a mouthful of words. G_{OL}-plane has been added in the wording to emphasize that we are using an analysis based on Eq. (7.2a). The real question lies in what safety margin we should impose on a given system. This question leads to the definitions of gain and phase margins, which constitute the basis of the general relative stability criteria for closed-loop systems.

When we make a Nyquist plot, we usually just map the positive-imaginary axis from $\omega = 0$ to infinity, as opposed to the entire axis starting from negative infinity. If a system is unstable, the resulting plot will contribute only π to the $(-1, 0)$ point as opposed to

---

[9] The formal explanation is in the *Web Support*. For a quick idea, our result is based on writing $G_{OL}(j\omega) = -1$. One simple thinking of instability is that if we feed back a sinusoidal wave, it will undergo a $-180°$ phase shift at the summing point of a negative-feedback loop. If the amplitude of the wave is less than one after passing through the summing point, it will die out. However, if the amplitude is larger than one, the oscillations will grow.

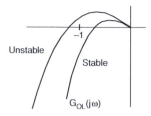

Figure 8.3. Illustration of the stable versus unstable possibilities under the Nyquist stability criterion.

2π – what encirclement really means. However, just mapping the positive-imaginary axis is sufficient to observe whether the plot may encircle the $(-1, 0)$ point.

8.3.2. Gain and Phase Margins

Once we understand the origin of Nyquist stability criterion, putting it to use is easy. Suppose we have a closed-loop system with the characteristic equation $1 + G_cG_p = 0$. With the point $(-1, 0)$ as a reference and the $G_c(j\omega)G_p(j\omega)$ curve on a Nyquist plot, we can establish a relative measure on how safe we are – that is, how far we are from the $(-1, 0)$ point. There are two possibilities. They are shown in Fig. 8.4, together with their interpretations on a Bode plot.

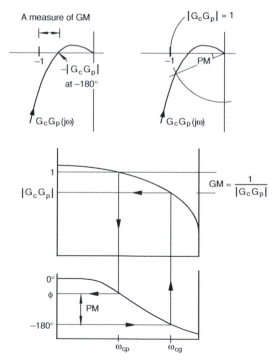

Figure 8.4. Interpretation of the gain and the phase margins based on the Nyquist stability criterion with Nyquist and Bode plots.

(1) On the negative-real axis ($-180°$), find the "distance" of $|G_cG_p|$ from $(-1,0)$. This is the **gain margin** (GM). The formal definition is

$$\text{GM} = \frac{1}{|G_c(j\omega_{cg})G_p(j\omega_{cg})|}, \tag{8.21}$$

where $G_c(j\omega)G_p(j\omega)$ is evaluated at the point at which it has a phase lag of $-180°$. The particular frequency at this point is the **gain crossover frequency**, $\omega = \omega_{cg}$. The smaller the magnitude of G_cG_p at $-180°$, the larger the GM and the "safer we are."

(2) Find the frequency where the magnitude $|G_c(j\omega)G_p(j\omega)|$ is 1. This particular frequency is the **phase crossover frequency**, $\omega = \omega_{cp}$. We then find the angle between G_cG_p and $-180°$. This is the **phase margin** (PM). The formal definition is

$$\text{PM} = \phi - (-180°) = \phi + 180°, \tag{8.22}$$

where the phase lag ϕ (a negative value) is measured at the point where $G_c(j\omega_{cp})G_p(j\omega_{cp})$ has a magnitude of one. The larger the angle, the larger the PM and the "safer we are."

For most control systems, we usually take a GM between 1.7 and 2 and a PM between $30°$ and $45°$ as the design specifications.

The Nyquist stability criterion can be applied to Bode plots. In fact, the calculation in which the Bode plot is used is much easier. To obtain the GM, we find the value of $|G_cG_p|$ that corresponds to a phase lag of $-180°$. To find the PM, we look up the phase lag corresponding to when $|G_cG_p|$ is 1.

Once again, recall that simple first- and second-order systems with no positive zeros are always stable, and no matter what a computer program may return, the GM and the PM have no meaning. Also, the GM crossover frequency is the same as the so-called ultimate frequency when we do the $s = j\omega$ substitution in the closed-loop characteristic equation or when we measure the value with the Ziegler–Nichols ultimate-cycle method.

The GM and the PM are used in the next section for controller design. Before that, let's plot different controller transfer functions and infer their properties in frequency-response analysis. Generally speaking, any function that introduces additional phase lag or magnitude tends to be destabilizing, and the effect is frequency dependent.

We skip the proportional controller, which is just $G_c = K_c$. Again, do the plots by using sample numbers with MATLAB as you read the examples.

Example 8.10: Derive the magnitude and the phase lag of the transfer function of a PI controller.

We could make the substitution $s = j\omega$ into $G_c(s) = K_c[1 + (1/\tau_I s)]$. However, we can obtain the result immediately if we see that the function is a product of an integrator and a first-order lead:

$$G_c(s) = K_c \frac{1}{\tau_I s}(\tau_I s + 1).$$

Thus

$$|G_c(j\omega)| = K_c \frac{1}{\omega \tau_I}\sqrt{(1 + \omega^2\tau_I^2)}, \qquad \angle G_c(j\omega) = -90° + \tan^{-1}(\omega\tau_I).$$

To do a demonstration plot, we may try the following MATLAB statements:

```
kc=1; % Just some arbitrary numbers
taui=2;
G=tf(kc*[taui 1],[taui 0]);
figure(1), bode(G);
figure(2), nyquist(G);
```

On the magnitude plot, the low-frequency asymptote is a line with slope -1. The high-frequency asymptote is a horizontal line at K_c. The phase-angle plot starts at $-90°$ at very low frequencies and approaches $0°$ in the high-frequency limit. On the polar plot, the $G_c(j\omega)$ locus is a vertical line that approaches from negative infinity at $\omega = 0$. At infinity frequency, it is at the K_c point on the real axis.

Integral control adds an additional phase lag ($-90°$) at low frequencies below the corner frequency $1/\tau_I$. A larger value of the integral time constant will limit the frequency range in which the controller introduces phase lag. This is one reason why choosing a large τ_I tends to be more stable than choosing a system with a small τ_I.[10]

Example 8.11: Derive the magnitude and the phase lag of the transfer function of an ideal PD controller.

The result is that of a first-order lead, as in Example 8.4. From $G_c(s) = K_c(1 + \tau_D s)$, we have, after $s = j\omega$ substitution,

$$G_c(j\omega) = K_c(1 + j\omega\tau_D),$$

and thus

$$|G_c(j\omega)| = K_c\sqrt{\left(1 + \omega^2\tau_D^2\right)}, \qquad \angle G_c(j\omega) = \tan^{-1}(\omega\tau_D).$$

On the magnitude plot, the low-frequency asymptote is a horizontal line at K_c. The high-frequency asymptote has a slope of $+1$. The phase-angle plot starts at $0°$ at very low frequencies and approaches $90°$ in the high-frequency limit. On the polar plot, the $G_c(j\omega)$ locus is a vertical line that starts at the point K_c on the real axis and approaches infinity. Based on the phase-angle plot, the PD controller provides a phase lead and thus a stabilizing effect. At the same time, the higher magnitude at higher-frequency ranges will amplify noises. There is a practical limit as to how fast a response a PD controller can handle.

The MATLAB statements are essentially the same as the first-order lead function:

```
kc=1; % Just some numbers we pick arbitrarily
taud=2;
G=tf(kc*[taud 1],1);
figure(1), bode(G);
figure(2), nyquist(G);
```

---

[10] Furthermore, we may want to choose τ_I such that $1/\tau_I$ is smaller than the corner frequency associated with the slowest open-loop pole of the process function. In this way, we help to stabilize the system by reducing the phase lag that is due to the integration before the phase lag of the process function "kicks in." However, integral control will not be effective if τ_I is too large, and there will be a design trade-off when we work with very slow processes. We will test this idea in Homework Problem II.38.

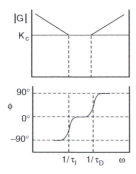

Figure E8.12. Only high- and low-frequency asymptotes are shown here. Fill in the rest with the help of MATLAB.

Example 8.12: Derive the magnitude and the phase lag of the controller transfer function

$$G_c(s) = K_c \left(\frac{1 + \tau_I s}{\tau_I s} \right)(1 + \tau_D s),$$

which is a PID controller with ideal derivative action and in the so-called interacting form. We look at this function because it is a product of an integrator and two first-order leads, and we can identify the high- and the low-frequency asymptotes easily. It is not identical to the ideal (noninteracting) PID controller, but the numerical results are very similar.

First, we need to generate the plots. Use Fig. E8.12 to help interpret the MATLAB-generated Bode plot[11]:

```
kc=1;
taui=4;
taud=1;
Gi=tf(kc*[taui 1],[taui 0]);
Gd=tf([taud 1],1);
G=Gi*Gd;
bode(G);
```

By choosing $\tau_D < \tau_I$ (i.e., corner frequencies $1/\tau_D > 1/\tau_I$), we find that the magnitude plot has a notch shape. How sharp it is will depend on the relative values of the corner frequencies. The low-frequency asymptote below $1/\tau_I$ has a slope of -1. The high-frequency asymptote above $1/\tau_D$ has a slope of $+1$. The phase-angle plot starts at $-90°$, rises to $0°$ after the corner frequency $1/\tau_I$, and finally reaches $90°$ at the high-frequency limit.

Relatively speaking, a PID controller behaves like a PI controller at low frequencies, whereas it is more like a PD controller at high frequencies. The controller is most desirable in the midrange, where it has the features of both PI and PD controllers. Also, in the notch region, the controller function has the lowest magnitude and allows for a larger GM for the system.

---

[11] If you want to see a plot for an ideal PID controller, use

```
G=tf(kc*[taui*taud taui 1],[taui 0]);
```

Example 8.13: Derive the magnitude and the phase lag of the transfer functions of **phase-lead** and **phase-lag compensators**. In many electromechanical control systems, the controller G_c is built with relatively simple R-C circuits and takes the form of a lead–lag element:

$$G_c(s) = K \frac{(s + z_0)}{(s + p_0)}.$$

Here, z_0 and p_0 are just two positive numbers. There are obviously two possibilities: case (a) $z_0 > p_0$, and case (b) $z_0 < p_0$. Sketch the magnitude and the phase-lag plots of G_c for both cases. Identify which case is the phase-lead and which case is the phase-lag compensation. What types of classical controllers may phase-lead and phase-lag compensations resemble?

We may look at the controller transfer function in the time-constant form:

$$G_c(s) = \left(K \frac{z_0}{p_0} \right) \frac{(s/z_0 + 1)}{(s/p_0 + 1)},$$

where we could further write $K_c = K z_0 / p_0$, but we will use the pole-zero form. Either way, we should see that the corner frequencies are at $\omega = z_0$ and p_0. To make a Bode plot, we theoretically should do the $s = j\omega$ substitution, but we can write the magnitude and the phase angle immediately if we recognize that the function is a product of a first-order lead and a first-order lag. Hence, making use of Examples 8.2 and 8.4, we can write

$$|G_c(j\omega)| = K \sqrt{\omega^2 + z_0^2} \frac{1}{\sqrt{\omega^2 + p_0^2}},$$

$$\angle G_c(j\omega) = \tan^{-1}(\omega/z_0) + \tan^{-1}(-\omega/p_0).$$

Figure E8.13 is a rough hand sketch with the high- and low-frequency asymptotes. It is meant to help interpret the MATLAB plots that we will generate next.

(a) With $z_0 > p_0$, the phase angle is always negative, and this is the phase-lag compensator. The MATLAB statements to get an illustrative plot are

```
k=1;
zo=4;    % Try repeat with various choices of zo and po
po=1;
G=zpk(-zo,-po,k);
bode(G);
```

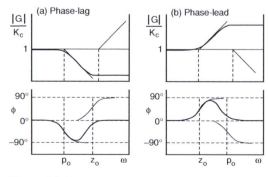

Figure E8.13.

The shape of the magnitude plot resembles that of a PI controller, but with an upper limit on the low-frequency asymptote. We can infer that the phase-lag compensator could be more stabilizing than a PI controller with very slow systems.[12] The notch-shaped phase angle plot of the phase-lag compensator is quite different from that of a PI controller. The phase lag starts at $0°$ versus $-90°$ for a PI controller. From a stability point of view, a phase-lag compensator is preferred to a PI controller. On the other hand, without an integrating function, the phase-lag compensator cannot eliminate offset.

(b) With $z_0 < p_0$, the phase-angle is positive, and this is the phase-lead compensator. First, we need an illustrative plot. Sample MATLAB statements to use are

```
zo=1;
po=4;
G=zpk(-zo,-po,1);
bode(G);
```

The nice feature of the phase-lead compensator, and for that matter a real PD controller, is that it limits the high-frequency magnitude. In contrast, an ideal PD controller has no upper limit and would amplify high-frequency input noises much more significantly.

Example 8.14: Designing phase-lead and phase-lag compensators. Consider a simple unity feedback loop with the characteristic equation $1 + G_c G_p = 0$ and with a first-order process function $G_p = [K_p/(\tau_p s + 1)]$. What are the design considerations if we use either a phase-lead or a phase-lag compensator? Consider the consequences by using Bode and root-locus plots.

With $K_c = K z_0/p_0$, the closed-loop transfer function is

$$\frac{C}{R} = \frac{G_c G_p}{1 + G_c G_p} = \frac{K_c K_p (s/z_0 + 1)}{(s/p_0 + 1)(\tau_p s + 1) + K_c K_p (s/z_0 + 1)},$$

and after one more algebraic step, we will see that the system steady-state gain is $K_c K_p/(1 + K_c K_p)$, which means that there will be an offset whether we use a phase-lead or a phase-lag compensator.

From the characteristic polynomial, it is probable that we will get either an overdamped or an underdamped system response, depending on how we design the controller. The choice is not clear from the algebra, and this is where the root-locus plot comes in handy. From the perspective of a root-locus plot, we can immediately make the decision that, no matter what, both z_0 and p_0 should be larger than the value of $1/\tau_p$ in G_p. That's how we may "steer" the closed-loop poles away from the imaginary axis for better system response. (If we know our root locus, we should know that this system is always stable.)

(a) Let's first consider a phase-lead compensator, $z_0 < p_0$. We first construct the Bode and the root-locus plots that represent a system containing the compensator and a first-order process:

---

[12] From the perspective of a root-locus plot, a phase-lag compensator adds a large open-loop zero and a relatively small open-loop pole. And for a phase-lead compensator, we are adding a large open-loop pole. When $p_0 \gg z_0$ (or $1/p_0 \ll 1/z_0$), we can also look at the phase-lead compensator as the real PD controller. How to design their locations, of course, depends on the particular process that we need to control. We will see that in Example 8.14.

```
Kp=1; % Arbitrary numbers for the process function
taup=1;
Gp=tf(Kp,[taup 1]);

zo=2; % Phase-lead, zo < po
po=4;
Gc=zpk(-zo,-po,1)
figure(1), bode(Gc*Gp)
figure(2), rlocus(Gc*Gp)
```

The root-locus plot resembles that of a real PD controller. The system remains overdamped with no complex closed-loop poles. One root locus runs from the "real PD pole" $-p_0$ to negative infinity. The other is from $-\tau_p$ to $-z_0$, which limits the fastest possible system response. How large z_0, and thus K_c, can be depends on the real physical hardware and process.

On the Bode plot, the corner frequencies are, in increasing order, $1/\tau_p$, z_0, and p_0. The two magnitude high frequency asymptotes intersecting $1/\tau_p$ and p_0 are those of first-order lags. The high frequency asymptote at $\omega = z_0$ is that of a first-order lead. The largest phase lag of the system is $-90°$ at very high frequencies. The system is always stable, as displayed by the root-locus plot.

(b) With a phase-lag compensator, $z_0 > p_0$, we can use these statements:

```
% Gp remains the same as in part (a)
zo=4;
po=2;
Gc=zpk(-zo,-po,1)
figure(1), bode(Gc*Gp)
figure(2), rlocus(Gc*Gp)
```

The shape of the root-locus plot resembles that of a PI controller, except of course we do not have an open-loop pole at the origin anymore. The root loci approach one another from $-\tau_p$ and $-p_0$, then break away from the real axis to form a circle that breaks in to the left of the open-loop zero at $-z_0$. One locus approaches negative infinity and the other goes toward $-z_0$. We may design the controller with an approach similar to that in Example 7.7.

On the Bode plot, the corner frequencies are, in increasing order, $1/\tau_p$, p_0, and z_0. The individual term frequency asymptotes have the same properties as in part (a). The largest phase lag of the system is larger than $-90°$ just past $\omega = p_0$, but still much less than $-180°$. This system is always stable.

8.4. Controller Design

The concept of GMs and PMs derived from the Nyquist criterion provides a general relative stability criterion. Frequency-response graphical tools such as Bode, Nyquist, and Nichols plots can all be used in ensuring that a control system is stable. As in root-locus plots, we can vary only one parameter at a time, and the common practice is to vary the proportional gain.

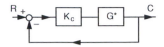

Figure 8.5. A simple unity feedback system.

8.4.1. How Do We Calculate Proportional Gain without Trial and Error?

This is a big question when we use, for example, a Bode plot. Let's presume that we have a closed-loop system in which we know "everything" but the proportional gain, and we write the closed-loop characteristic equation as

$$1 + G_{OL} = 1 + K_c G^* = 0,$$

where $G_{OL} = G_c G_a G_p G_m$. We further rewrite the function as $K_c G^*$ to indicate that we would like to find K_c (Fig. 8.5). The notation G^* is more than just the product of $G_a G_p G_m$; G^* includes the integral and the derivative terms if we use a PID controller.

With the straight textbook definition and explanation, the GM and the PM of the closed-loop system apply to only the magnitude and the phase-lag plots by use of the entire open-loop function, $|G_{OL}|$ and $\angle G_{OL}$. It means that we need to know the value for the proportional gain K_c. Of course, we do not, and the whole affair appears to be a trial-and-error calculation. The question is whether we can calculate K_c without guessing. The answer is yes. The next question is whether we can get the answer from the plots of $|G^*|$ and $\angle G^*$. This answer is also yes.

From the definition of GM, we have

$$GM = \frac{1}{|G_{OL}(j\omega_{cg})|} \quad \text{or} \quad GM = \frac{1}{K_c |G^*(j\omega_{cg})|}, \tag{8.23}$$

where the magnitudes are evaluated at the gain crossover frequency ω_{cg} as defined by the $-180°$ phase lag. We can find $|G^*(j\omega_{cg})|$ simply by a Bode plot of G^* itself. The key is that the phase angle of K_c is zero, $\angle G^*$ is identical to $\angle G_{OL}$, and both $|G^*|$ and $|G_{OL}|$ have the same ω_{cg}.[13]

From the definition of GM, GM $= 1$ at marginal stability. Hence, with GM $= 1$ in Eq. (8.23), we can evaluate $K_{cu} = 1/|G^*(j\omega_{cg})|$. Equation (8.23) can alternatively be stated as

$$GM = \frac{K_{cu}}{K_c}. \tag{8.24}$$

Once we know the value of K_{cu}, we can calculate the K_c for a given GM in a problem. Typically, we select GM, say, 1.7, to find the proportional gain.

We can also use the Bode plot of G^* to do PM calculations. From the textbook definition, we are supposed to find the phase angle $\phi = \angle G_{OL}$, where $|G_{OL}| = 1$. If the PM is 45°, ϕ should be $-135°$. It appears that we need to know K_c beforehand to calculate G_{OL}, but we do not.

We use the fact that $\angle G^*$ is identical to $\angle G_{OL}$ and ω_{cp} is the same in both plots, so we can go backward. On the G^* Bode plot, we can find the value of $|G^*(j\omega_{cp})|$ that corresponds to

---

[13] We can use MATLAB to do the Bode plot of G^* and use the `margin()` function, which will return the "GM" of G^*, but we now know it is really $1/|G^*(j\omega_{cg})|$ or, following Eq. (8.24), also the value for the ultimate gain K_{cu}.

a phase lag of, say, $-135°$. Now $|G_{OL}| = K_c|G^*|$ and because $|G_{OL}| = 1$ at the PM, we can find $K_c = 1/|G^*(j\omega_{cp})|$ that will provide a phase margin of $45°$ without trial and error.

How do we know the answer is correct? Just "plug" K_c back into G_{OL} and repeat the Bode plot by using G_{OL}. It does not take that much time to check with MATLAB. Now we are finally ready for some examples. Again, run MATLAB to confirm the results while you read them.

Example 7.2C: Let's revisit Example 7.2 with the closed-loop characteristic equation:

$$1 + K_c \frac{1}{(s+3)(s+2)(s+1)} = 0.$$

If we want to design a PI controller, how should we proceed with frequency-response methods? Let's presume that the unit of the time constants is in minutes.

The first step is to find the ultimate gain. With the given third-order process transfer function, we use the following MATLAB commands,

```
p=poly([-1 -2 -3]);
G=tf(1,p);
margin(G);
```

MATLAB returns K_{cu} as 35.6 dB (at $\omega_{cg} = 3.3$ rad/min),[14] which we easily recalculate as 60 ($10^{35.6/20}$).

Note that the low-frequency asymptote of the magnitude plot is not 1 (0 dB). Why? That's because the transfer function is not in the time-constant form. If we factor the function accordingly, we should expect a low-frequency asymptote of $1/6$ (-15.6 dB).

If we take a GM of 1.7 in Eq. (8.24), we should use a proportional gain of $K_c = 60/1.7 = 35.3$. This is the case if we use only a proportional controller as in Example 6.2. We repeat the Bode plot and margin calculation with $K_c = 35.3$:

```
G=tf(35.3,p);
margin(G);
```

Now we should find that the GM is indeed 1.7 (4.6 dB, $10^{4.6/20}$) and the PM is $18.8°$, which is a bit low according to the design rule of thumb.

We have yet to tackle the PI controller. There are, of course, different ways to find a good integral time constant. With frequency response, we have the handy tool of the Ziegler–Nichols ultimate-cycle tuning relations. So with $K_{cu} = 60$ and $\omega_{cg} = 3.3$ rad/min, we find by referring to Table 6.1 that if we use only a proportional controller, we should use $K_c = 30$, and if we use a PI controller, we should use $K_c = 27.3$ and $\tau_I = 1.58$ min.[15]

Using the Ziegler–Nichols tuning parameters, we repeat the proportional controller system Bode plot:

```
G=tf(30,p);
margin(G);
```

We should find a GM of 2 (6 dB) and a PM of $25.4°$, which is definitely a bit more conservative than the 1.7 GM result.

[14] MATLAB always considers time to be in seconds, but this should not concern us as long as we keep our own time units consistent.

[15] All these calculations are done with the M-file recipe.m from the *Web Support*.

With the PI controller, we use the following statements:

```
kc=27.3; taui=1.58;
Gc=tf(kc*[taui 1],[taui 0]);
G=tf(1,p);    % p was defined at the beginning of this example
margin(Gc*G);
```

We should find a GM of 1.47 (3.34 dB) and a PM of 12.3°. Both margins are a bit small. If we do a root-locus plot on each case and with the help of `rlocfind()` in MATLAB , we should find that the corresponding closed-loop poles of these results are indeed quite close to the imaginary axis.

Where do we go from here? We may stay with the design or we may increase the margins. We also can use MATLAB to simulate the closed-loop time-domain response and from the underdamped time-response curve, estimate numerically the effective damping ratio and other quantities such as percentage of OS. If the time-domain response does not meet our specification, we will have to tune the values of K_c or τ_I.

If we want to increase the margin, we either have to reduce the value of K_c or increase τ_I. One possibility is to keep $\tau_I = 1.58$ min and repeat the Bode plot calculation to find a new K_c that may provide a gain margin of, say, 2 (6 dB), as in the case of using only the proportional controller. To do so, we first need to find the new ultimate gain by using the PI controller:

```
kc=1; taui=1.58;
Gc=tf(kc*[taui 1],[taui 0]);
margin(Gc*G);                 % G remains as above
```

MATLAB should return $K_{cu} = 40.2$ (32.1 dB). Thus, following Eq. (8.24), we need to use $K_c = 40.2/2 = 20.1$ to meet the GM specification of 2. You can double check the result yourself with `kc=20.1`, `taui=1.58`. If so, you should find that the phase margin is now 23° – a bit low, but we probably can accept that. After this, we may proceed to the time-domain response calculations.

The ideas in this example can be applied to a PID controller. Yes, controller design is indeed an iterative process. A computer is a handy tool, but we still need to know what we are doing.

Example 7.2D: Back in the last example with a proportional controller, a gain margin of 1.7 created a system with a very small PM. What proportional gain should we use to achieve a PM of at least 45°?

Following the explanation after Eq. (8.24), we should calculate the phase angle of $G(j\omega) = [(j\omega + 3)(j\omega + 2)(j\omega + 1)]^{-1}$. Of course, we'd rather use bode() in MATLAB:

```
p=poly([-1 -2 -3]);
G=tf(1,p);
[mag,phase,w]=bode(G);
mag=mag(1,:);           % MATLAB returns a three-dimensional
                        % array[16]
```

---

[16] The MATLAB function `bode()` returns the actual magnitude even though the documentation says decibels. This is a detail that we can check with the function `freqresp()` as explained on the *Web Support*, especially with future upgrades of the software.

```
phase=phase(1,:);
tmp=[w';mag;phase]'  % Be careful with the primes here
```

From the tmp matrix of frequency, magnitude, and phase angle, we find that at $\omega = 1.74$ rad/min, $|G| = 0.054$ and $\angle G = -131.4°$, which provides a PM of 48.6°. Also, at $\omega = 2.21$ rad/min, $|G| = 0.037$ and $\angle G = -150°$, which provides a PM of 30°. (We will need to do an interpolation if the problem statement dictates, say, a PM of exactly 45°.)

To achieve a PM of 48.6°, we use a proportional gain of $K_c = 1/0.054 = 18.5$. We can repeat the Bode plot calculation with MATLAB by using $K_c = 18.5$ and the statements

```
G=tf(18.5,p);
margin(G);
```

The result confirms that the system has a PM of 48.6° and a GM of 3.2 (10.2 dB), a much more conservative design than a GM of 1.7. If we choose to use $K_c = 1/0.037 = 26.96$, a MATLAB calculation should confirm a PM of 30° and a GM of 2.2.

Example 7.4A: This time, let's revisit Example 7.4, which is a system with dead time. We would like to know how to start designing a PI controller. The closed-loop characteristic equation with a proportional controller is (again assuming the time unit is in minutes)

$$1 + K_c \frac{0.8e^{-2s}}{5s + 1} = 0.$$

The first step is to find the ultimate gain. Following Example 8.6, we can add easily the extra phase lag that is due to the dead-time function:

```
G=tf(0.8,[5 1]);
tdead=2;
[Mag,Phase,freq]=bode(G);
Mag=Mag(1,:);
Phase=Phase(1,:) - ((180/pi)*tdead*freq');
[Gm,Pm,Wcg,Wcp]=margin(Mag,Phase,freq)
```

We should find $K_{cu} = 5.72$ at $\omega_{cg} = 0.893$ rad/min, which is exactly what we found in Example 7.4, but which takes a lot more work. If we take a GM of 1.7, we should use a proportional gain of $K_c = 5.72/1.7 = 3.36$. We use

```
G=tf(3.36*0.8,[5 1]);
```

and repeat the calculation. We should find the corresponding PM to be 54.6°, which is plenty.

These are the controller settings if we again use the Ziegler–Nichols tuning relations (or really recipe.m): with $K_{cu} = 5.73$ and $\omega_{cg} = 0.895$ rad/min, we should use a proportional controller with $K_c = 2.87$, and if we use a PI controller, we should use $K_c = 2.61$ and $\tau_I = 5.85$ min. The tasks of checking the GM, PM, and the time-domain response are left as a Review Problem.

Example 5.7D: We can now finally wrap up the dye-mixing problem that we left in Example 5.7C in Chap. 6.

(a) The control system can be unstable if we place the photodetector too far downstream. To cut down on the algebra, we look into the problem with a slight simplification. At this point, we use only a proportional controller. Because the regulating valve is so much faster than the mixing process, we retain only the mixing first-order lag to obtain the approximate closed-loop characteristic equation:

$$1 + \frac{K_c K_v K_p K_m e^{-t_d s}}{(\tau_p s + 1)} = 0.$$

Again for illustration purpose, we supposedly have chosen K_c such that $K_c K_v K_p K_m = 5$, and τ_p is the mixing-process time constant. Find, without trial and error and without further approximation, the maximum distance L that the photodetector can be placed downstream such that the system remains stable. (There are two ways to get the answer. The idea of using magnitude and phase angle and the Nyquist criterion is by far the less confusing method and less prone to algebraic mistakes.)

(b) To factor in some safety margin, we install the photodetector at half the maximum distance that we found in part (a). With the same proportional gain and the same approximation used in part (a), what is the GM?

(c) Now that we can install the photodetector at a well-chosen distance, we can put the dynamics of the regulating valve back into our closed-loop analysis. What is the critical proportional gain when we include the first-order lag that is due to the regulating valve? And what is the proportional gain if we want a GM of 1.7?

(d) Finally we come to the controller design. All we know is that customers may be fussy about the color of their jeans. Of course, we also want to avoid the need to dispose of off-spec dye solutions unnecessarily. Despite these concerns, an old plant engineer mentioned that the actual dye tank downstream is huge and an OS of as much as 20 to 25% is acceptable as long as the mixing system settles down "quickly." So select your choice of controller, performance specification, and controller gains. Double check your final design with time-domain simulation and frequency-response analysis to see that we have the proper controller design.

(a) Let's use the abbreviation $G_{OL} = G_c G_v G_p G_m$, and thus the magnitude and phase-angle equations, from Example 5.7, are

$$|G_{OL}| = \left| \frac{5}{\tau_p s + 1} \right| |e^{-t_d s}|, \qquad \angle G_{OL} = \tan^{-1}(-\omega \tau_p) - t_d \omega,$$

where $\tau_p = 4s$. At crossover ω_{cg}, $\angle G_{OL} = -180°$ and GM $= 1$, meaning that $|G_{OL}| = 1$. Because the magnitude of the dead-time transfer function is unity, the magnitude equation is simply

$$1 = \frac{5}{\sqrt{1 + 4^2 \omega_{cg}^2}} \quad \text{or } \omega_{cg} = 1.12s^{-1}.$$

With the crossover frequency known, we now use the phase-angle equation to find the dead time:

$$-180° = \tan^{-1}(-4 \times 1.12) - t_d(1.12)(180/\pi) \quad \text{or } t_d = 1.45s.$$

Note the unit conversion to angles in the phase equation. With our arbitrary choice of proportional gain such that $K_c K_v K_p K_m = 5$, a dead time of $t_d = 1.45s$ is associated with GM $= 1$.

(b) Refer back to Example 5.7 in Chap. 5. The average fluid velocity is 400 cm/s. Thus the photodetector is located $(1.45)(400) = 580$ cm downstream from the mixer. To reduce the distance by half means that we now install the sensor at a location $580/2 = 290$ cm downstream. The reduced transport lag is now $1.45/2 = 0.725s$.

To find the new GM, we need to, in theory, reverse the calculation sequence. We first use the phase equation to find the new crossover frequency ω_{cg}. Then we use the magnitude equation to find the new $|G_{OL}|$, and the new GM is of course $1/|G_{OL}|$. However, because we now know the values of t_d, τ_p, and $K_c K_v K_p K_m$, we might as well use MATLAB. These are the statements:

```
k=5;
tdead=0.725;
taup=4; % The large time constant; dominant pole is at 1/4
G=tf(k,[taup 1]);

freq=logspace(-1,1)';
[Mag,Phase]=bode(G,freq);
Mag=Mag(1,:);
Phase=Phase(1,:) - ((180/pi)*tdead*freq');
[Gm,Pm,Wcg,Wcp]=margin(Mag,Phase,freq)
```

We should find that the new gain margin is 1.86.

(c) We now include the regulating valve and still use a proportional controller. The closed-loop equation is

$$1 + K_c\left(\frac{0.8}{0.2s+1}\right)\left(\frac{0.6}{4s+1}\right)2.6e^{-0.725s} = 0.$$

With MATLAB, the statements are

```
k=0.8*0.6*2.6;
G=tf(k, conv([0.2 1],[4 1]));
[Mag,Phase]=bode(G,freq);
Mag=Mag(1,:);
Phase=Phase(1,:) - ((180/pi)*tdead*freq');
[Gm,Pm,Wcg,Wcp]=margin(Mag,Phase,freq)
```

We should find that the ultimate gain is $K_{c,u} = 6.42$ at the crossover frequency of $\omega_{cg} = 1.86\,\text{s}^{-1}$. To keep to a gain margin of 1.7, we need to reduce the proportional gain to $K_c = 6.42/1.7 = 3.77$.

(d) With $K_{c,u} = 6.42$ and $\omega_{cg} = 1.86\,\text{s}^{-1}$, we can use the Ziegler–Nichols ultimate-gain tuning relations (with recipe.m) to find the following values for different objectives:

	K_c	τ_I	τ_D
Quarter decay	3.8	1.7	0.42
Little overshoot	2.1	1.7	1.1
No overshoot	1.3	1.7	1.1

If we repeat the time-response simulations as in Example 5.7C in Chap. 6, we should find that the settings for the quarter decay lead to a 51% OS (roughly a 0.26 DR), the little OS settings have a 27% OS, and the so-called no OS settings still have ~8% OS.

Finally, there is no unique solution for the final design. The problem statement leaves us with a lot of latitude, especially when we have seen from the time-response simulations that many combinations of controller settings give us similar closed-loop responses. In process engineering, we do not always have to be fastidious with the exact time-response specifications and hence values of the controller settings. Many real-life systems can provide acceptable performances within a certain range of response characteristics. As for controller settings, a good chance is that we have to perform field tuning to account for anything ranging from inaccurate process identification to shifting operating conditions of nonlinear processes.

For the present problem, and from all the settings provided by the different methods, we may select $\tau_I = 3s$ and $\tau_D = 0.5s$. We next tune the proportional gain to give us the desired response. The closed-loop equation with an ideal PID controller is now

$$1 + K_c \left(1 + \frac{1}{\tau_I s} + \tau_D s \right) \frac{1.248}{(0.2s + 1)(4s + 1)} e^{-0.725s} = 0.$$

First, we need MATLAB to find the ultimate gain:

```
taui=3;
taud=0.5;
gc=tf([taui*taud (taui+taud) 1],[taui 0]); % ideal PID without
                                           % the Kc
tdead=0.725;
k=0.8*0.6*2.6;
G=tf(k, conv([0.2 1],[4 1]));
[Mag,Phase]=bode(gc*G,freq);
Mag=Mag(1,:);
Phase=Phase(1,:) - ((180/pi)*tdead*freq');
[Gm,Pm,Wcg,Wcp]=margin(Mag,Phase,freq)
```

We should find the ultimate gain to be $K_{cu} = 5.87$. And at a gain margin of 1.7, we need to use a proportional gain of $K_c = 5.87/1.7 = 3.45$. A time-response simulation shows that the system, with respect to a unit-step change in the set point, has an OS of 23%. This tuning is slightly less oscillatory than if we had chosen $\tau_I = 3$ s and $\tau_D = 0.3$ s, as suggested by ITAE (Example 5.7A). In this case, $K_{cu} = 6.79$ and $K_c = 4$, which is closer to the K_c from the Ziegler–Nichols tuning. Again, confirm these results in the Review Problems.

8.4.2. A Final Word: Can Frequency-Response Methods Replace Root Locus?

No. These methods complement each other. Very often a design decision is made only after analyses with both methods.

The root-locus method gives us a good indication of the transient response of a system and the effect of varying the controller gain. However, we need a relatively accurate model for the analysis, not to mention that root locus does not handle dead time as well.

Frequency methods can give us the relative stability (the GMs and PMs). In addition, we could construct the Bode plot with experimental data by using a sinusoidal or pulse input, i.e., the subsequent design does not need a (theoretical) model. If we do have a model, the data can be used to verify the model. However, there are systems that have more than one crossover frequency on the Bode plot (the magnitude and the phase lag do not decrease monotonically with frequency), and it would be hard to judge which is the appropriate one with the Bode plot alone.

Review Problems

(1) Derive Eqs. (8.19) and (8.20). Use MATLAB to plot the resonant frequency and maximum magnitude as functions of the damping ratio with $K = 1$.

(2) What are the low- and the high-frequency asymptotes of the minimum phase function $(s + z)/(s + p)$ versus the simplest nonminimum-phase function $(s - z)/(s + p)$ in a Bode plot?

(3) What is the bandwidth of a second-order function?

(4) We used $\tau_D < \tau_I$ in Example 8.12. What if we use $\tau_D > \tau_I$ in our PID controller design? What if we use a real PID controller?

(5) Sketch the Bode plots for $G(s) = s^n$, with $n = \pm 1, \pm 2, \ldots$, etc.

(6) In Example 8.12, we used the interacting form of a PID controller. Derive the magnitude and the phase-angle equations for the ideal noninteracting PID controller. (It is called noninteracting because the three controller modes are simply added together.) See that this function will have the same frequency asymptotes.

(7) Finish the controller calculations in Example 5.7D.

Hints:

(1) The plotting statements can be

```
z = 0.05:0.01:0.7;
wr = sqrt(1-2*z.*z);
dum = sqrt(1-z.*z);
Mp = 1./(2*z.*dum);
plot(z,wr, z,Mp);
```

(2) Try to perform the Bode plots with MATLAB. The magnitude plots are identical. The phase angle of the nonminimum-phase example will go from $0°$ to $-180°$, whereas you'd see a minimum of the phase angle in the minimum-phase function. Thus for a transfer function that is minimum phase, we may identify the function from simply the magnitude plot. But we cannot do the same if the function is nonminimum phase.

(3) We need to find the frequency ω_b when the magnitude drops from the low frequency asymptote by $1/\sqrt{2}$. From the magnitude equation in Example 8.3, we need to solve

$$\left(1 - \tau^2 \omega_b^2\right)^2 + (2\zeta\tau\omega_b)^2 = 2$$

If we now "solve" this equation using $\tau^2\omega^2$ as a variable, we should find

$$\tau^2\omega_b^2 = 1 - 2\zeta_2 + \sqrt{4\zeta^2(\zeta^2 - 1) + 2}$$

and the final from with ω_b explicitly on the LHS is one small step away.

(4) Sketching with MATLAB should help.

(5) $G(j\omega) = j^n \omega^n$. This function is real if n is even, imaginary if n is odd. Also, $|G| = \omega^n$, and the phase angle of $G(j\omega)$ is $\tan^{-1}(0)$ when n is even and is $\tan^{-1}(\infty)$ when n is odd.

(6) Substitution of $s = j\omega$ into $G_c(s) = K_c[1 + (1/\tau_I s) + \tau_D s]$ gives

$$G_c(j\omega) = K_c\left(1 + \frac{1}{j\omega\tau_I} + j\omega\tau_D\right) = K_c\left(1 + j\frac{\tau_I \tau_D \omega^2 - 1}{\omega\tau_I}\right),$$

and thus

$$|G_c(j\omega)| = K_c\sqrt{1 + \left(\tau_D\omega - \frac{1}{\omega\tau_I}\right)^2},$$

$$\angle G_c(j\omega) = \tan^{-1}\left(\tau_D\omega - \frac{1}{\omega\tau_I}\right).$$

The magnitude equation has slopes of -1 and $+1$ at very low and very high frequencies. In the phase-angle equation, the two limits are $-90°$ and $+90°$, as in Example 8.12. Furthermore, from the phase-angle equation of the ideal controller, the "trough" center should be located at the frequency $\omega = (\tau_I \tau_D)^{-1/2}$. The polar plot of the ideal PID controller is like combining the images of a PI and an ideal PD controller – a vertical line at K_c that extends from negative infinity at $\omega = 0$ toward positive infinity at extremely high frequencies.

(7) The MATLAB statements and plots are provided on the *Web Support*.

9

Design of State-Space Systems

We now return to the use of state-space representation that was introduced in Chap. 4. As you may have guessed, we want to design control systems based on state-space analysis. A state feedback controller is very different from the classical PID controller. Our treatment remains introductory, and we will stay with linear or linearized SISO systems. Nevertheless, the topics here should enlighten (!) us as to what modern control is all about.

What Are We Up to?

- Evaluating the controllability and observability of a system.
- Designing pole placement of state feedback systems. Applying Ackermann's formula.
- Designing with full-state and reduced-order observers (estimators).

9.1. Controllability and Observability

Before we formulate a state-space system, we need to raise two important questions. One is whether the choice of inputs (the manipulated variables) may lead to changes in the states, and the second is whether we can evaluate all the states based on the observed output. These are what we call the controllability and the observability problems.

9.1.1. Controllability

A system is said to be completely state controllable if there exists an input $u(t)$ that can drive the system from any given initial state $x_0(t_0 = 0)$ to any other desired state $x(t)$. To derive the controllability criterion, let us restate the linear system and its solution from Eqs. (4.1), (4.2), and (4.10):

$$\dot{x} = Ax + Bu, \tag{9.1}$$

$$y = Cx, \tag{9.2}$$

$$x(t) = e^{At}x(0) + \int_0^t e^{-A(t-\tau)}Bu(\tau)\,d\tau. \tag{9.3}$$

With our definition of controllability, there is no loss of generality if we choose to have $\mathbf{x}(t) = 0$, i.e., moving the system to the origin. Thus Eq. (9.3) becomes

$$\mathbf{x}(0) = -\int_0^t e^{-\mathbf{A}\tau}\mathbf{B}u(\tau)\,d\tau. \tag{9.4}$$

We next use Eq. (4.15), i.e., the fact that we can expand the matrix exponential function as a closed-form series:

$$e^{\mathbf{A}t} = \alpha_0(t)\mathbf{I} + \alpha_1(t)\mathbf{A} + \alpha_2(t)\mathbf{A}^2 + \cdots + \alpha_{n-1}(t)\mathbf{A}^{n-1}. \tag{9.5}$$

Substitution of Eq. (9.5) into Eq. (9.4) gives

$$\mathbf{x}(0) = -\sum_{k=0}^{n-1}\mathbf{A}^k\mathbf{B}\int_0^t \alpha_k(\tau)u(\tau)\,d\tau.$$

We now hide the ugly mess by defining the $(n \times 1)$ vector β with elements

$$\beta_k(\tau) = \int_0^t \alpha_k(\tau)u(\tau)\,d\tau,$$

and Eq. (9.4) appears as

$$\mathbf{x}(0) = -\sum_{k=0}^{n-1}\mathbf{A}^k\mathbf{B}\beta_k = -[\mathbf{B}\,|\,\mathbf{AB}\,|\,\mathbf{A}^2\mathbf{B}\,|\cdots|\,\mathbf{A}^{n-1}\mathbf{B}]\begin{bmatrix}\beta_0\\\beta_1\\\vdots\\\beta_{n-1}\end{bmatrix}. \tag{9.6}$$

If Eq. (9.6) is to be satisfied, the $(n \times n)$ matrix $[\mathbf{B}\ \mathbf{AB}\cdots\mathbf{A}^{n-1}\mathbf{B}]$ must be of rank n. This is a necessary and sufficient condition for controllability. Hence we can state that a system is completely controllable if and only if the controllability matrix

$$\mathbf{C}_0 = [\mathbf{B}\ \mathbf{AB}\ \mathbf{A}^2\mathbf{B}\ \cdots\ \mathbf{A}^{n-1}\mathbf{B}] \tag{9.7}$$

is of rank n.

The controllability condition is the same even when we have multiple inputs, $\mathbf{u}$. If we have r inputs, then $\mathbf{u}$ is $(r \times 1)$, $\mathbf{B}$ is $(n \times r)$, each of the β_k is $(r \times 1)$, β is $(nr \times 1)$, and $\mathbf{C}_0$ is $(n \times nr)$.

When we have multiple outputs $\mathbf{y}$, we want to control the output rather than the states. Complete state controllability is neither necessary nor sufficient for actual output controllability. With the output $\mathbf{y} = \mathbf{C}\mathbf{x}$ and the result in Eq. (9.6), we can *infer* that the **output controllability matrix** is

$$\mathbf{C}_0 = [\mathbf{CB}\ \mathbf{CAB}\ \mathbf{CA}^2\mathbf{B}\ \cdots\ \mathbf{CA}^{n-1}\mathbf{B}]. \tag{9.8}$$

If we have m outputs, $\mathbf{y}$ is $(m \times 1)$ and $\mathbf{C}$ is $(m \times n)$. If we also have r inputs, then the output controllability matrix is $(m \times nr)$. From our interpretation of Eq. (9.6), we can also infer that to have complete output controllability, the matrix in Eq. (9.8) must have rank m.

9.1.2. Observability

The linear time-invariant system in Eqs. (9.1) and (9.2) is completely observable if every initial state $\mathbf{x}(0)$ can be determined from the output $y(t)$ over a finite time interval. The

Design of State-Space Systems

concept of observability is useful because in a given system, not all of the state variables are accessible for direct measurement. We will need to estimate the unmeasurable state variables from the output in order to construct the control signal.

Because our focus is to establish the link between y and $\mathbf{x}$, or observability, it suffices to consider only the unforced problem:

$$\dot{\mathbf{x}} = \mathbf{Ax}, \tag{9.9}$$

$$y = \mathbf{Cx}. \tag{9.10}$$

Substitution of the solution of Eq. (9.9) into Eq. (9.10) gives

$$y(t) = \mathbf{C}e^{\mathbf{A}t}\mathbf{x}(0).$$

We again take that we can expand the exponential function as in Eq. (9.5). Thus we have

$$y(t) = \sum_{k=0}^{n-1} \alpha_k(t)\mathbf{C}\mathbf{A}^k\mathbf{x}(0) = [\alpha_0\, \alpha_1\, \cdots\, \alpha_{n-1}] \begin{bmatrix} \mathbf{C} \\ \mathbf{CA} \\ \vdots \\ \mathbf{CA}^{n-1} \end{bmatrix} \mathbf{x}(0). \tag{9.11}$$

With the same reasoning that we applied to Eq. (9.6), we can infer that to have complete observability, the observability matrix[1]

$$\mathbf{O}_b = \begin{bmatrix} \mathbf{C} \\ \mathbf{CA} \\ \vdots \\ \mathbf{CA}^{n-1} \end{bmatrix} \tag{9.12}$$

must be of rank n. When we have m outputs, $\mathbf{y}$ is $(m \times 1)$, $\mathbf{C}$ is $(m \times n)$, and $\mathbf{O}_b$ is $(mn \times n)$.

Example 9.1: Consider a third-order model:

$$\mathbf{A} = \begin{bmatrix} 0 & 1 & 0 \\ 0 & 0 & 1 \\ -6 & -11 & -6 \end{bmatrix}, \quad \mathbf{B} = \begin{bmatrix} 0 \\ 0 \\ 1 \end{bmatrix}, \quad \mathbf{C} = [1\ 0\ 0],$$

which is the controllable canonical form of the problem in Example 4.9. Construct the controllability and the observability matrices.

To compute the controllability matrix, we can use the MATLAB function `ctrb()`:

```
A=[0 1 0; 0 0 1; -6 -11 -6];
B=[0; 0; 1];
Co=ctrb(A,B)
```

Or we can use the definition itself:

```
Co=[B A*B A^2*B]
```

[1] Controllability and observability are dual concepts. With $\mathbf{C} = \mathbf{B}^T$ and $\mathbf{A} = \mathbf{A}^T$, we can see that $\mathbf{O}_b = \mathbf{C}_0^T$.

Either way, we should obtain

$$\mathbf{C}_0 = \begin{bmatrix} 0 & 0 & 1 \\ 0 & 1 & -6 \\ 1 & -6 & 25 \end{bmatrix},$$

which has a rank of 3, and the model is completely state controllable.

Similarly, we can use the MATLAB function `obsv()` for the observability matrix:

```
C=[1 0 0];
Ob=obsv(A,C)
```

Or we can use the definition

```
Ob=[C; C*A; C*A^2]
```

We should find that `Ob` is the identity matrix, which of course, is of rank 3.

Example 4.8A: We now revisit the fermentor example 4.8. Our question is whether we can control the cell mass and glucose concentration by adjusting only D.

From Eq. (E4.38) in Example 4.8, we have

$$\mathbf{A} = \begin{bmatrix} 0 & C_1\mu' \\ -\dfrac{\mu}{Y} & -\dfrac{C_1}{Y}\mu' - \mu \end{bmatrix}, \quad \mathbf{B} = \begin{bmatrix} -C_1 \\ \dfrac{C_1}{Y} \end{bmatrix}.$$

First, we evaluate

$$\mathbf{AB} = \begin{bmatrix} 0 & C_1\mu' \\ -\dfrac{\mu}{Y} & -\dfrac{C_1}{Y}\mu' - \mu \end{bmatrix} \begin{bmatrix} -C_1 \\ \dfrac{C_1}{y} \end{bmatrix} = \begin{bmatrix} \dfrac{C_1^2\mu'}{Y} \\ -\dfrac{C_1^2\mu}{Y^2} \end{bmatrix}.$$

The controllability matrix is

$$\mathbf{C}_0 = [\mathbf{B}\ \mathbf{AB}] = \begin{bmatrix} -C_1 & \dfrac{C_1^2\mu'}{Y} \\ \dfrac{C_1}{Y} & -\dfrac{C_1^2\mu'}{Y^2} \end{bmatrix}.$$

The determinant of $\mathbf{C}_0$ is 0 and the rank of $\mathbf{C}_0$ is 1. Hence, both cell mass and substrate cannot be controlled simultaneously by just varying D. The answer is quite obvious with just a bit of intuition. If we insist on using D as the only input, we can control either C_1 or C_2, but not both quantities. To effectively regulate both C_1 and C_2, we must implement a system with two inputs. An obvious solution is to adjust the glucose feed concentration (C_{20}) as well as the total flow rate (dilution rate D).

Now we will see what happens with two inputs. Compared with Eq. (E4.38), here $\mathbf{A}$ remains the same, and $\mathbf{B}$ in Eq. (E4.47) is now a (2×2) matrix with a rank of 2. Hence the controllability matrix $\mathbf{C}_0 = [\mathbf{B}\ \mathbf{AB}]$ is a (2×4) matrix and it must have a rank of 2 (as at least $\mathbf{B}$ is), and both C_1 and C_2 are controllable.

Design of State-Space Systems

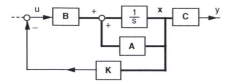

Figure 9.1. Closed-loop system with state feedback.

9.2.   Pole-Placement Design

9.2.1. Pole Placement and Ackermann's Formula

When we used root locus for controller design in Chap. 7, we chose a dominant pole (or a conjugate pair if complex). With state-space representation, we have the mathematical tool to choose all the closed-loop poles. To begin, we restate the state-space model in Eqs. (4.1) and (4.2):

$$\dot{\mathbf{x}} = \mathbf{A}\mathbf{x} + \mathbf{B}u, \tag{9.13}$$

$$y = \mathbf{C}\mathbf{x}. \tag{9.14}$$

With a control system, the input u is now the manipulated variable that is driven by the control signal (Fig. 9.1). For the moment, we consider only the regulator problem and omit changes in the set point. We state the simple control law that depends on full state feedback as

$$u(t) = -\mathbf{K}\mathbf{x} = -K_1 x_1(t) - K_2 x_2(t) \cdots -K_n x_n(t), \tag{9.15}$$

where $\mathbf{K}$ is the **state feedback gain** ($1 \times n$) vector. In this formulation, the feedback information requires $\mathbf{x}(t)$, meaning that we must be able to measure all the state variables.

We now substitute Eq. (9.15) into Eq. (9.13) to arrive at the system equation

$$\dot{\mathbf{x}} = (\mathbf{A} - \mathbf{B}\mathbf{K})\mathbf{x}. \tag{9.16}$$

The eigenvalues of the system matrix $\mathbf{A} - \mathbf{B}\mathbf{K}$ are called the regulator poles. What we want is to find $\mathbf{K}$ such that it satisfies how we select all the eigenvalues (or where we put all the closed-loop poles).

To do that easily, we first need to put our model [Eq. (9.13)] in the controllable canonical form as in Eq. (4.19):

$$\dot{\mathbf{x}} = \begin{bmatrix} 0 & 1 & 0 & \cdots & 0 \\ 0 & 0 & 1 & \cdots & 0 \\ \vdots & & & & \vdots \\ 0 & 0 & 0 & \cdots & 1 \\ -a_0 & -a_1 & -a_2 & \cdots & -a_{n-1} \end{bmatrix} \mathbf{x} + \begin{bmatrix} 0 \\ 0 \\ \vdots \\ 0 \\ 1 \end{bmatrix} u. \tag{9.17}$$

After substituting for u using Eq. (9.15), we find that the system matrix in Eq. (9.16) is

$$\mathbf{A} - \mathbf{BK} = \begin{bmatrix} 0 & 1 & 0 & \cdots & 0 \\ 0 & 0 & 1 & \cdots & 0 \\ \vdots & & & & \vdots \\ 0 & 0 & 0 & \cdots & 1 \\ -a_0 & -a_1 & -a_2 & \cdots & -a_{n-1} \end{bmatrix} - \begin{bmatrix} 0 \\ 0 \\ \vdots \\ 0 \\ 1 \end{bmatrix} [K_1 \ K_2 \ \cdots \ K_n]$$

or

$$\mathbf{A} - \mathbf{BK} = \begin{bmatrix} 0 & 1 & 0 & \cdots & 0 \\ 0 & 0 & 1 & \cdots & 0 \\ \vdots & & & & \vdots \\ 0 & 0 & 0 & \cdots & 1 \\ -a_0 - K_1 & -a_1 - K_2 & -a_2 - K_3 & \cdots & -a_{n-1} - K_n \end{bmatrix}. \tag{9.18}$$

As in Eq. (4.21), the closed-loop characteristic equation $|s\mathbf{I} - \mathbf{A} + \mathbf{BK}| = 0$ will appear as

$$s^n + (a_{n-1} + K_n)s^{n-1} + \cdots + (a_1 + K_2)s + (a_0 + K_1) = 0. \tag{9.19}$$

We next return to our assertion that we can choose all our closed-loop poles, or in terms of eigenvalues, $\lambda_1, \lambda_2, \ldots, \lambda_n$. This desired closed-loop characteristic equation is

$$(s - \lambda_1)(s - \lambda_2) \cdots (s - \lambda_n) = s^n + \alpha_{n-1}s^{n-1} + \cdots + \alpha_1 s + \alpha_0 = 0, \tag{9.20}$$

where we compute the coefficients α_i by expanding the terms on the LHS. By matching the coefficients of like powers of s in Eqs. (9.19) and (9.20), we obtain

$$a_0 + K_1 = \alpha_0,$$
$$a_1 + K_2 = \alpha_1,$$
$$\cdots$$
$$a_{n-1} + K_n = \alpha_{n-1}.$$

Thus in general, we can calculate all the state feedback gains in $\mathbf{K}$ by

$$K_i = \alpha_{i-1} - a_{i-1}, \quad i = 1, 2, \ldots, n. \tag{9.21}$$

This is the result of full state feedback pole-placement design. If the system is completely state controllable, we can compute the state gain vector $\mathbf{K}$ to meet our selection of all the closed-loop poles (eigenvalues) through the coefficients α_i.

There are other methods in pole-placement design. One of them is **Ackermann's formula**. The derivation of Eq. (9.21) is predicated on the fact that we have put Eq. (9.13) in the controllable canonical form. Ackermann's formula requires only that system (9.13) be completely state controllable. If so, we can evaluate the state feedback gain as [2]

$$\mathbf{K} = [0 \ 0 \ \cdots \ 1][\mathbf{B} \ \mathbf{AB} \ \cdots \ \mathbf{A}^{n-1}\mathbf{B}]^{-1}\alpha_c(\mathbf{A}), \tag{9.22}$$

---

[2] Roughly, Ackermann's formula arises from the application of the Cayley–Hamilton theorem to Eq. (9.20). The details of the derivation are in the *Web Support*.

where

$$\alpha_c(\mathbf{A}) = \mathbf{A}^n + \alpha_{n-1}\mathbf{A}^{n-1} + \cdots + \alpha_1\mathbf{A} + \alpha_0\mathbf{I} \qquad (9.23)$$

is the polynomial derived from the desired eigenvalues as in Eq. (9.20), except now $\alpha_c(\mathbf{A})$ is an $(n \times n)$ matrix.

9.2.2. Servo Systems

We now reintroduce the change in reference, $r(t)$. We will stay with analyzing a SISO system. By a proper choice in the indexing of the state variables, we select $x_1 = y$. In a feedback loop, the input to the process model may take the form

$$u(t) = K_r r(t) - \mathbf{K}\mathbf{x}(t),$$

where K_r is some gain associated with the change in the reference and $\mathbf{K}$ is the state feedback gain as defined in Eq. (9.15). One of the approaches that we can take is to choose $K_r = K_1$ such that $u(t)$ is

$$u(t) = K_1[r(t) - x_1(t)] - K_2 x_2(t) - \cdots - K_n x_n(t), \qquad (9.24)$$

where we may recognize that $r(t) - x_1(t)$ is the error $e(t)$.

The system equation is now

$$\dot{\mathbf{x}} = \mathbf{A}\mathbf{x} + \mathbf{B}[K_1 r - \mathbf{K}\mathbf{x}]$$

or

$$\dot{\mathbf{x}} = (\mathbf{A} - \mathbf{B}\mathbf{K})\mathbf{x} + \mathbf{B}K_1 r. \qquad (9.25)$$

The system matrix and thus design procedures remain the same as in the regulator problem in Eq. (9.16).[3]

9.2.3. Servo Systems with Integral Control

You may have noticed that nothing that we have covered so far does integral control as in a PID controller. To implement integral action, we need to add one state variable, as in

---

[3] The system must be asymptotically stable. At the new steady state (as $t \to \infty$), we have

$$0 = (\mathbf{A} - \mathbf{B}\mathbf{K})\mathbf{x}(\infty) + \mathbf{B}K_1 r(\infty)$$

and subtracting this equation from Eq. (9.25), we have

$$\dot{\mathbf{x}} = (\mathbf{A} - \mathbf{B}\mathbf{K})[\mathbf{x} - \mathbf{x}(\infty)] + \mathbf{B}K_1[r - r(\infty)].$$

If we define $\mathbf{e} = \mathbf{x} - \mathbf{x}(\infty)$, and also $r(t)$ as a step function such that r is really a constant for $t > 0$, the equation is simply

$$\dot{\mathbf{e}} = (\mathbf{A} - \mathbf{B}\mathbf{K})\mathbf{e}.$$

Not only is this equation identical to the form in Eq. (9.16), but we also can interpret the analysis as equivalent to a problem in which we want to find $\mathbf{K}$ such that the steady-state error $\mathbf{e}(t)$ approaches zero as quickly as possible.

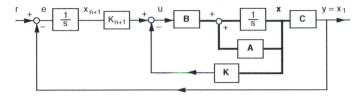

Figure 9.2. State feedback with integral control.

Fig. 9.2. Here, we integrate the error $[r(t) - x_1(t)]$ to generate the new variable x_{n+1}. This quantity is multiplied by the additional feedback gain K_{n+1} before being added to the rest of the feedback data.

The input to the process model now takes the form

$$u(t) = K_{n+1} x_{n+1}(t) - \mathbf{K}\mathbf{x}(t). \tag{9.26}$$

The differential equation for x_{n+1} is

$$\dot{x}_{n+1} = r(t) - \mathbf{C}\mathbf{x}. \tag{9.27}$$

We have written $x_1 = y = \mathbf{C}\mathbf{x}$ just so that we can package this equation in matrix form in the next step. Substituting Eq. (9.26) into state model (9.13), together with Eq. (9.27), we can write this $(n + 1)$ system as

$$\begin{bmatrix} \dot{\mathbf{x}} \\ \dot{x}_{n+1} \end{bmatrix} = \begin{bmatrix} \mathbf{A} - \mathbf{B}\mathbf{K} & \mathbf{B}K_{n+1} \\ -\mathbf{C} & 0 \end{bmatrix} \begin{bmatrix} \mathbf{x} \\ x_{n+1} \end{bmatrix} + \begin{bmatrix} 0 \\ 1 \end{bmatrix} r. \tag{9.28}$$

In terms of dimensions, $(\mathbf{A} - \mathbf{B}\mathbf{K})$, $\mathbf{B}$, and $\mathbf{C}$ remain, respectively, $(n \times n)$, $(n \times 1)$, and $(1 \times n)$. We can interpret the system matrix as

$$\begin{bmatrix} \mathbf{A} - \mathbf{B}\mathbf{K} & \mathbf{B}K_{n+1} \\ -\mathbf{C} & 0 \end{bmatrix} = \begin{bmatrix} \mathbf{A} & 0 \\ -\mathbf{C} & 0 \end{bmatrix} - \begin{bmatrix} \mathbf{B} \\ 0 \end{bmatrix} [\mathbf{K} \ -K_{n+1}] = \hat{\mathbf{A}} - \hat{\mathbf{B}}\hat{\mathbf{K}}, \tag{9.29}$$

where now our task is to find the $(n + 1)$ state feedback gains:

$$\hat{\mathbf{K}} = [\mathbf{K} \ -K_{n+1}]. \tag{9.30}$$

With Eq. (9.29), we can view the characteristic equation of the system as

$$|s\mathbf{I} - \hat{\mathbf{A}} + \hat{\mathbf{B}}\hat{\mathbf{K}}| = 0, \tag{9.31}$$

which is in the familiar form of the problem in Eq. (9.16). Thus we can make use of the pole-placement techniques in Subsection 9.2.1.

Example 9.2: Consider the second-order model in Example 9.1. What are the state feedback gains if we specify that the closed-loop poles are to be at $-3 \pm 3j$ and -6?

With the given model in the controllable canonical form, we can use Eq. (9.21). The MATLAB statements are

```
A=[0 1 0; 0 0 1; -6 -11 -6]; % Should find
p1=poly(A)                    % [1 6 11 6], coefficients a_i in
                              % Eq. (9.19)
```

```
P=[-3+3j -3-3j -6];
p2=poly(P)                        % [1 12 54 108], coefficients αᵢ in
                                  % Eq.(9.20)
p2-p1                             % [0 6 43 102], Kᵢ as in Eq.(9.21)
```

To obtain the state feedback gains with Eq. (9.21), we should subtract the coefficients of the polynomial p1 from p2, starting with the last constant coefficient. The result is, indeed,

$$\mathbf{K} = (K_1, K_2, K_3) = (108 - 6,\ 54 - 11,\ 12 - 6) = (102,\ 43,\ 6).$$

Check 1: The same result can be obtained with the MATLAB function `acker()`, which uses Ackermann's formula. The statements are

```
B=[0; 0; 1];
acker(A,B,P)                      % Should return [102 43 6]
```

Check 2. We can do the Ackermann's formula step by step. The statements are

```
M=[B A*B A^2*B];                  % controllability matrix
ac=polyvalm(p2,A);                % Eq.(9.23)
[0 0 1]* inv(M)* ac               % Eq.(9.22)
```

To evaluate the matrix polynomial in Eq. (9.23), we use the MATLAB function `poly-valm()`, which applies the coefficients in p2 to the matrix A.

Example 4.7B: Let us revisit the two CSTR-in-series problems in Example 4.7. Use the inlet concentration as the input variable and check that the system is controllable and observable. Find the state feedback gain such that the reactor system is very slightly underdamped with a damping ratio of 0.8, which is equivalent to approximately a 1.5% OS.
From Eq. (E4.27) of Example 4.7, the model is

$$\frac{d}{dt} \begin{bmatrix} c_1 \\ c_2 \end{bmatrix} = \begin{bmatrix} -5 & 0 \\ 2 & -4 \end{bmatrix} \begin{bmatrix} c_1 \\ c_2 \end{bmatrix} + \begin{bmatrix} 4 \\ 0 \end{bmatrix} c_0$$

and C_2 is the only output. We can construct the model and check the controllability and observability with

```
A=[-5 0; 2 -4];
B=[4; 0];
C=[0 1];
D=0;
rank(ctrb(A,B))  % should find rank = 2
rank(obsv(A,C))  % for both matrices
```

Both the controllability and the observability matrices are of rank 2. Hence the system is controllable and observable.
To achieve a damping ratio of 0.8, we can find that the closed-loop poles must be at $-4.5 \pm 3.38\,j$ (by using a combination of what we learned in Example 7.5 and Fig. 2.5), but

we can cheat with MATLAB and use root-locus plots!

```
[q,p]=ss2tf(A,B,C,D);      % converts state-space to transfer
                           % function⁴
Gp=tf(q,p);
rlocus(Gp)
sgrid(0.8,1)
[kc,P]=rlocfind(Gp)        % should find kc = 1.46
```

We now apply the closed-loop poles P directly to Ackermann's formula:

```
K=acker(A,B,P)             % should find K = [0 1.46]
```

The state-space state feedback gain (K_2) related to the output variable C_2 is the same as the proportional gain obtained with root locus. Given any set of closed-loop poles, we can find the state feedback gain of a controllable system by using state-space pole-placement methods. The use of root locus is not necessary, but it is a handy tool that we can take advantage of.

Example 4.7C: Add integral action to the system in Example 4.7B so we can eliminate the steady-state error.

To find the new state feedback gain is a matter of applying Eq. (9.29) and Ackermann's formula. The hard part is to make an intelligent decision on the choice of closed-loop poles. Following the lead of Example 4.7B, we use root-locus plots to help us. With the understanding that we have two open-loop poles at -4 and -5, a reasonable choice of the integral time constant is $1/3$ min. With the open-loop zero at -3, the reactor system is always stable, the dominant closed-loop pole is real, and the reactor system will not suffer from excessive oscillation.

Hence our first step is to use root locus to find the closed-loop poles of a PI control system with a damping ratio of 0.8. The MATLAB statements to continue with Example 4.7B are

```
kc=1; taui=1/3;
Gc=tf(kc*[taui 1],[taui 0]);
rlocus (Gc*Gp);             % Gp is from Example 4.7B
sgrid(0.8,1)
[kc, P]=rlocfind(Gc*Gp)    % should find proportional gain kc=1.66
```

The closed-loop poles P are roughly at -2.15 and $-3.43 \pm 2.62j$, which we apply immediately to Ackermann's formula by using $\hat{\mathbf{A}}$ and $\hat{\mathbf{B}}$ in Eq. (9.29):

```
Ah=[A zeros(2,1); -C 0];   % Eq.(9.29)
Bh=[B; 0];
Kh=acker(Ah,Bh,P)         % should find Kh = [0 1.66 -4.99]
```

The state feedback gain including integral control $\hat{\mathbf{K}}$ is [0 1.66 −4.99]. Unlike the simple proportional gain, we cannot expect that $K_{n+1} = 4.99$ would resemble the integral time

---

⁴ Another way here is to make use of the analytical result in Example 4.7:

```
Gp=zpk([],[-5 -4],8);      % transfer function C2/Co taken from
                           % Eq.(E4.30a)
```

constant in classical PI control. For doing the time-domain simulation, the task is similar to the hints that are provided for Example 7.5B in the Review Problems. The actual statements are also provided on the *Web Support*.

Example 7.5B: Consider the second-order system in Example 7.5. What are the state feedback gains if we specify that the closed-loop poles are to be $-0.375 \pm 0.382j$ as determined in Example 7.5A?

The problem posed in Examples 7.5 and 7.5A is not in the controllable canonical form (unless we do the transform ourselves). Thus we use Ackermann's formula. The MATLAB statements are

```
G=tf(1,conv([2 1],[4 1]));      % Make the state-space object
                                % from the transfer function
S=ss(G);
scale=S.c(2);                   % Rescale MATLAB model matrices
S.c=S.c/scale; S.b=S.b*scale;
P=[-0.375+0.382j -0.375-0.382j] % Define the closed-loop poles
k=acker(S.a,S.b,P)             % Calculate the feedback gains
```

MATLAB will return the vector [0 1.29], meaning that $K_1 = 0$ and $K_2 = 1.29$, which was the proportional gain obtained in Example 7.5A. Because $K_1 = 0$, we feed back only the controlled variable as analogous to proportional control. In this simple example, the state-space system is virtually the classical system with a proportional controller.

A note of caution is necessary when we let MATLAB generate the state-space model from a transfer function. The vector **C** (from S.c) is [0 0.5], which means that the indexing is reversed such that x_2 is the output variable and x_1 is the derivative of x_2. Second, **C** is not [0 1], and hence we have to rescale the matrices **B** and **C**. These two points are further covered in MATLAB Session 4.

9.3. State Estimation Design

9.3.1. State Estimator

The pole-placement design is predicated on the feedback of *all* the state variables **x** (Fig. 9.1). Under many circumstances, this may not be true. We have to estimate unmeasurable state variables or signals that are too noisy to be measured accurately. One approach to work around this problem is to estimate the state vector with a model. The algorithm that performs this estimation is called the **state observer** or the **state estimator**. The estimated state $\tilde{\mathbf{x}}$ is then used as the feedback signal in a control system (Fig. 9.3). A full-order state observer

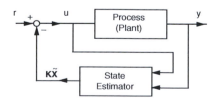

Figure 9.3. Concept of using a state estimator to generate an estimated state feedback signal.

9.3. State Estimation Design

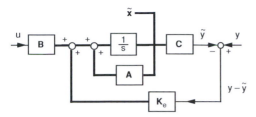

Figure 9.4. A probable model for a state estimator.

estimates all the states even when some of them are measured. A reduced-order observer does the smart thing and skips these measurable states.

The next task is to seek a model for the observer. We stay with a SISO system, but the concept can be extended to multiple outputs. The estimate should embody the dynamics of the plant (process). Thus one probable model, as shown in Fig. 9.4, is to assume that the state estimator has the same structure as the plant model, as in Eqs. (9.13) and (9.14) or Fig. 9.1. The estimator also has the *identical* plant matrices $\mathbf{A}$ and $\mathbf{B}$. However, one major difference is the addition of the estimation error, $y - \tilde{y}$, in the computation of the estimated state $\tilde{\mathbf{x}}$.

The estimated state variables based on Fig. 9.4 can be described by (details in Review Problems)

$$\dot{\tilde{\mathbf{x}}} = \mathbf{A}\tilde{\mathbf{x}} + \mathbf{B}u + \mathbf{K}_e(y - \mathbf{C}\tilde{\mathbf{x}})$$
$$= (\mathbf{A} - \mathbf{K}_e\mathbf{C})\tilde{\mathbf{x}} + \mathbf{B}u + \mathbf{K}_e y. \tag{9.32}$$

Here, $\tilde{y} = \mathbf{C}\tilde{\mathbf{x}}$ has been used in writing the error in the estimation of the output, $(y - \tilde{y})$. The $(n \times 1)$ observer gain vector $\mathbf{K}_e$ does a weighting on how the error affects each estimate. In the next two subsections, we will apply the state estimator in Eq. (9.32) to a state feedback system and see how we can formulate the problem such that the error $(y - \tilde{y})$ can become zero.

9.3.2. Full-Order State Estimator System

A system that makes use of the state estimator is shown in Fig. 9.5, where, for the moment, changes in the reference are omitted. What we need is the set of equations that describes this regulator system with state estimation.

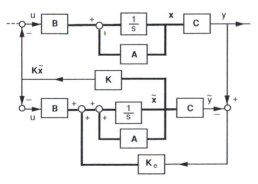

Figure 9.5. A regulator system with a controller–estimator.

By itself, the estimator in Eq. (9.32) has the characteristic equation

$$|s\mathbf{I} - \mathbf{A} + \mathbf{K}_e\mathbf{C}| = 0. \tag{9.33}$$

Our intention is to use the estimated states to provide feedback information:

$$u = -\mathbf{K}\tilde{\mathbf{x}}. \tag{9.34}$$

The state-space model of Eq. (9.13) now appears as

$$\dot{\mathbf{x}} = \mathbf{A}\mathbf{x} + \mathbf{B}u = \mathbf{A}\mathbf{x} - \mathbf{B}\mathbf{K}\tilde{\mathbf{x}}. \tag{9.35}$$

If we substitute $y = \mathbf{C}\mathbf{x}$ into Eq. (9.32), we can integrate Eqs. (9.32) and (9.35) simultaneously to compute $\mathbf{x}(t)$ and $\tilde{\mathbf{x}}(t)$. In matrix form, this set of $2n$ equations can be written as

$$\frac{d}{dt}\begin{bmatrix} \mathbf{x} \\ \tilde{\mathbf{x}} \end{bmatrix} = \begin{bmatrix} \mathbf{A} & -\mathbf{B}\mathbf{K} \\ \hline \mathbf{K}_e\mathbf{C} & \mathbf{A} - \mathbf{K}_e\mathbf{C} - \mathbf{B}\mathbf{K} \end{bmatrix}\begin{bmatrix} \mathbf{x} \\ \tilde{\mathbf{x}} \end{bmatrix}. \tag{9.36}$$

9.3.3. Estimator Design

With Eq. (9.36), it is not obvious how $\mathbf{K}_e$ affects the choices of $\mathbf{K}$. We now derive a form of Eq. (9.36) that is based on the error of the estimation and is easier for us to make a statement on its properties. We define the state error vector as

$$\mathbf{e}(t) = \mathbf{x}(t) - \tilde{\mathbf{x}}(t). \tag{9.37}$$

Subtracting Eq. (9.32) from Eq. (9.35) and using $y = \mathbf{C}\mathbf{x}$, we should find that

$$(\dot{\mathbf{x}} - \dot{\tilde{\mathbf{x}}}) = (\mathbf{A} - \mathbf{K}_e\mathbf{C})(\mathbf{x} - \tilde{\mathbf{x}}) \quad \text{or} \quad \dot{\mathbf{e}} = (\mathbf{A} - \mathbf{K}_e\mathbf{C})\mathbf{e}. \tag{9.38}$$

This error equation has the same characteristic equation as the estimator in Eq. (9.33). The goal is to choose eigenvalues of the estimator such that the error decays away quickly. We may note that the form of Eq. (9.38) is the same as that of the regulator problem. Thus we should be able to use the tools of pole placement for the estimator design. In fact, we can apply, without derivation, a modified form of Ackermann's formula to evaluate

$$\mathbf{K}_e = \alpha_e(\mathbf{A})\begin{bmatrix} \mathbf{C} \\ \mathbf{C}\mathbf{A} \\ \vdots \\ \mathbf{C}\mathbf{A}^{n-1} \end{bmatrix}^{-1}\begin{bmatrix} 0 \\ 0 \\ \vdots \\ 1 \end{bmatrix}, \tag{9.39}$$

where, as analogous to Eq. (9.20),

$$\alpha_e(s) = s^n + \alpha_{n-1}s^{n-1} + \cdots + \alpha_1 s + \alpha_0 \tag{9.40}$$

is the polynomial derived from our own chosen estimator eigenvalues. Equation (9.39) is different from Eq. (9.22) because we are now solving the *dual* problem for the $(n \times 1)$ vector $\mathbf{K}_e$.

Next, we can replace $\tilde{\mathbf{x}}$ in Eq. (9.35) with the definition of the error vector, and the equation becomes

$$\dot{\mathbf{x}} = \mathbf{A}\mathbf{x} - \mathbf{B}\mathbf{K}(\mathbf{x} - \mathbf{e}). \tag{9.41}$$

Equations (9.38) and (9.41) can be put in matrix form as

$$\begin{bmatrix} \tilde{\mathbf{x}} \\ \dot{\mathbf{e}} \end{bmatrix} = \begin{bmatrix} \mathbf{A} - \mathbf{BK} & \mathbf{BK} \\ \mathbf{0} & \mathbf{A} - \mathbf{K}_e\mathbf{C} \end{bmatrix} \begin{bmatrix} \mathbf{x} \\ \mathbf{e} \end{bmatrix}. \tag{9.42}$$

Now it is clear that the characteristic equation of the controller–estimator system is

$$|s\mathbf{I} - \mathbf{A} + \mathbf{BK}||s\mathbf{I} - \mathbf{A} + \mathbf{K}_e\mathbf{C}| = 0. \tag{9.43}$$

We have the very important result that choices for the eigenvalues for the pole-placement design and the observer design can be made independently. Generally, we want the observer response to be two to five times faster than the system response. We should not have to worry about saturation as the entire observer is software based, but we do have to consider noise and sensitivity problems.

Example 9.3: Consider the second-order model in Example 9.1, which was used in Example 9.2 to calculate the state feedback gains. What is the observer gain vector $\mathbf{K}_e$ if we specify that the estimator error should have eigenvalues -9 repeated thrice?

With eigenvalues selected at -9, we have chosen the estimator to be faster than the state feedback, and all the errors are to decay exponentially. We make use of the Ackermann's formula in Eq. (9.39) for observer gains. The MATLAB statements are

```
A=[0 1 0; 0 0 1; -6 -11 -6];   % Define the model
B=[0; 0; 1];
C=[1 0 0];

pe=poly([-9 -9 -9]);           % Make estimator polynomial (9.40)
ae=polyvalm(pe,A);
Ob=[C; C*A; C*A^2];
Ke=ae*inv(Ob)*[0; 0; 1]        % Eq. (9.39)
```

We should find that $\mathbf{K}_e = (21, 106, -144)$. The estimator calculations are purely mathematical, and the values of the observer gains can be negative. Furthermore, we can check that the system of equations in Eq. (9.42) has the correct eigenvalues as suggested by Eq. (4.43).

```
K=[102 43 6];  % Feedback gains calculated from Example 9.2
A11=A-B*K;     % Submatrices in Eq.(9.42)
A12=B*K;
A21=zeros(3,3);
A22=A-Ke*C;
BIGA=[A11 A12; A21 A22];
eig(BIGA)
```

Indeed, we should find that the big matrix `BIGA` has eigenvalues of $-3 \pm 3j$, -6, and -9 repeated three times.

9.3.4. Reduced-Order Estimator

We should not have to estimate variables that we can measure. It is logical to design a **reduced-order estimator** that estimates only the states that cannot be measured or are too noisy to be measured accurately. Following our introductory practice, we consider only one

Design of State-Space Systems

measured output. The following development assumes that we have selected x_1 to be the measured variable. Hence the output is

$$y = \mathbf{C}\mathbf{x} = [1 \ 0 \ \cdots \ 0]\mathbf{x}. \tag{9.44}$$

Next, we partition the state vector as

$$\mathbf{x} = \begin{bmatrix} x_1 \\ \mathbf{x}_e \end{bmatrix}, \tag{9.45}$$

where $\mathbf{x}_e = [x_2 \ \cdots \ x_n]$ contains the $(n-1)$ states that have to be estimated. State model equation (9.13) is partitioned accordingly as

$$\begin{bmatrix} \dot{x}_1 \\ \dot{\mathbf{x}}_e \end{bmatrix} = \begin{bmatrix} a_{11} & \mathbf{A}_{1e} \\ \mathbf{A}_{e1} & \mathbf{A}_{ee} \end{bmatrix} \begin{bmatrix} x_1 \\ \mathbf{x}_e \end{bmatrix} + \begin{bmatrix} b_1 \\ \mathbf{B}_e \end{bmatrix} u, \tag{9.46}$$

where the dimensions of $\mathbf{A}_{1e}$, $\mathbf{A}_{e1}$, and $\mathbf{A}_{ee}$ are, respectively, $(1 \times n - 1)$, $(n - 1 \times 1)$, and $(n - 1 \times n - 1)$, and that of $\mathbf{B}_e$ is $(n - 1 \times 1)$.

The next task is to make use of the full state estimator equations. Before that, we have to remold Eq. (9.46) as if it were a full state problem. This exercise requires some careful bookkeeping of notations. Let's take the first row in Eq. (9.46) and make it constitute the output equation. Thus we make a slight rearrangement:

$$\dot{x}_1 - a_{11}x_1 - b_1 u = \mathbf{A}_{1e}\mathbf{x}_e$$

such that it takes the form of $y = \mathbf{C}\mathbf{x}$. We repeat with the second row of (9.46) and put it as

$$\dot{\mathbf{x}}_e = \mathbf{A}_{ee}\mathbf{x}_e + (\mathbf{A}_{e1}x_1 + \mathbf{B}_e u)$$

such that it can be compared with $\dot{\mathbf{x}} = A\mathbf{x} + Bu$.

The next step is to take the full state estimator in Eq. (9.32),

$$\dot{\tilde{\mathbf{x}}} = (\mathbf{A} - \mathbf{K}_e\mathbf{C})\tilde{\mathbf{x}} + \mathbf{B}u + \mathbf{K}_e y,$$

and substitute term by term by using the reduced-order model equations.[5] The result is, finally,

$$\dot{\tilde{\mathbf{x}}}_e = (\mathbf{A}_{ee} - \mathbf{K}_{er}\mathbf{A}_{1e})\tilde{\mathbf{x}}_e + (\mathbf{A}_{e1}x_1 + \mathbf{B}_e u) + \mathbf{K}_{er}(\dot{x}_1 - a_{11}x_1 - b_1 u), \tag{9.47}$$

which is the reduced-order equivalent of Eq. (9.32). Note that in this equation $x_1 = y$.

---

[5] The matching of terms for reduced-order substitution in Eq. (9.31) to derive Eqs. (9.47)–(9.49):

Full-order state estimator	Reduced-order state estimator
$\tilde{\mathbf{x}}$	$\tilde{\mathbf{x}}_e$
y	$\dot{x}_1 - a_{11}x_1 - b_1 u$
$\mathbf{C}$	$\mathbf{A}_{1e}$
$\mathbf{A}$	$\mathbf{A}_{ee}$
$\mathbf{K}_e(n \times 1)$	$\mathbf{K}_{er}(n - 1 \times 1)$
$\mathbf{B}u$	$\mathbf{A}_{e1}x_1 + \mathbf{B}_e u$

The computation of the $(n-1)$ weighting factors in $\mathbf{K}_{er}$ can be based on the equivalent form of Eq. (9.38). Again, doing the substitution for the notations, we find that the error estimate becomes

$$\dot{\mathbf{e}} = (\mathbf{A}_{ee} - \mathbf{K}_{er}\mathbf{A}_{1e})\mathbf{e}, \tag{9.48}$$

which means that Ackermann's formula in Eq. (9.39) now takes the form

$$\mathbf{K}_{er} = \alpha_e(\mathbf{A}_{ee}) \begin{bmatrix} \mathbf{A}_{1e} \\ \mathbf{A}_{1e}\mathbf{A}_{ee} \\ \vdots \\ \mathbf{A}_{1e}\mathbf{A}_{ee}^{n-1} \end{bmatrix}^{-1} \begin{bmatrix} 0 \\ 0 \\ \vdots \\ 1 \end{bmatrix}. \tag{9.49}$$

We are not quite done yet. If we use Eq. (9.47) to compute $\tilde{\mathbf{x}}_e$, it requires taking the derivative of x_1, an exercise that can easily amplify noise. Therefore we want a modified form that allows us to replace this derivative. To begin, we define a new variable

$$\tilde{\mathbf{x}}_{e1} = \tilde{\mathbf{x}}_e - \mathbf{K}_{er}x_1. \tag{9.50}$$

This variable is substituted into Eq. (9.47) to give

$$(\dot{\tilde{\mathbf{x}}}_{e1} + \mathbf{K}_{er}\dot{x}_1) = (\mathbf{A}_{ee} - \mathbf{K}_{er}\mathbf{A}_{1e})(\tilde{\mathbf{x}}_{e1} + \mathbf{K}_{er}x_1)$$
$$+ (\mathbf{A}_{e1}x_1 + \mathbf{B}_e u) + \mathbf{K}_{er}(\dot{x}_1 - a_{11}x_1 - b_1 u).$$

After cancellation of the derivative term, we have

$$\dot{\tilde{\mathbf{x}}}_{e1} = (\mathbf{A}_{ee} - \mathbf{K}_{er}\mathbf{A}_{1e})\tilde{\mathbf{x}}_{e1}$$
$$+ (\mathbf{A}_{ee}\mathbf{K}_{er} - \mathbf{K}_{er}\mathbf{A}_{1e}\mathbf{K}_{er} + \mathbf{A}_{e1} - \mathbf{K}_{er}a_{11})x_1 + (\mathbf{B}_e - \mathbf{K}_{er}b_1)u. \tag{9.51}$$

This differential equation is used to compute $\tilde{\mathbf{x}}_{e1}$, which then is used to calculate $\tilde{\mathbf{x}}_e$ with Eq. (9.50). With the estimated states, we can compute the feedback to the state-space model as

$$u = -\begin{bmatrix} K_1 & \mathbf{K}_{1e}^T \end{bmatrix} \begin{bmatrix} x_1 \\ \tilde{\mathbf{x}}_e \end{bmatrix}. \tag{9.52}$$

The application of Eqs. (9.50)–(9.52) is a bit involved and best illustrated as shown in Fig. 9.6.

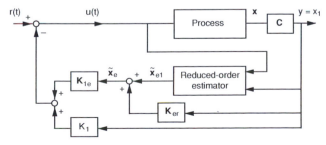

Figure 9.6. State feedback with reduced-order estimator.

Example 9.4: Consider the estimator in Example 9.3: What is the reduced-order observer gain vector K_{er} if we specify that the estimator error should have eigenvalues of -9 repeated twice?

We can use Eq. (9.49), and the MATLAB statements are

```
A=[0 1 0; 0 0 1; -6 -11 -6];
N=size(A,1);
a11=A(1,1);                    % Extract matrix partitions as in
                               % Eq.(9.46)
A1e=A(1,2:N);
Ae1=A(2:N,1);
Aee=A(2:N,2:N);

pe=poly([-9 -9]);             % Make estimator polynomial
ae=polyvalm(pe,Aee);
Ob=[A1e; A1e*Aee];
Ker=ae*inv(Ob)*[0; 1]   % Eq.(9.49) for n=2
```

We should find that $K_{er} = (12\ -2)$.

After all this fancy mathematics, a word of caution is needed. It is extremely dangerous to apply the state estimate as presented in this chapter. Why? The first hint is in Eq. (9.32). We have assumed perfect knowledge of the plant matrices. Of course, we rarely do. Furthermore, we have omitted actual terms for disturbances, noises, and errors in measurements. Despite these drawbacks, material in this chapter provides the groundwork to attack serious problems in modern control.

Review Problems

(1) For the second-order transfer function,

$$\frac{Y}{U} = \frac{1}{s^2 + 2\zeta\omega_n s + \omega_n^2},$$

derive the controllable canonical form. If the desired poles of a closed-loop system are to be placed at λ_1 and λ_2, what should be the state feedback gains?

(2) Presume we do not know what the estimator should be other than that it has the form

$$\dot{\tilde{x}} = F\tilde{x} + Gu + Hy.$$

Find Eq. (9.32).

(3) Do the time-response simulation in Example 7.5B. We found that the state-space system has a steady-state error. Implement integral control and find the new state feedback gain vector. Perform a time-response simulation to confirm the result.

(4) With respect to Fig. R9.4, what is the transfer function equivalent to the controller–estimator system in Eq. (9.32)?

9.3. State Estimation Design

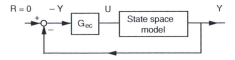

Figure R9.4.

Hints:

(1) The controllable canonical form was derived in Example 4.1. The characteristic polynomial of $(s\mathbf{I} - \mathbf{A} + \mathbf{BK})$ should be

$$s^2 + (2\zeta\omega_n + K_2)s + (\omega_n^2 + K_1) = 0.$$

The characteristic polynomial of desired poles is

$$s^2 + (\lambda_1 + \lambda_2)s + \lambda_1\lambda_2 = 0.$$

Thus

$$K_1 = \lambda_1\lambda_2 - \omega_n^2, \qquad K_2 = (\lambda_1 + \lambda_2) - 2\zeta\omega_n.$$

(2) The Laplace transform of the given equation is

$$s\tilde{\mathbf{X}} = \mathbf{F}\tilde{\mathbf{X}} + \mathbf{G}U + \mathbf{H}Y.$$

Substituting $Y = \mathbf{CX}$, we have

$$\tilde{\mathbf{X}} = (s\mathbf{I} - \mathbf{F})^{-1}[\mathbf{G}U + \mathbf{HCX}].$$

We further substitute for $\mathbf{X} = (s\mathbf{I} - \mathbf{A})^{-1}\mathbf{B}U$ with the simple state-space model to give

$$\tilde{\mathbf{X}} = (s\mathbf{I} - \mathbf{F})^{-1}[\mathbf{G} + \mathbf{HC}(s\mathbf{I} - \mathbf{A})^{-1}\mathbf{B}]U.$$

What we want is to dictate that the transfer function of this estimator is the same as that of the state-space model:

$$(s\mathbf{I} - \mathbf{F})^{-1}[\mathbf{G} + \mathbf{HC}(s\mathbf{I} - \mathbf{A})^{-1}\mathbf{B}] = (s\mathbf{I} - \mathbf{A})^{-1}\mathbf{B}.$$

Moving the second term to the RHS and factoring out the $(s\mathbf{I} - \mathbf{A})^{-1}\mathbf{B}$ gives

$$(s\mathbf{I} - \mathbf{F})^{-1}\mathbf{G} = [\mathbf{I} - (s\mathbf{I} - \mathbf{F})^{-1}\mathbf{HC}](s\mathbf{I} - \mathbf{A})^{-1}\mathbf{B}.$$

Thus we can multiply $(s\mathbf{I} - \mathbf{F})$ to both sides to have

$$\mathbf{G} = [(s\mathbf{I} - \mathbf{F}) - \mathbf{HC}](s\mathbf{I} - \mathbf{A})^{-1}\mathbf{B}.$$

Finally,

$$[(s\mathbf{I} - \mathbf{F}) - \mathbf{HC}]^{-1}\mathbf{G} = (s\mathbf{I} - \mathbf{A})^{-1}\mathbf{B}.$$

Comparing term by term, we have

$$\mathbf{F} + \mathbf{HC} = \mathbf{A}, \quad \text{or} \quad \mathbf{F} = \mathbf{A} - \mathbf{HC}$$
$$\mathbf{G} = \mathbf{B}.$$

This result is what we need in Eq. (9.32) if we also set $\mathbf{H} = \mathbf{K}_e$.

(3) For the time-response simulation, we also duplicate the classical control design for comparision. The statements are

```
G=tf(1,conv([2 1],[4 1]));
S=ss(G);                          % MATLAB uses reverse
                                  % indexing
scale=S.c(2);                     % Need to rescale B and
                                  % C too
S.c=S.c/scale;
S.b=S.b*scale;
P=[-0.375+0.382j -0.375-0.382j];  % Define the closed-loop
                                  % poles
K=acker(S.a,S.b,P)
% Compute the system matrices for plotting
A = S.a - S.b*K                   % Makes system matrix,
                                  % Eq.(9.25)
B = S.b*K(2)
C = S.c
D=0;
step(A,B,C,D)
hold % to add the classical design result
Gcl=feedback(1.29*G,1);
% Kc=1.29 was the proportional gain obtained in
% Example 7.5A
step(Gcl)
```

To eliminate offset, we need Subsection 9.2.3. With an added state that is due to integration, we have to add one more closed-loop pole. We choose it to be -1, sufficiently faster than the real part of the complex poles. The statements are

```
G=tf(1,conv([2 1],[4 1]));
S=ss(G); % Generates the matrices S.a, S.b, S.c, S.d
Ah=[S.a zeros(2,1); -S.c 0]       % A-head in Eq.(9.29)
Bh=[S.b; 0]                       % B-head
P=[-0.375+0.382j -0.375-0.382j -1]; % Add a faster pole
                                  % at -1
kh=acker(Ah,Bh,P)                 % K-head in (9.29)
```

We should find $\hat{\mathbf{K}} = [2\ 3.6\ -2.3]$. To do the time-response simulation, we can use

```
Asys=Ah-Bh*kh;    % System matrix(9.29)
Bsys=[0; 0; 1];   % Follows Eq.(9.28)
Csys=[S.c 0];
step(Asys, Bsys,Csys,0)
```

(4) For the estimator, y is the input and u is the output. With $u = -\mathbf{K}\tilde{\mathbf{x}}$, the Laplace transform of Eq. (9.32) is

$$[s\mathbf{I} - \mathbf{A} + \mathbf{K}_e\mathbf{C} + \mathbf{BK}]\tilde{\mathbf{X}}(s) = \mathbf{K}_e Y(s)$$

or

$$\tilde{\mathbf{X}}(s) = [s\mathbf{I} - \mathbf{A} + \mathbf{K}_e\mathbf{C} + \mathbf{BK}]^{-1}\mathbf{K}_e Y(s).$$

We now substitute $\tilde{\mathbf{X}}$ back into the Laplace transform of $u = -\mathbf{K}\tilde{\mathbf{x}}$ to obtain

$$U(s) = -\mathbf{K}[s\mathbf{I} - \mathbf{A} + \mathbf{K}_e\mathbf{C} + \mathbf{BK}]^{-1}\mathbf{K}_e Y(s) = -G_{ec}(s)Y(s).$$

10

Multiloop Systems

T here are many advanced strategies in classical control systems. Only a limited selection of examples is presented in this chapter. We start with cascade control, which is a simple introduction to a multiloop, but essentially SISO, system. We continue with feedforward and ratio control. The idea behind ratio control is simple, and it applies quite well to the furnace problem that is used as an illustration. Finally, a multiple-input multiple-output (MIMO) system is addressed with a simple blending problem as illustration, and the problem is used to look into issues of interaction and decoupling. These techniques build on what we have learned in classical control theories.

What Are We Up to?

- Applying classical controller analysis to cascade control, feedforward control, feedforward–feedback control, ratio control, and the Smith predictor for time-delay compensation
- Analyzing a MIMO system with relative gain array and assessing the pairing of manipulated and controlled variables
- Attempting to decouple and eliminate the interactions in a two-input two-output system

10.1. Cascade Control

A common design found in process engineering is cascade control. This is a strategy that allows us to handle load changes more effectively with respect to the manipulated variable.

To illustrate the idea, we consider the temperature control of a gas furnace, which is used to heat up a cold process stream. The fuel-gas flow rate is the manipulated variable, and its flow is subject to fluctuations that are due to upstream pressure variations.

In a simple single-loop system, we measure the outlet temperature, and the temperature controller (TC) sends its signal to the regulating valve. If there is fluctuation in the fuel-gas flow rate, this simple system will not counter the disturbance until the controller senses that the temperature of the furnace has deviated from the set point (T^s).

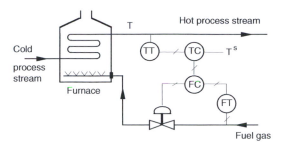

Figure 10.1. Cascade control of the temperature of a furnace, taken to be the same as that of the outlet process stream. The temperature controller (TC) does not actuate the regulating valve directly; it sends its signal to a secondary flow-rate control loop, which in turn ensures that the desired fuel-gas flow is deliverd.

A **cascade control system** can be designed to handle a fuel-gas disturbance more effectively (Fig. 10.1). In this case, a **secondary loop** (also called the *slave loop*) is used to adjust the regulating valve and thus manipulate the fuel-gas flow rate. The TC (the master or primary controller) sends its signal, in terms of the desired flow rate, to the secondary flow control loop – in essence, the signal is the set point of the secondary flow controller (FC).

In the secondary loop, the FC compares the desired fuel-gas flow rate with the measured flow rate from the flow transducer (FT) and adjusts the regulating valve accordingly. This inner flow control loop can respond immediately to fluctuations in the fuel-gas flow to ensure that the proper amount of fuel is delivered.

To be effective, the secondary loop must have a faster response time (smaller time constant) than the outer loop. Generally, we use as high a proportional gain as feasible. In control jargon, we say that the inner loop is tuned very tightly.

We can use a block diagram to describe Fig. 10.1. Cascade control adds an inner control loop with secondary controller function G_{c_2} [Fig. 10.2(a)]. This implementation of cascade

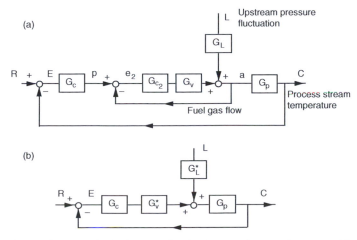

Figure 10.2. (a) Block diagram of a simple cascade control system with reference to the furnace problem, (b) reduced block diagram of a cascade control system.

control requires two controllers and two measured variables (fuel-gas flow and furnace temperature). The furnace temperature is the controlled variable, and the fuel-gas flow rate remains the only manipulated variable.

For cleaner algebra, we omit the measurement transfer functions, taking $G_{m_1} = G_{m_2} = 1$. A disturbance, such as upstream pressure, that specifically leads to changes in the fuel-gas flow rate is now drawn to be part of the secondary flow control loop. (A disturbance such as change in the process stream inlet temperature, which is not part of the secondary loop, would still be drawn in its usual location, as in Section 5.2).

We now reduce the block diagram. The first step is to close the inner loop so the system becomes a standard feedback loop [Fig. 10.2(b)]. With hindsight, the result should be intuitively obvious. For now, we take the slow route. Using the lowercase letter locations in Fig. 10.2(a), we write the algebraic equations

$$e_2 = p - a,$$
$$a = G_{c_2} G_v e_2 + G_L L.$$

Substitution of e_2 leads to

$$a = G_{c_2} G_v (p - a) + G_L L,$$

and the result after rearrangement is a form that allows us to draw Fig. 10.2(b):

$$a = \left(\frac{G_{c_2} G_v}{1 + G_{c_2} G_v} \right) p + \left(\frac{G_L}{1 + G_{c_2} G_v} \right) L = G_v^* p + G_L^* L,$$

where

$$G_v^* = \left(\frac{G_{c_2} G_v}{1 + G_{c_2} G_v} \right), \qquad G_L^* = \left(\frac{G_L}{1 + G_{c_2} G_v} \right). \tag{10.1}$$

The remaining task to derive the closed-loop transfer functions is routine. Again, slowly, we can write the relation in Fig. 10.2(b) as

$$C = G_L^* G_p L + G_c G_v^* G_p E,$$

and substituting $E = R - C$, we have, after rearrangement,

$$C = \left(\frac{G_c G_v^* G_p}{1 + G_c G_v^* G_p} \right) R + \left(\frac{G_p G_L^*}{1 + G_c G_v^* G_p} \right) L. \tag{10.2}$$

The closed-loop characteristic polynomial of this cascade system is

$$1 + G_c G_v^* G_p = 0. \tag{10.3}$$

If we now substitute G_v^* from Eq. (10.1), the characteristic polynomial takes the form[1]

$$1 + G_{c_2} G_v + G_c G_{c_2} G_v G_p = 0. \tag{10.3a}$$

---

[1] If we remove the secondary loop, this characteristic equation should reduce to that of a conventional feedback system equation. It is not obvious from Eq. (10.3) because our derivation has taken the measurement function G_{m_2} to be unity. If we had included G_{m_2} in a more detailed analysis, we could get the single-loop result by setting $G_{c_2} = 1$ and $G_{m_2} = 0$.

So far, we know that the secondary loop helps to reduce disturbance in the manipulated variable. If we design the control loop properly, we should also accomplish a faster response in the actuating element: the regulating valve. To go one step further, cascade control can even help to make the entire system more stable. These points may not be intuitive. A simple example is used to illustrate these features.

Example 10.1: Consider a simple cascade system as shown in Fig. 10.2(a) with a PI controller in the primary loop and a proportional controller in the slave loop. For simplicity, consider first-order functions:

$$G_p = \frac{0.8}{2s + 1}, \quad G_v = \frac{0.5}{s + 1}, \quad G_L = \frac{0.75}{s + 1}.$$

(a) How can the proper choice of K_{c_2} of the controller in the slave loop help to improve the actuator performance and eliminate disturbance in the manipulated variable (e.g., fuel-gas flow in the furnace temperature control)?

If we substitute $G_{c_2} = K_{c_2}$ and $G_v = [K_v/(\tau_v s + 1)]$ into G_v^* in Eq. (10.1), we should find

$$G_v^* = \left[\frac{K_{c_2} K_v}{(\tau_v s + 1) + K_{c_2} K_v}\right] = \frac{K_v^*}{\tau_v^* s + 1}, \tag{E10.1}$$

where

$$K_v^* = \left[\frac{K_{c_2} K_v}{1 + K_{C_2} K_v}\right], \quad \tau_v^* = \left[\frac{\tau_v}{1 + K_{c_2} K_v}\right]. \tag{E10.2}$$

Similarly, substituting $G_L = [K_L/(\tau_v s + 1)]$ into G_L^* should give

$$K_L^* = \left(\frac{K_L}{1 + K_{c_2} K_v}\right). \tag{E10.3}$$

Thus as the proportional gain K_{c_2} becomes larger, K_v^* approaches unity gain, meaning that there is a more effective change in the manipulated variable, and K_L^* approaches zero, meaning that the manipulated variable is becoming less sensitive to changes in the load. Furthermore, the effective actuator time constant τ_v^* will become smaller, meaning a faster response.

(b) The slave loop affords us a faster response with respect to the actuator. What is the proportional gain K_{c_2} if we want the slave-loop time constant τ_v^* to be only one tenth of the original time constant τ_v in G_v?

From the problem statement, $K_v = 0.5$ and $\tau_v = 1$ s. Thus $\tau_v^* = 0.1$ s, and substitution of these values into τ_v^* of Eqs. (E10.2) gives

$$0.1 = \left(\frac{1}{1 + 0.5 \, K_{c_2}}\right), \quad \text{or } K_{c_2} = 18.$$

The steady-state gain is

$$K_v^* = \frac{(18)(0.5)}{1 + (18)(0.5)} = 0.9.$$

The slave loop will have a 10% offset with respect to desired set-point changes in the secondary controller.

Multiloop Systems

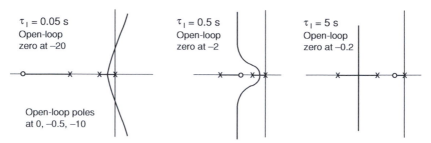

$\tau_1 = 0.05$ s
Open-loop
zero at –20

Open-loop poles
at 0, –0.5, –10

$\tau_1 = 0.5$ s
Open-loop
zero at –2

$\tau_1 = 5$ s
Open-loop
zero at –0.2

Figure E10.1.

(c) So far, we have used only proportional control in the slave loop. We certainly expect offset in this inner loop. Why do we stay with proportional control here?

The modest 10% offset that we have in the slave loop is acceptable under many circumstances. As long as we have integral action in the outer loop, the primary controller can make necessary adjustments in its output and ensure that there is no steady-state error in the controlled variable (e.g., the furnace temperature).

(d) Now we tackle the entire closed-loop system with the primary PI controller. Our task here is to choose the proper integral time constant among the given values of 0.05, 0.5, and 5s. We can tolerate underdamped response but absolutely not a system that can become unstable. Of course, we want a system response that is as fast as we can make it, i.e., with a proper choice of proportional gain. Select and explain your choice of the integral time constant.

Among other methods, root locus is the most instructive in this case. With a PI primary controller and numerical values, Eq. (10.3) becomes

$$1 + K_c \left(\frac{\tau_I s + 1}{\tau_I s} \right) \left(\frac{0.9}{0.1s + 1} \right) \left(\frac{0.8}{2s + 1} \right) = 0.$$

With MATLAB, we can easily prepare the root-locus plots of this equation for the cases of $\tau_I = 0.05$, 0.5, and 5s. (You should do it yourself. Only a rough sketch is shown in Fig. E10.1. Help can be found in the Review Problems.)

From the root-locus plots, it is clear that the system may become unstable when $\tau_I = 0.05s$. The system is always stable when $\tau_I = 5s$, but the speed of the system response is limited by the dominant pole between the origin and −0.2. The proper choice is $\tau_I = 0.5s$, in which case the system is always stable but the closed-loop poles can move farther away, loosely speaking, from the origin.

(e) Take the case *without* cascade control and, using the integral time constant that you have selected in part (d), determine the range of proportional gain that you can use (without the cascade controller and secondary loop) to maintain a stable system. How is this different from when we use cascade control?

With the choice of $\tau_I = 0.5s$, but without the inner loop or the secondary controller, the closed-loop equation is

$$1 + G_c G_v G_p = 1 + K_c \left(\frac{0.5s + 1}{0.5s} \right) \left(\frac{0.5}{s + 1} \right) \left(\frac{0.8}{2s + 1} \right) = 0,$$

which can be expanded to

$$s^3 + 1.5s^2 + (0.5 + 0.2K_c)s + 0.4K_c = 0.$$

With the Routh–Hurwitz analysis in Chap. 7, we should find that, to have a stable system, we must keep $K_c < 7.5$. (You fill in the intermediate steps in the Review Problems. Other techniques such as root locus, direct substitution, or frequency response in Chap. 8 should arrive at the same result.)

With cascade control, we know from part (d) that the system is always stable. Nevertheless, we can write the closed-loop characteristic equation

$$1 + K_c \left(\frac{0.5s + 1}{0.5s} \right) \left(\frac{0.9}{0.1s + 1} \right) \left(\frac{0.8}{2s + 1} \right) = 0$$

or

$$0.1s^3 + 1.05s^2 + (0.5 + 0.36K_c)s + 0.72K_c = 0$$

A Routh–Hurwitz analysis can confirm that. The key point is that, with cascade control, the system becomes more stable and allows us to use a larger proportional gain in the primary controller. The main reason is the much faster response (smaller time constant) of the actuator in the inner loop.[2]

10.2. Feedforward Control

To counter probable disturbances, we can take an even more proactive approach than cascade control and use feedforward control. The idea is that if we can make measurements of disturbance changes, we can use this information and our knowledge of the process model to make proper adjustments in the manipulated variable *before* the disturbance has a chance to affect the controlled variable.

We will continue with the gas furnace to illustrate **feedforward control**. For simplicity, let's make the assumption that changes in the furnace temperature (T) can be effected by changes in the fuel-gas flow rate (F_{fuel}) and the cold process stream flow rate (F_s). Other variables, such as the process stream temperature, are constant.

In Section 10.1, the fuel-gas flow rate is the manipulated variable (M) and cascade control is used to handle its fluctuations. Now we also consider changes in the cold process stream flow rate as another disturbance (L). Let's presume further that we have derived diligently from heat and mass balances the corresponding transfer functions, G_L and G_p, and we have the process model

$$C = G_L L + G_p M, \tag{10.4}$$

where we have used the general notation C as the controlled variable in place of furnace temperature T.

---

[2] If you are skeptical of this statement, try to do the Bode plots of the systems with and without cascade control and think about the consequence of changing the break frequency (or bandwidth) of the valve function. If you do not pick up the hint, the answer can be found on the *Web Support* in the details of Example 10.1.

We want the controlled variable to track set-point changes (R) precisely, so we substitute the ideal case $C = R$, and rearrange Eq. (10.4) to

$$M = \frac{1}{G_p} R - \frac{G_L}{G_p} L. \qquad (10.5)$$

This equation provides us with a model-based rule as to how the manipulated variable should be adjusted when we either change the set point or are faced with a change in the load variable. Equation (10.5) is the basis of what is called *dynamic* feedforward control because Eq. (10.4) has to be derived from a time-domain differential equation (a transient model).[3]

In Eq. (10.5), $1/G_p$ is the **set-point tracking controller**. This is what we need if we install only a feedforward controller, which in reality, we seldom do.[4] Under most circumstances, the change in set point is handled by a feedback control loop, and we need to implement only the second term of Eq. (10.5). The transfer function $-G_L/G_p$ is the **feedforward controller** (or the *disturbance rejection controller*). In terms of disturbance rejection, we may also see how the feedforward controller arises if we command $C = 0$ (i.e., no change), and write Eq. (10.4) as

$$0 = G_L L + G_p M.$$

To see how we implement a feedforward controller, we now turn to a block diagram (Fig. 10.3).[5] For the moment, we omit the feedback path from our general picture. With the expectation that we will introduce a feedback loop, we will not implement the set-point tracking term in Eq. (10.5). The implementation of feedforword control requires measurement of the load variable.

If there is more than one load variable, we theoretically could implement a feedforward controller on each one. However, that may not be good engineering. Unless there is a compelling reason, we should select the variable that either undergoes the most severe fluctuation or has the strongest impact on the controlled variable.

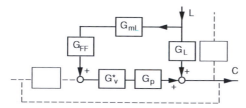

Figure 10.3. A feedforward control system on a major load variable with measurement function G_{ML} and feedforward controller G_{FF}.

---

[3] In contrast, we could have done the derivation by using steady-state models. In such a case, we would arrive at the design equation for a *steady-state* feedforward controller. We'll skip this analysis. As will be shown in Eq. (10.9), we can identify this steady-state part from the dynamic approach.

[4] The set-point tracking controller not only becomes redundant as soon as we add feedback control, but it also unnecessarily ties the feedforward controller into the closed-loop characteristic equation.

[5] If the transfer functions G_L and G_p are based on a simple process model, we know quite well that they should have the same characteristic polynomial. Thus the term $-G_L/G_p$ is nothing but a ratio of the steady-state gains, $-K_L/K_p$.

Here, we use L to denote the major load variable and its corresponding transfer function is G_L. We measure the load variable with a sensor, G_{mL}, which transmits its signal to the feedforward controller G_{FF}. The feedforward controller then sends its decision to manipulate the actuating element, or valve, G_v. In the block diagram, the actuator transfer function is denoted by G_v^*. The idea is that cascade control may be implemented with the actuator, G_v, as we have derived in Eq. (10.1). We simply use G_v^* to reduce clutter in the diagram.

With the feedforward and load path shown, the corresponding algebraic representation is

$$C = [G_L + G_{mL} G_{FF} G_v^* G_p] L. \tag{10.6}$$

The ideal feedforward controller should allow us to make proper adjustment in the actuator to achieve perfect rejection of load changes. To have $C = 0$, the theoretical feedforward controller function is

$$G_{FF} = -\frac{G_L}{G_{mL} G_v^* G_p}, \tag{10.7}$$

which is a slightly more complete version of what we derived in Eq. (10.5).

Before we blindly try to program Eq. (10.7) into a computer, we need to recognize certain pitfalls. If we write out the transfer functions in Eq. (10.7), we should find that G_{FF} is not physically realizable – the polynomial in the numerator has a higher order than the one in the denominator.[6]

If we approximate the composite function $G_{mL} G_v^* G_p$ as a first-order function with dead time, $K e^{-\theta s}/(\tau s + 1)$, Eq. (10.7) would appear as

$$G_{FF} = -\frac{K_L}{K} \frac{\tau s + 1}{\tau_p s + 1} e^{\theta s}.$$

Now the dead time appears as a positive exponent or an advance in time. We cannot foresee the future, and this idea is not probable either.[7]

The consequence is that the most simple implementation of a feedforward controller, especially with off-the-shelf hardware, is a lead–lag element with a gain:

$$G_{FF} = K_{FF} \frac{\tau_{FLD} s + 1}{\tau_{FLG} s + 1}. \tag{10.8}$$

---

[6] If we go by the book, G_L and G_p are the transfer functions to a process and their dynamic terms (characteristic polynomial) in Eq. (10.7) must cancel out. The feedforward transfer function would be reduced to something that looks like $(-K_L/K_{mL} K_v^* K_p)(\tau_{mL} s + 1)(\tau_v^* s + 1)$ whereas the denominator is just 1. In the simplest case, in which the responses of the transmitter and the valve are extremely fast such that we can ignore their dynamics, the feedforward function consists of only the steady-state gains, as in Eq. (10.9).

[7] If the load transfer function in Eq. (10.7) had also been approximated as a first-order function with dead time, say, of the form $K_L e^{-t_d s}/(\tau_p s + 1)$, the feedforward controller would appear as

$$G_{FF} = -\frac{K_L}{K} \frac{\tau s + 1}{\tau_p s + 1} e^{-(t_d - \theta)s}.$$

Now, if $t_d > \theta$, it is possible for the feedforward controller to incorporate dead-time compensation. The situation in which we may find that the load function dead time is larger than that in the feedforward path of $G_m G_v^* G_p$ is not obvious from our simplified block diagram. Such a circumstance arises when we deal with more complex processing equipment consisting of several units (i.e., multicapacity process) and the disturbance enters farther upstream than where the controlled and manipulated variables are located.

Based on Eq. (10.7), the gain of this feedforward controller is

$$K_{FF} = -\frac{K_L}{K_{mL} K_v^* K_p}. \tag{10.9}$$

This is the **steady-state compensator**. The lead–lag element with lead time constant τ_{FLD} and lag time constant τ_{FLG} is the **dynamic compensator**. Any dead-time in the transfer functions in Eq. (10.7) is omitted in this implementation.

When we tune the feedforward controller, we may take, as a first approximation, τ_{FLD} as the sum of the time constants τ_m and τ_v^*. Analogous to the "real" derivative control function, we can choose the lag time constant to be a tenth smaller, $\tau_{FLG} \approx 0.1\,\tau_{FLD}$. If the dynamics of the measurement device is extremely fast, $G_m = K_{mL}$, and if we have cascade control, the time constant τ_v^* is also small, and we may not need the lead–lag element in the feedforward controller. Just the use of the steady-state compensator K_{FF} may suffice. In any event, the feedforward controller must be tuned with computer simulations and subsequently field tests.

10.3. Feedforward–Feedback Control

Because we do not have the precise model function G_p embedded in the feedforward controller function in Eq. (10.8), we cannot expect perfect rejection of disturbances. In fact, feedforward control is never used by itself; it is implemented in conjunction with a feedback loop to provide the so-called feedback trim [Fig. 10.4(a)]. The feedback loop handles (1) measurement errors, (2) errors in the feedforward function, (3) changes in unmeasured load variables, such as the inlet process stream temperature in the furnace that one single feedforward loop cannot handle, and of course, (4) set-point changes.

Our next task is to find the closed-loop transfer functions of this feedforward–feedback system. Among other methods, we should see that we can "move" the $G_v^* G_p$ term, as shown in Fig. 10.4(b). (You can double check with algebra.) After this step, the rest is routine. We can almost write the final result immediately. Anyway, we should see that

$$C = [G_L + G_{mL} G_{FF} G_v^* G_p]L + [G_c G_v^* G_p]E,$$

$$E = R - G_m C.$$

After substitution for E and rearrangement, we arrive at

$$C = \frac{G_L + G_{mL} G_{FF} G_v^* G_p}{1 + G_m G_c G_v^* G_p} L + \frac{G_c G_v^* G_p}{1 + G_m G_c G_v^* G_p} R. \tag{10.10}$$

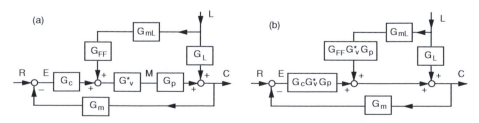

Figure 10.4. (a) A feedforward-feedback control system, (b) the diagram after $G_v^* G_p$ is moved.

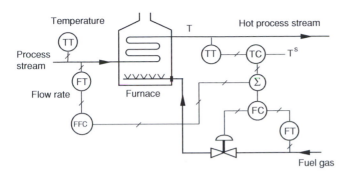

Figure E10.2.

If we do not have cascade control, G_v^* is simply G_v. If we are using cascade control, we can substitute Eq. (10.1) for G_v^*, but we'll skip this messy algebraic step. The key point is that the closed-loop characteristic polynomial is

$$1 + G_m G_c G_v^* G_p = 0, \tag{10.11}$$

and the feedforward controller G_{FF} does not affect the system stability.

Example 10.2: Consider the temperature control of a gas furnace used in heating a process stream. The probable disturbances are in the process stream temperature and flow rate and the fuel-gas flow rate. Draw the schematic diagram of the furnace temperature control system and show how feedforward, feedback, and cascade controls can all be implemented together to handle load changes.

The design in Fig. E10.2 is based on our discussion of cascade control. The fuel-gas flow is the manipulated variable, and so we handle disturbance in the fuel-gas flow with a flow controller (FC) in a slave loop. This secondary loop remains the same as the G_v^* function in Eq. (10.1), where the secondary transfer function is denoted by G_{c_2}.

Of the other two load variables, we choose the process stream flow rate as the major disturbance. The flow transducer FT sends the signal to the feedforward controller (FFC, transfer function G_{FF}). A summer ($\sum$) combines the signals from both the feedforward and the feedback controllers, and its output becomes the set point for the secondary fuel-gas flow rate controller (FC).

The handling of disturbance in the inlet process stream temperature is passive. Any changes in this load variable will affect the furnace temperature. The change in furnace temperature is measured by the outlet temperature transducer (TT) and sent to the feedback temperature controller (TC). The primary controller then acts accordingly to reduce the deviation in the furnace temperature.

10.4. Ratio Control

We are not entirely finished with the furnace. There is one more piece missing from the whole picture – the air flow rate. We need to ensure sufficient air flow for efficient combustion. The regulation of air flow is particularly important in the reduction of air pollutant emission.

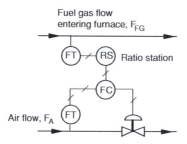

Figure 10.5. Simple ratio control of the air flow rate.

To regulate the air flow rate with respect to the fuel-gas flow rate, we can use **ratio control**. Figure 10.5 illustrates one of the simplest implementations of this strategy. Let's say the air to fuel-gas flow rates must be kept at some constant ratio,

$$R = \frac{F_A}{F_{FG}}. \tag{10.12}$$

What we can do easily is to measure the fuel-gas flow rate, multiply the value by R in the so-called ratio station, and send the signal as the set point to the air flow controller. The calculation can be based on actual flow rates rather than on deviation variables.

A more sophisticated implementation is **full metering control** (Fig. 10.6). In this case, we send the signals from the fuel-gas controller (FC in the fuel-gas loop) and the air flow transmitter (FT) to the ratio controller (RC), which takes the desired flow ratio (R) as the set point. This controller calculates the proper air flow rate, which in turn becomes the set point to the air flow controller (FC in the air flow loop). If we take away the secondary flow control loops on both the fuel-gas and the air flow rates, what we have is called parallel positioning control. In this simpler case, of course, the performance of the furnace is subject to fluctuations in fuel- and air-supply lines.

We are skipping the equations and details because the air flow regulation should not affect the stability and system analysis of the fuel-gas controller and ratio control is best implemented with Simulink in simulation and design projects.

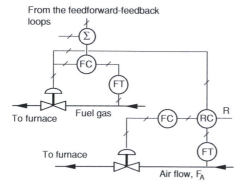

Figure 10.6. Full metering ratio control of fuel and air flows.

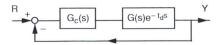

Figure 10.7. System with inherent dead time.

10.5. Time-Delay Compensation – The Smith Predictor

There are different schemes to handle systems with a large dead time. One of them is the Smith predictor. It is not the most effective technique, but it provides a good thought process.

Consider a unit feedback system with a time delay in its process function (Fig. 10.7). The characteristic polynomial is

$$1 + G_c(s)G(s)e^{-t_d s} = 0. \tag{10.13}$$

We know from frequency-response analysis that time lag introduces extra phase lag, reduces the gain margin, and is a significant source of instability. This is mainly because the feedback information is outdated.

If we have a model for the process, i.e., we know $G(s)$ and t_d, we can predict what may happen and feed back this estimation. From the way the dead-time compensator (or predictor) is written (Fig. 10.8), we can interpret the transfer function as follows. Assuming that we know the process model, we feed back the "output" calculation based on this model. We also have to subtract out the "actual" calculated time-delay output information.

Now the error E also includes the feedback information from the dead-time compensator:

$$E = R - Y - E[G_c G(1 - e^{-t_d s})],$$

and substituting

$$Y = G_c G e^{-t_d s} E,$$

we have

$$E = R - E[G_c G e^{-t_d s} + G_c G(1 - e^{-t_d s})],$$

where the exponential terms cancel out and we are left with simply

$$E = R - EG_c G. \tag{10.14}$$

The time-delay effect is canceled out, and this equation at the summing point is equivalent to a system without dead time (in which the forward path is $C = G_c GE$). With simple

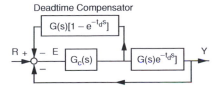

Figure 10.8. Implementation of the Smith predictor.

209

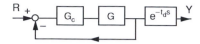

Figure 10.9. An interpretation of the compensator effect.

block-diagram algebra, we can also show that the closed-loop characteristic polynomial with the Smith predictor is simply

$$1 + G_c G = 0. \tag{10.15}$$

The time delay is removed. With the delay compensator included, we can now use a larger proportional gain without becoming unstable. Going back to the fact that the feedback information is $G_c G R$, we can also interpret the compensator effect as in Fig. 10.9. The Smith predictor is essentially making use of state feedback as opposed to output feedback.

Just like feedforward control (or any other model-based control), we have perfect compensation only if we know the precise process model. Otherwise, the effectiveness of the compensator (or predictor) is diminished. Assume that we have an imperfect model approximation $H(s)$ and dead-time estimation θ ($H \neq G$ and $\theta \neq t_d$); the feedback information is now

$$Y + [H(1 - e^{-\theta s})]G_c R = [G_c G e^{-t_d s} + G_c H(1 - e^{-\theta s})]R$$
$$= G_c[G e^{-t_d s} + H(1 - e^{-\theta s})]R,$$

where the RHS becomes $G_c G R$ if and only if $H = G$ and $\theta = t_d$. Note that the time-delay term is an exponential function. Error in the estimation of the dead time is more detrimental than error in the estimation of the process function G.

Because few things are exact in this world, we most likely have errors in the estimation of the process and the dead time. Therefore we only have partial dead-time compensation, and we must be conservative in picking controller gains based on the characteristic polynomial $1 + G_c G = 0$.

In a chemical plant, time delay is usually a result of a transport lag in the pipe flow. If the flow rate is fairly constant, the use of the Smith predictor is acceptable. If the flow rate varies for whatever reason, this compensation method will not be effective.

10.6. Multiple-Input Multiple-Output Control

In this section, we analyze a **MIMO** system. There are valuable insights that can be gained from using the classical transfer function approach. One decision that we need to appreciate is the proper pairing of manipulated and controlled variables. To do that, we also need to know how strong the interaction is among different variables.

The key points will be illustrated with a blending process. Here, we mix two streams with mass flow rates m_1 and m_2, and both the total flow rate F and the composition x of a solute A are to be controlled (Fig. 10.10). With simple intuition, we know changes in both m_1 and m_2 will affect F and x. We can describe the relations with the block diagram in Fig. 10.11, where interactions are represented by the two yet to be derived transfer functions G_{12} and G_{21}.

Given Fig. 10.11 and classical control theory, we can infer the structure of the control system, which is shown in Fig. 10.12. That is, we use two controllers and two feedback

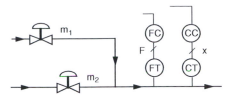

Figure 10.10. A blending system with manipulated and controlled variable pairings yet to be determined.

loops, in which, for simplicity, the measurement and the actuator functions have been omitted.

Simple reasoning can illustrate now interactions may arise in Fig. 10.12. If a disturbance (not shown in diagram) moves C_1 away from its reference R_1, the controller G_{c_1} will respond to the error and alter M_1 accordingly. The change in M_1, however, will also affect C_2 by means of G_{21}. Hence C_2 is forced to change and deviate from R_2. Now the second controller G_{c_2} kicks in and adjusts M_2, which in turn also affects C_1 by means of G_{12}.

With this case the system may eventually settle, but it is just as likely that the system in Fig. 10.12 will spiral out of control. It is clear that loop interactions can destabilize a control system, and tuning controllers in a MIMO system can be difficult. One logical thing that we can do is to reduce loop interactions by proper pairing of manipulated and controlled variables. This is the focus of the analysis in the following subsections.

10.6.1. MIMO Transfer Functions

We now derive the transfer functions of the MIMO system. This sets the stage for the more detailed analysis that follows. The transfer functions in Fig. 10.11 depend on the process that we have to control, and we derive them in the next subsection for the blending process. Here, we consider the general system shown in Fig. 10.12.

With the understanding that the errors are $E_1 = R_1 - C_1$ and $E_2 = R_2 - C_2$ in Fig. 10.12, we can write immediately

$$M_1 = G_{c_1}(R_1 - C_1), \tag{10.16}$$

$$M_2 = G_{c_2}(R_2 - C_2). \tag{10.17}$$

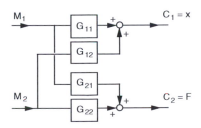

Figure 10.11. Block diagram of an interacting 2×2 process, with the output x and F referring to the blending problem.

211

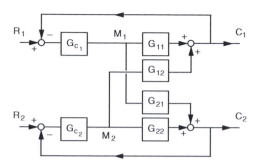

Figure 10.12. Block diagram of a 2×2 servo system. The pairing of the manipulated and controlled variables is not necessarily the same as that shown in Fig. 10.11.

The process (also in Fig. 10.11) can be written as

$$C_1 = G_{11}M_1 + G_{12}M_2, \tag{10.18}$$

$$C_2 = G_{21}M_1 + G_{22}M_2. \tag{10.19}$$

Substituting for M_1 and M_2 by using Eqs. (10.16) and (10.17), and factoring C_1 and C_2 to the left, we find that Eqs. (10.18) and (10.19) become

$$\left(1 + G_{11}G_{c_1}\right)C_1 + G_{12}G_{c_2}C_2 = G_{11}G_{c_1}R_1 + G_{12}G_{c_2}R_2, \tag{10.20}$$

$$G_{21}G_{c_1}C_1 + \left(1 + G_{22}G_{c_2}\right)C_2 = G_{21}G_{c_1}R_1 + G_{22}G_{c_2}R_2. \tag{10.21}$$

Making use of Kramer's rule, we should identify (derive!) the *system characteristic equation*,

$$p(s) = \left(1 + G_{11}G_{c_1}\right)\left(1 + G_{22}G_{c_2}\right) - G_{12}G_{21}G_{c_1}G_{c_2} = 0, \tag{10.22}$$

which, of course, is what governs the dynamics and the stability of the system. We may recognize that when either $G_{12} = 0$ or $G_{21} = 0$, the interaction term is zero.[8] In either case, the system characteristics analysis can be reduced to those of two single-loop systems:

$$1 + G_{11}G_{c_1} = 0, \qquad 1 + G_{22}G_{c_2} = 0.$$

Now back to finding the transfer functions with interaction. To make the algebra appear a bit cleaner, we consider the following two cases. When $R_2 = 0$, we can derive, from Eqs. (10.20) and (10.21),

$$\frac{C_1}{R_1} = \frac{G_{11}G_{c_1} + G_{c_1}G_{c_2}(G_{11}G_{22} - G_{12}G_{21})}{p(s)}. \tag{10.23}$$

When $R_1 = 0$, we can find

$$\frac{C_1}{R_2} = \frac{G_{12}G_{c_2}}{p(s)}. \tag{10.24}$$

---

[8] When $G_{12} = G_{21} = 0$, the system is decoupled and behaves identically to two single loops. When either $G_{12} = 0$ or $G_{21} = 0$, the situation is referred to as one-way interaction, which is sufficient to eliminate recursive interactions between the two loops. In such a case, one of the loops is not affected by the second whereas it becomes a source of disturbance to this second loop.

If both references change simultaneously, we just need to add their effects in Eqs. (10.23) and (10.24) together. (What about C_2? You'll get to try that in the Review Problems.)

It is apparent from Eq. (10.22) that, with interaction, the controller design of the MIMO system is different from a SISO system. One logical question is under what circumstances may we make use of SISO designs as an approximation? Or in other words, can we tell if the interaction may be weak? This takes us to the next two subsections.

10.6.2. Process Gain Matrix

We come back to derive the process transfer functions for the blending problem.[9] The total mass flow balance is

$$F = m_1 + m_2, \tag{10.25}$$

where F is the total flow rate after blending, and m_1 and m_2 are the two inlet flows that we manipulate. The mass balance for a solute A (without using the subscript A explicitly) is

$$x F = x_1 m_1 + x_2 m_2, \tag{10.26}$$

where x is the mass fraction of A after blending and x_1 and x_2 are the mass fractions of A in the two inlet streams. We want to find the transfer functions as shown in Fig. 10.11:

$$\begin{bmatrix} X(s) \\ F(s) \end{bmatrix} = \begin{bmatrix} G_{11}(s) & G_{12}(s) \\ G_{21}(s) & G_{22}(s) \end{bmatrix} \begin{bmatrix} M_1(s) \\ M_2(s) \end{bmatrix}. \tag{10.27}$$

We take stream m_1 to be pure solute A and stream m_2 to be pure solvent. In this example, $x_1 = 1$ and $x_2 = 0$, and Eq. (10.26) is simplified to

$$x = \frac{m_1}{F} = \frac{m_1}{m_1 + m_2}. \tag{10.28}$$

Because x_i and m_i are functions of time, we need to linearize Eq. (10.26). A first-order Taylor expansion of x is

$$x \approx \left(\frac{m_1}{F}\right)_s + \left[\frac{m_2}{(m_1 + m_2)^2}\right]_s (m_1 - m_{1,s}) - \left[\frac{m_1}{(m_1 + m_2)^2}\right]_s (m_2 - m_{2,s}),$$

where the subscript s to the brackets denotes terms evaluated at steady state. The first term on the RHS is really the value of x at steady state, x_s, which can be moved to the LHS to make the deviation variable in x. With that, we take the Laplace transform to obtain the transfer functions of the deviation variables:

$$X(s) = G_{11}(s)M_1(s) + G_{12}(s)M_2(s), \tag{10.29}$$

where

$$G_{11}(s) = \left[\frac{m_2}{(m_1 + m_2)^2}\right]_s = K_{11}, \qquad G_{12}(s) = -\left[\frac{m_1}{(m_1 + m_2)^2}\right]_s = K_{12}. \tag{10.30}$$

---

[9] Because the point is to illustrate the analysis of interactions, we are using only steady-state balances, and it should not be a surprise that the transfer functions end up being only steady-state gains in Eq. (10.32). For a general dynamic problem in which we have to work with the transfer functions $G_{ij}(s)$, we can still apply the results here by making use of the steady-state gains of the transfer functions.

The transfer functions are constants, and hence we denote them with the gains K_{11} and K_{12}. If the solvent flow rate m_2 increases, the solute will be diluted. Hence K_{12} is negative.

The functions G_{21} and G_{22} are much easier. From Eq. (10.25), we can see immediately that

$$G_{21}(s) = K_{21} = 1, \qquad G_{22}(s) = K_{22} = 1. \tag{10.31}$$

Thus, in this problem, the process transfer function matrix of Eq. (10.27) can be written in terms of the steady-state **gain matrix:**

$$\begin{bmatrix} X(s) \\ F(s) \end{bmatrix} = \begin{bmatrix} K_{11} & K_{12} \\ K_{21} & K_{22} \end{bmatrix} \begin{bmatrix} M_1(s) \\ M_2(s) \end{bmatrix}. \tag{10.32}$$

In more general terms, we replace the LHS of Eq. (10.32) with a controlled variable vector:

$$\mathbf{C}(s) = \mathbf{KM}(s), \tag{10.33}$$

where $\mathbf{C} = [\mathbf{X} \ \mathbf{F}]^T$. If there is a policy such that the manipulated variables can regulate the controlled variables, we must be able to find an inverse of the gain matrix such that

$$\mathbf{M}(s) = \mathbf{K}^{-1}\mathbf{C}(s). \tag{10.34}$$

Example 10.3: If $m_1 = 0.1$ g/s and $m_2 = 10$ g/s, what is the process gain matrix? What are the interpretations?

Making use of Eqs. (10.30), we can calculate $K_{11} = 9.8 \times 10^{-2}$ and $K_{12} = -9.8 \times 10^{-4}$. With Eqs. (10.31), the process gain matrix is

$$\mathbf{K} = \begin{bmatrix} 9.8 \times 10^{-2} & -9.8 \times 10^{-4} \\ 1 & 1 \end{bmatrix}.$$

Under the circumstances of the particular set of numbers given, changing either m_1 or m_2 has a stronger effect on the total flow rate F than on x. With respect to the composition x, changing the solute flow m_1 has a much stronger effect than changing the solvent flow. The situation very much resembles a one-way interaction.

We may question other obvious examples of the process gain matrix. The sweetest is an identity matrix, meaning no interaction among the manipulated and controlled variables. Here is a quick summary of several simple possibilities[10]:

$\mathbf{K} = \begin{bmatrix} 1 & 0 \\ 0 & 1 \end{bmatrix}$: No interaction. Controller design is like that of single-loop systems.

$\mathbf{K} = \begin{bmatrix} 1 & \delta \\ \delta & 1 \end{bmatrix}$: Strong interaction if δ is close to 1; weak interaction if $\delta \ll 1$.

$\mathbf{K} = \begin{bmatrix} 1 & 1 \\ 0 & 1 \end{bmatrix}, \begin{bmatrix} 1 & 0 \\ 1 & 1 \end{bmatrix}$: One-way interaction

---

[10] There is more to "looking" at $\mathbf{K}$. We can, for example, make use of its singular value and condition number, which should be deferred to a second course in control.

10.6.3. Relative Gain Array

You may not find observing the process gain matrix satisfactory. That takes us to the **relative gain array (RGA)**, which can provide for a more quantitative assessment of the effect of changing a manipulated variable on different controlled variables. We start with the blending problem before coming back to the general definition.

For the blending process, the relative gain parameter of the effect of m_1 on x is defined as

$$\lambda_{x,m_1} = \frac{\partial x/\partial m_1|_{m_2}}{\partial x/\partial m_1|_F}. \tag{10.35}$$

It is the ratio of the partial derivative evaluated under two different circumstances. On top, we look at the effect of m_1 while holding m_2 constant. The calculation represents an open-loop experiment, and the value is referred to as an open-loop gain. In the denominator, the total flow rate, the other controlled variable, is held constant. Because we are varying (in theory) m_1, F can be constant only if we have a closed-loop with perfect control involving m_2. The partial derivative in the denominator is referred to as some closed-loop gain.

How do we interpret the relative gain? The idea is that if m_2 does not interfere with m_1, the derivative in the denominator should not be affected by the closed-loop involving m_2, and its value should be the same as the open-loop value in the numerator. In other words, if there is no interaction, $\lambda_{x,m_1} = 1$.

Example 10.4: Evaluate the RGA matrix for the blending problem.
The complete RGA matrix for the 2×2 blending problem is defined as

$$\Lambda = \begin{bmatrix} \lambda_{x,m_1} & \lambda_{x,m_2} \\ \lambda_{F,m_1} & \lambda_{F,m_2} \end{bmatrix}. \tag{E10.4}$$

For the first element, we use Eq. (10.28) to find

$$\frac{\partial x}{\partial m_1}\bigg|_{m_2} = \frac{m_2}{(m_1 + m_2)^2}, \qquad \frac{\partial x}{\partial m_1}\bigg|_F = \frac{1}{F} = \frac{1}{m_1 + m_2}.$$

Hence, with the definition in Eq. (10.35),

$$\lambda_{x,m_1} = \frac{m_2}{m_1 + m_2} = 1 - x. \tag{E10.5}$$

We proceed to find the other three elements (see Review Problems), and we have the RGA for the blending problem:

$$\Lambda = \begin{bmatrix} 1 - x & x \\ x & 1 - x \end{bmatrix}. \tag{E10.6}$$

There are several general points regarding this problem that should be noted, i.e., without proving them formally here. The sum of all the entries in each row and each column of the RGA Λ is 1. Thus, in the case of a 2×2 problem, all we need is to evaluate one element. Furthermore, the calculation is based on only open-loop information. In Example 10.4, the derivation is based on Eqs. (10.25) and (10.26).

We can now state the general definition of the relative gain array Λ. For the element relating the ith controlled variable to the jth manipulated variable,

$$\lambda_{i,j} = \frac{\partial c_i / \partial m_j|_{m_{k,k \neq j}}}{\partial c_i / \partial m_j|_{c_{k,k \neq i}}}, \tag{10.36}$$

where the (open-loop) gain in the numerator is evaluated with all other manipulated variables held constant and all the loops open (no loops!). The (closed-loop) gain in the denominator is evaluated with all the loops – other than the ith loop – closed. The value of this so-called closed-loop gain reflects the effect from other closed loops *and* the open loop between m_j and c_i.

The RGA can be derived in terms of the process steady-state gains. Making use of gain matrix equation (10.32), we can find (not that hard; see Review Problems)

$$\lambda_{x,m_1} = \frac{1}{1 - \frac{K_{12}K_{21}}{K_{11}K_{22}}}, \tag{10.37}$$

which can be considered a more general form of Eq. (E10.5) and hence (E10.6).[11]

The next question comes back to the meaning of the RGA and how that may influence our decision in pairing manipulated with controlled variables. Here is the simple interpretation, making use of Eqs. (10.36) and (10.37):

$\lambda_{i,j} = 1$: Requires $K_{12}K_{21} = 0$. "Open-loop" gain is the same as the "closed-loop" gain. The controlled variable (or loop) i is not subject to interaction from other manipulated variables (or other loops). Of course, we know nothing about whether other manipulated variables may interact and affect other controlled variables. Nevertheless, pairing the ith controlled variable to the jth manipulated variable is desirable.

$\lambda_{i,j} = 0$: The open-loop gain is zero. The manipulated variable j has no effect on the controlled variable i. Of course m_j may still influence other controlled variables (by means of one-way interaction). Either way, it makes no sense to pair m_j with c_i in a control loop.

$0 < \lambda_{i,j} < 1$: No doubt there are interactions from other loops, and from Eq. (10.37), some of the process gains must have opposite signs (or act in different directions). When $\lambda_{i,j} = 0.5$, we can interpret that the effect of the interactions is identical to the open-loop gain – recall the statement after Eq. (10.36). When $\lambda_{i,j} > 0.5$, the interaction is less than the main effect of m_j on c_i. However, when $\lambda_{i,j} < 0.5$, the interactive effects predominate and we want to avoid pairing m_j with c_i.

$\lambda_{i,j} > 1$: There are interactions from other loops as well, but now with all the process gains having the same sign. Avoid pairing m_j with c_i if $\lambda_{i,j}$ is much larger than 1.

$\lambda_{i,j} < 0$: We can infer by using Eq. (10.36) that the open-loop and the closed-loop gains have different signs or opposing effects. The overall influence of the other loops is in opposition to the main effect of m_j on c_i. Moreover, from Eq. (10.37), the interactive product $K_{12}K_{21}$ must be larger than the direct terms $K_{11}K_{22}$. Undesirable interaction is strong. The overall multiloop system may become unstable easily if we open up one of its loops. We definitely should avoid pairing m_j with c_i.

---

[11] For your information, RGA can be computed as the so-called Hadamard product, $\lambda_{ij} = K_{ij}K_{ji}^{-1}$, which is the element-by-element product of the gain matrix $\mathbf{K}$ and the transpose of its inverse. You can confirm this by repeating the examples with MATLAB calculations.

To sum up, the key is to pair the manipulated and the controlled variables such that the relative gain parameter is positive and as close to one as possible.

Example 10.5: If $m_1 = 0.1$ g/s and $m_2 = 10$ g/s, what is the proper pairing of manipulated and controlled variables? What if $m_1 = 9$ g/s and $m_2 = 1$ g/s?

In the first case, in which m_1 is very small, it is like a dosing problem. From Eq. (10.28), $x = 0.0099$. Because $x \ll 1$, λ_{x,m_1} is very close to 1 by Eq. (E10.5). Thus interaction is not significant if we pair x with m_1 and F with m_2. Physically, we essentially manipulate the total flow with the large solvent flow m_2 and tune the solute concentration by manipulating m_1.

In the second case, $x = 0.9$. Now $\lambda_{x,m_1} = 0.1$ by Eq. (E10.5). Because $\lambda_{x,m_1} \ll 1$, we do not want to pair x with m_1. Instead, we pair x with m_2 and F with m_1. Now we regulate the total flow with the larger solute flow m_1 and tune the concentration with the solvent m_2.

10.7. Decoupling of Interacting Systems

After proper pairing of manipulated and controlled variables, we still have to design and tune the controllers. The simplest approach is to tune each loop individually and conservatively while the other loop is in manual mode. At a more sophisticated level, we may try to decouple the loops mathematically into two noninteracting SISO systems with which we can apply single-loop tuning procedures. Several examples applicable to a 2×2 system are offered here.

10.7.1. Alternative Definition of Manipulated Variables

We seek choices of manipulated variables that may decouple the system. A simple possibility is to pick them to be the same as the controlled variables. In the blending problem, the two new manipulated variables can be defined as[12]

$$\mu_1 = F, \tag{10.38}$$
$$\mu_2 = x. \tag{10.39}$$

Once the controller (a computer) evaluates these two manipulated variables, it also computes on the fly the actual signals necessary for the two mass flow rates m_1 and m_2. The computation follows directly balance equations (10.25) and (10.28). Figure 10.13 is a schematic diagram of how this idea may be implemented.

10.7.2. Decoupler Functions

In this subsection, we add the so-called **decoupler functions** to a 2×2 system. Our starting point is Fig. 10.12. The closed-loop system equations can be written in matrix form, virtually by visual observation of the block diagram, as

$$\begin{bmatrix} C_1 \\ C_2 \end{bmatrix} = \begin{bmatrix} G_{11} & G_{12} \\ G_{21} & G_{22} \end{bmatrix} \begin{bmatrix} G_{c_1} & 0 \\ 0 & G_{c_2} \end{bmatrix} \begin{bmatrix} R_1 - C_1 \\ R_2 - C_2 \end{bmatrix}. \tag{10.40}$$

---

[12] The blending problem can be reduced to one-way interaction if we use m_1 instead of x as the new manipulated variable μ_2. We will do that in the Review Problems.

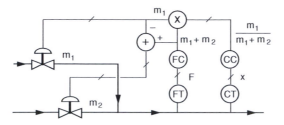

Figure 10.13. A decoupled control scheme. The controller outputs are the manipulated variables in Eqs. (10.38) and (10.39), and they are rewritten based on their definitions in Eqs. (10.25) and (10.28).

In matrix form, this equation looks deceptively simple, but if we expand the algebra, we should arrive at Eqs. (10.20) and (10.21) again.

In a system with interactions, G_{12} and G_{21} are not zero, but we can manipulate the controller signal such that the system appears (mathematically) to be decoupled. So let's try to transform the controller output with a matrix **D**, which will contain our decoupling functions. The manipulated variables are now

$$\begin{bmatrix} M_1 \\ M_2 \end{bmatrix} = \begin{bmatrix} d_{11} & d_{12} \\ d_{21} & d_{22} \end{bmatrix} \begin{bmatrix} G_{c_1} & 0 \\ 0 & G_{c_2} \end{bmatrix} \begin{bmatrix} R_1 - C_1 \\ R_2 - C_2 \end{bmatrix},$$

and the system equations become

$$\begin{bmatrix} C_1 \\ C_2 \end{bmatrix} = \begin{bmatrix} G_{11} & G_{12} \\ G_{21} & G_{22} \end{bmatrix} \begin{bmatrix} d_{11} & d_{12} \\ d_{21} & d_{22} \end{bmatrix} \begin{bmatrix} G_{c_1} & 0 \\ 0 & G_{c_2} \end{bmatrix} \begin{bmatrix} R_1 - C_1 \\ R_2 - C_2 \end{bmatrix} = \mathbf{GDG}_c \begin{bmatrix} R_1 - C_1 \\ R_2 - C_2 \end{bmatrix}.$$

$$(10.41)$$

To decouple the system equations, we require that $\mathbf{GDG}_c$ be a diagonal matrix. Define $\mathbf{G}_0 = \mathbf{GDG}_c$, and the previous step can be solved for **C**:

$$\begin{bmatrix} C_1 \\ C_2 \end{bmatrix} = [\mathbf{I} + \mathbf{G}_0]^{-1} \mathbf{G}_0 \begin{bmatrix} R_1 \\ R_2 \end{bmatrix}. \tag{10.42}$$

Because $\mathbf{G}_0$ is diagonal, the matrix $[\mathbf{I} + \mathbf{G}_0]^{-1}\mathbf{G}_0$ is also diagonal, and happily, we have two decoupled equations in Eq. (10.42).

Now we have to find **D**. Because $\mathbf{G}_c$ is already diagonal, we require that **GD** be diagonal:

$$\begin{bmatrix} G_{11} & G_{12} \\ G_{21} & G_{22} \end{bmatrix} \begin{bmatrix} d_{11} & d_{12} \\ d_{21} & d_{22} \end{bmatrix} = \begin{bmatrix} H_1 & 0 \\ 0 & H_2 \end{bmatrix}. \tag{10.43}$$

With a little bit of algebra to match term by term, we should find (see Review Problems)

$$d_{11} = \frac{G_{22}H_1}{G_{11}G_{22} - G_{12}G_{21}}, \quad d_{22} = \frac{G_{11}H_2}{G_{11}G_{22} - G_{12}G_{21}}, \tag{10.44}$$

$$d_{21} = \frac{-G_{21}}{G_{22}}d_{11}, \quad d_{12} = \frac{-G_{12}}{G_{11}}d_{22}. \tag{10.45}$$

We have six unknowns (four d_{ij} and two H_i) but only four equations. We have to make two (arbitrary) decisions. One possibility is to choose (or define)

$$H_1 = \frac{G_{11}G_{22} - G_{12}G_{21}}{G_{22}}, \qquad H_2 = \frac{G_{11}G_{22} - G_{12}G_{21}}{G_{11}} \tag{10.46}$$

such that d_{11} and d_{22} become 1. (We can also think in terms of choosing both $d_{11} = d_{22} = 1$ and then derive the relations for H_1 and H_2.) It follows that

$$d_{21} = \frac{-G_{21}}{G_{22}}, \qquad d_{12} = \frac{-G_{12}}{G_{11}}. \tag{10.47}$$

Now the closed-loop equations are

$$\begin{bmatrix} C_1 \\ C_2 \end{bmatrix} = \begin{bmatrix} H_1 & 0 \\ 0 & H_2 \end{bmatrix} \begin{bmatrix} G_{c_1} & 0 \\ 0 & G_{c_2} \end{bmatrix} \begin{bmatrix} R_1 - C_1 \\ R_2 - C_2 \end{bmatrix} = \begin{bmatrix} H_1 G_{c_1} & 0 \\ 0 & H_2 G_{c_2} \end{bmatrix} \begin{bmatrix} R_1 - C_1 \\ R_2 - C_2 \end{bmatrix}, \tag{10.48}$$

from which we can easily write, for each row of the matrices,

$$\frac{C_1}{R_1} = \frac{G_{c_1} H_1}{1 + G_{c_1} H_1}, \qquad \frac{C_2}{R_2} = \frac{G_{c_2} H_2}{1 + G_{c_2} H_2}, \tag{10.49}$$

and the design can be based on the two characteristic equations

$$1 + G_{c_1} H_1 = 0, \qquad 1 + G_{c_2} H_2 = 0. \tag{10.50}$$

Recall from Eqs. (10.46) that H_1 and H_2 are defined entirely by the four plant functions G_{ij}. This is another example of model-based control. With the definitions of H_1 and H_2 given in Eqs. (10.46), the calculations are best performed with a computer.

10.7.3. Feedforward Decoupling Functions

A simpler approach is to use only two decoupler functions and implement them as if they were feedforward controllers that may reduce the disturbance arising from loop interaction. As implemented in Fig. 10.14, we use the function D_{12} to "foresee" and reduce the interaction that is due to G_{12}. Likewise, D_{21} is used to address the interaction that is due to G_{21}. To find these two decoupling functions, we focus on how to cancel the interaction at the points identified as a and b in Fig. 10.14.

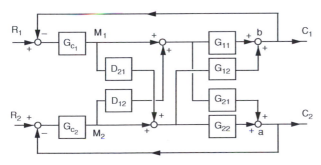

Figure 10.14. A decoupling scheme in which two feedforward-like decoupler functions are used.

Let's pick point a first. If the signal from M_1 through G_{21} can be canceled by the compensation through D_{21}, we can write

$$G_{21}M_1 + G_{22}D_{21}M_1 = 0.$$

Cancel out M_1 and we have

$$D_{21} = -G_{21}/G_{22}. \qquad (10.51)$$

Similarly, if D_{12} can cancel the effect of G_{12} at point b, we have

$$G_{12}M_2 + G_{11}D_{12}M_2 = 0$$

or

$$D_{12} = -G_{12}/G_{11}. \qquad (10.52)$$

We may note that Eqs. (10.51) and (10.52) are the same as d_{21} and d_{12} in Eqs. (10.47). The strategy of implementing D_{12} and D_{21} is similar to the discussion of feedforward controllers in Section 10.2, and typically we remove the time-delay terms and apply a lead–lag compensator as in Eq. (10.8). If the time constant of the first-order lead is similar to the time constant of the first-order lag, then we just need a steady-state compensator.

Example 10.6: A classic example of a MIMO problem is a distillation column.[13] From open-loop step tests, the following transfer functions are obtained:

$$\begin{bmatrix} X_D(s) \\ X_B(s) \end{bmatrix} = \begin{bmatrix} \dfrac{0.07e^{-3s}}{12s+1} & \dfrac{-0.05e^{-s}}{15s+1} \\ \dfrac{0.1e^{-4s}}{11s+1} & \dfrac{-0.15e^{-2s}}{10s+1} \end{bmatrix} \begin{bmatrix} L(s) \\ V(s) \end{bmatrix}.$$

In this model, x_D and x_B are the distillate and the bottom compositions, respectively; L is the reflux flow rate, and V is the boil-up rate. Design a 2×2 MIMO system with PI controllers and decouplers as in Fig. 10.14.

Before we design the MIMO system, we need to check the paring of variables. The steady-state gain matrix is

$$\mathbf{K} = \begin{bmatrix} 0.07 & -0.05 \\ 0.1 & -0.15 \end{bmatrix}.$$

With Eqs. (10.37) and (E10.6), the RGA is

$$\mathbf{\Lambda} = \begin{bmatrix} 1.91 & -0.91 \\ -0.91 & 1.91 \end{bmatrix}.$$

The relative gain parameter $\lambda_{x_D - L}$ is 1.91. It is not 1 but at least it is not negative. Physically, it also makes sense to manipulate the distillate composition with the more neighboring reflux

---

[13] Pardon me if you have not taken a course in separation processes yet, but you do not need to know what a distillation column is to read the example. In a simple-minded way, we can think of making moonshine. We have to boil a dilute alcohol solution at the bottom and we need a condenser at the top to catch the distillate. This is how we have the V and L manipulated variables. Furthermore, the transfer functions are what we obtain from doing an experiment, not from any theoretical derivation.

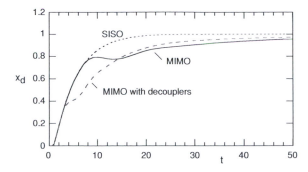

Figure E10.6.

flow. Therefore we pair $x_D - L$ and $x_B - V$. Next, with Eqs. (10.51) and (10.52), the two decoupling functions are

$$D_{12} = K_{d,12} \frac{12s + 1}{15s + 1}, \qquad D_{21} = K_{d,21} \frac{10s + 1}{11s + 1} \approx K_{d,21}.$$

To do the tuning, we can use the initial values $K_{d,12} = -0.05/0.07 \approx -0.7$ and $K_{d,21} = -0.1/0.15 \approx -0.7$.

We will have to skip the details for the remainder of the exercise. You may try to generate a plot similar to that of Fig. E10.6 in the Review Problems.

This is roughly how it is done. All the simulations are performed with Simulink. First, we use G_{11} and G_{22} as the first-order with dead-time functions and apply them to the ITAE tuning relations in Table 6.1. With that, we have the PI controller settings of two SISO systems. The single-loop response to a unit-step change in the set point of x_D is labeled SISO in Fig. E10.6. We retain the ITAE controller settings and apply them to a Simulink block diagram constructed as in Fig. 10.12. The result is labeled MIMO in the figure. Finally, we use Fig. 10.14 and the two decouplers, and the simulation result with the initial setting is labeled "MIMO with decouplers."

In this illustration, we do not have to detune the SISO controller settings. The interaction does not appear to be severely detrimental mainly because we have used the conservative ITAE settings. It would not be the case if we had tried Cohen–Coon relations. The decouplers also do not appear to be particularly effective. They reduce the oscillation, but also slow down the system response. The main reason is that the lead–lag compensators do not factor in the dead times in all the transfer functions.

Review Problems

(1) Derive Eqs. (10.3) and (10.3a) with measurement transfer functions G_{m_1} and G_{m_2} in the primary and the secondary loops. Confirm footnote 1 that this equation can be reduced to that of a single-loop system.

(2) Do the root-locus plots in Example 10.1(d). Confirm the stability analysis in Example 10.1(e).

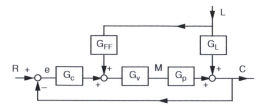

Figure R10.4.

(3) Draw the block diagram of the system in Example 10.2. Label the diagram with the proper variables.

(4) Attempt a numerical simulation of a feedforward–feedback system in Fig. R10.4. Consider the simplified block diagram with

$$G_v = \frac{0.5}{s+1}, \quad G_p = \frac{0.8}{2s+1}, \quad G_L = \frac{-0.4}{2s+1}.$$

(a) The load function has a negative gain. What does it mean?

(b) Omit for the moment the feedback loop and controller G_c, and consider only G_{FF} as defined in Eq. (10.8). Use MATLAB functions (or Simulink) to simulate the response in C when we impose a unit step change to L. Experiment with different values of the gain and the time constants in the lead–lag element.

(c) Consider a PI controller for the feedback loop with an integral time of 1.5 s; find the proportional gain such that the system has an underdamped behavior equivalent to a damping ratio of 0.7.

(d) With the feedback loop and PI controller in part (c), use MATLAB to simulate the response of C to a unit-step change in L. Repeat the different values of the feedforward controller as in part (b).

(5) Consider the simpler problem in Fig. R10.5 based on Fig. 10.12. If we implement only one feedback loop and one controller, how is the transfer function C_1/M_1 affected by the interaction?

(6) Derive the transfer functions C_2/R_1 and C_2/R_2 from Eqs. (10.20) and (10.21).

(7) Fill in the details and derive the RGA [Eq. (E10.6)] in Example 10.4.

(8) Derive Eq. (10.37).

(9) Show that we also can obtain Eq. (E10.6) by applying Eq. (10.37) to the blending problem.

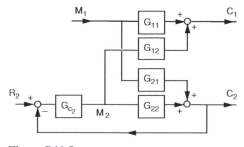

Figure R10.5.

(f) Repeat Subsection 10.7.1 by replacing the second manipulated variable in Eq. (10.39) with

$$\mu_2 = m_1.$$

Find the gain matrix and show that the relative gain parameter is 1. Show how this partially decoupling scheme can be implemented as analogous to Fig. 10.13.

(10) Derive Eqs. (10.44) and (10.45).

(11) Try to do the Simulink simulations in Example 10.6. If you need help, the Simulink file is on the *Web Support*.

Hints:

(2) The MATLAB statements can be

```
Part (d)
Gp=tf(0.8,[2 1]);
Gv=tf(0.9,[0.1 1]);   % With cascade control
taui=0.05;            % Just one example
Gi=tf([taui 1],[taui 0])
rlocus(Gi*Gv*Gp)

Part (e)
Gvo=tf(0.5,[1 1]);
rlocus(Gi*Gvo*Gp)
```

(4) (a) If L is the inlet process stream flow rate, the furnace temperature will decrease if the process flow increases.

(b) Use Eq. (10.9) and the comments that follow to select the parameters for the feedforward controller. Compare with the case when we do not have a feedforward controller by setting $K_{FF} = 0$. You should observe that the major compensation to the load change is contributed by the steady-state compensator.

(c) The proportional gain is ~ 1.4. The feedforward controller does not affect the system stability and we can design the controller G_c with only G_v, G_p, and the feedback loop. We have to use, for example, the root-locus method in Chap. 6 to do this part. Root locus can also help us to determine if $\tau_I = 1.5$ s is a good choice.

(d) You should find that the feedback loop takes over much of the burden in load changes. The system response is rather robust even with relatively large errors in the steady-state compensator.

(5)

$$C_2 = G_{22}G_{c_2}(R_2 - C_2) + G_{21}M_1,$$
$$C_1 = G_{11}M_1 + G_{12}G_{c_2}(R_2 - C_2).$$

Setting $R_2 = 0$,

$$C_2 = \frac{G_{21}}{1 + G_{c_2}G_{22}}M_1.$$

223

Substituting C_2 into the C_1 equation, we can find, after two algebraic steps,

$$C_1 = \left(G_{11} - \frac{G_{12}G_{21}G_{c_2}}{1 + G_{c_2}G_{22}}\right) M_1.$$

The second term in the parentheses is due to interaction.

(6) We apply Cramer's rule to find C_2 just as we had with C_1. The solution has the same characteristic polynomial in Eq. (10.22). The transfer functions are, with $R_1 = 0$,

$$\frac{C_2}{R_2} = \frac{G_{22}G_{c_2} + G_{c_1}G_{c_2}(G_{11}G_{22} - G_{12}G_{21})}{p(s)},$$

and with $R_2 = 0$,

$$\frac{C_2}{R_1} = \frac{G_{21}G_{c_1}}{p(s)}.$$

(7) We still use Eq. (10.28) as in Example 10.4. To find λ_{x,m_2},

$$\left.\frac{\partial x}{\partial m_2}\right|_{m_1} = \frac{-m_1}{(m_1 + m_2)^2}, \quad \left.\frac{\partial x}{\partial m_2}\right|_F = \frac{\partial}{\partial m_2}\left(\frac{F - m_2}{F}\right) = -\frac{1}{F},$$

$$\lambda_{x,m_2} = \frac{m_1}{m_1 + m_2} = x.$$

To find λ_{F,m_1},

$$\left.\frac{\partial F}{\partial m_1}\right|_{m_2} = 1, \text{ using Eq. (10.25)}, \quad \left.\frac{\partial F}{\partial m_1}\right|_x = \frac{1}{x}, \text{ using } F = m_1/x,$$

$$\lambda_{F,m_1} = x.$$

To find λ_{F,m_2}:

$$\left.\frac{\partial F}{\partial m_2}\right|_{m_1} = 1, \quad \left.\frac{\partial F}{\partial m_2}\right|_x = \frac{1}{1 - x}, \text{ using } xF = F - m_2, \quad F = m_2/(1 - x),$$

$$\lambda_{F,m_2} = 1 - x.$$

(8) We may just as well use Eq. (10.32) in its time-domain form:

$$\begin{bmatrix} x \\ F \end{bmatrix} = \begin{bmatrix} K_{11} & K_{12} \\ K_{21} & K_{22} \end{bmatrix} \begin{bmatrix} m_1 \\ m_2 \end{bmatrix},$$

where now x, F, m_1, and m_2 are deviation variables. From the first row, it is immediately obvious that

$$\left.\frac{\partial x}{\partial m_1}\right|_{m_2} = K_{11}.$$

We next substitute for m_2 by using the second row to get

$$x = K_{11}m_1 + K_{12}\frac{(F - K_{21}m_1)}{K_{22}}.$$

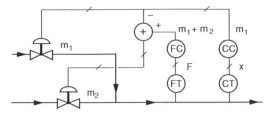

Figure R10.10.

Now we can find

$$\left.\frac{\partial x}{\partial m_1}\right|_F = K_{11} - \frac{K_{12}K_{21}}{K_{22}}.$$

From here on, getting Eq. (10.37) is a simple substitution step.
To derive Eq. (E10.6) by using K. This is just a matter of substituting the values of the K_{ij} from Eqs. (10.30) and (10.31) into Eq. (10.37). We should find once again that $\lambda_{x,m_1} = 1 - x$, as in Eq. (E10.5), and Eq. (E10.6) follows.

(9) We need to find how μ_1 and μ_2 affect F and x. With $\mu_1 = F$ and $\mu_2 = m_1$, we can rewrite the definition of $x = m_1/F$ as $x = \mu_1/\mu_2$. This is the form that we use to take a first-order Taylor expansion as we have done with the step after Eq. (10.28). The result in matrix form of the Laplace transform of the deviation variables is

$$\begin{bmatrix} F(s) \\ x(s) \end{bmatrix} = \begin{bmatrix} 1 & 0 \\ -\dfrac{\mu_2}{\mu_1^2}\bigg|_{ss} & \dfrac{1}{\mu_1}\bigg|_{ss} \end{bmatrix} \begin{bmatrix} \mu_1(s) \\ \mu_2(s) \end{bmatrix}.$$

By putting F in the first row, it is clear that we have a one-way interaction system. By Eq. (10.37), $\lambda = 1$. With $F = m_1 + m_2$ and m_1 as the output of the controllers, we can implement this scheme as in Fig. R10.10.

(10) We will find d_{11} and d_{21} as an illustration. The first column on the RHS of Eq. (10.43) is rewritten as two equations:

$$G_{11}d_{11} + G_{12}d_{21} = H_1,$$
$$G_{21}d_{11} + G_{22}d_{21} = 0.$$

Solving them simultaneously will lead to d_{11} and d_{21} in Eqs. (10.44) and (10.45). When $d_{11} = 1$ is chosen, Eqs. (10.44) can be rewritten as Eqs. (10.46).

MATLAB Tutorial Sessions

MATLAB is a formidable mathematics analysis package. An introduction to the most basic commands that are needed for our use is provided. No attempts are made to be comprehensive. We do not want a big, intimidating manual. The beauty of MATLAB is that we need to know only a tiny bit to get going and be productive. Once we get started, we can pick up new skills quickly with MATLAB's excellent on-line help features. We can learn only with hands-on work; these notes are written as a "walk-through" tutorial – you are expected to enter the commands into MATLAB as you read along.

Session 1. Important Basic Functions

For each session, the most important functions are put in a table at the beginning of the section for easy reference or review. The first one is on the basic commands and plotting. Try the commands as you read. You do not have to enter any text after the "%" sign. Any text that follows "%" is considered a comment and can be ignored. To save some paper, the results generated by MATLAB are omitted. If you need to see that for help, they are provided on the *Web Support*. There is also where any new MATLAB changes and upgrades are posted. Features in our tutorial sessions are based on MATLAB Version 6.1 and Control System Toolbox 5.1.

Important basic functions

General functions:

`cd`	Change subdirectory
`demo (intro)`	Launch the demo (introduction)
`dir (what)`	List of files in current directory (or only M-files)
`help, helpwin`	Help! Help window
`load`	Load workspace
`lookfor`	Keyword search
`print`	Print graph; can use pull-down menu
`quit`	Quit!
`save`	Save workspace
`who, whos`	List of variables in workspace

226

Calculation functions:

`conv`	Convolution function to multiply polynomials
`size, length`	Size of an array, length of a vector

Plotting functions:

`axis`	Override axis default of plot
`grid`	Add grid to plot
`hold`	Hold a figure to add more plots (curves)
`legend`	Add legend to plot
`plot`	Make plots
`text (gtext)`	Add text (graphical control) to plot
`title`	Add title to plot
`xlabel, ylabel`	Add axis labels to plot

M1.1. Some Basic MATLAB Commands

The following features are covered in this session:

- using help
- creating vectors, matrices, and polynomials
- simple matrix operations
- multiplying two polynomials with `conv()`

To begin, we can explore MATLAB by using its demonstrations. If you are new to MATLAB, it is highly recommended that you take a look at the introduction:

```
intro          % launch the introduction
demo           % launch the demo program
```

It is important to know that the MATLAB on-line help is excellent, and there are different ways to get that:

```
help           % old-fashioned help inside the Command Window
helpbrowser    % launch the help browser window; also available
               % from the Help pull-down menu and toolbar
```

We should make a habit of using the on-line help. The user interface of the help browser, which also works as a Web browser, is extremely intuitive, and it is highly recommended. When the `help` command is mentioned, that is just a general comment; it does not mean that you have to use the old-style help. To use help in the Command Window, turn the page display mode on first. Here's an example of seeking help on the print command with the old-style help:

```
more on        % turn the page mode on
help print
lookfor print  % general key-word search
which print    % list the path name of print.m
```

The help features and the Command Window interface tend to evolve quickly. For that reason, the *Web Support* is used to provide additional hints and tidbits so that we can quickly post the latest MATLAB changes. For now, a few more basic commands are introduced:

```
who          % list the variables that are currently defined
whos         % whos is a more detailed version of who
dir          % list the files in the current subdirectory
what         % list only the M-files
cd           % change the subdirectory
pwd          % list the present working directory
```

For fun, we can try

```
why
fix(clock)
```

MATLAB is most at home in dealing with arrays, which we refer to as matrices and vectors. They are all created by enclosing a set of numbers in brackets, [ ]. First, we define a row **vector** by entering, in the MATLAB Command Window,

```
x = [1 2 3 4 5 6 7 8 9 10]
```

If we add a semicolon at the end of a command, as in

```
x = [1 2 3 4 5 6 7 8 9 10];
```

we can suppress the display of the result. We can check what we have later by entering the name of the variable. To generate a column vector, we insert semicolons between numbers (a more specific example is given with a matrix below). The easier route is to take the transpose of x:

```
x = x'
```

Keep in mind that in MATLAB variables are case sensitive. Small letter x and capital X are two different variables.

We can also generate the row vector with the colon operator:

```
x = 1:10        % same as 1:1:10
y = 0:0.1:2     % just another example
```

The colon operator is very useful when we make longer vectors for plotting or calculations. With this syntax, the increment is squeezed between the beginning and the ending values of the vector and they are separated by colons. If the increment value is missing, the default increment is 1. Remember to add a semicolon at the end of each statement to suppress the display of a long string of numbers. This is skipped in the illustration just so you may see what is generated. When we do calculations based on vectors, MATLAB will vectorize the computation, which is much faster than if we write a loop construct as in the FOR loop in C or the DO loop in FORTRAN.

To create a **matrix**, we use a semicolon to separate the rows:

```
a = [1 2 3 ; 4 5 6 ; 7 8 9]
```

In place of the semicolons, we can also simply hit the return key as we generate the matrix.

There are circumstances in which we need the size of an array or the length of a vector. They can be found easily:

```
size(y)          % find the size of an array
length(y)        % find the length of a vector
```

In MATLAB, **polynomials** are stored exactly the same as vectors. Functions in MATLAB will interpret them properly if we follow the convention that a vector stores the coefficients of a polynomial in descending order – it begins with the highest-order term and *always* ends with a constant, even if it is zero. Some examples:

```
p1 =[1 -5 4]     % defines p1(s) = s^2  - 5*s + 4
p2 =[1 0 4]      % defines p2(s) = s^2 + 4
p3 =[1 -5 0]     % defines p3(s) = s^2 - 5*s
```

We can multiply two polynomials together easily with the convolution function `conv()`. For example, to expand $(s^2 - 5s + 4)(s^2 + 4)$, we can use

```
conv(p1,p2)      % this multiplies p1 by p2
```

or

```
conv([1 -5 4], [1 0 4])
```

MATLAB supports every imaginable way that we can manipulate vectors and matrices. We need to know only a few of them, and we will pick up these necessary ones along the way. For now, we'll do a couple of simple operations. With the vector x and matrix a that we've defined above, we can perform simple operations such as

```
y1 = 2*x         % multiplies x by a constant
y2 = sqrt(x)     % takes the square root of each element in x
b  = sqrt(a)     % takes the square root of each element in a
y3 = y1 + y2     % adds the two vectors
c  = a*b         % multiplies the two matrices
```

Note that all functions in MATLAB, such as `sqrt()`, are smart enough that they accept scalars, vectors, and, where appropriate, matrices.[1]

When we operate on an *element-by-element* basis, we need to add a period before the operator. Examples based on the two square matrices a and b:

```
d  = a.^3        % takes the cube of each element
a3 = a^3         % versus the cube of the matrix
e  = a.*b        % multiplies each element a(i,j)*b(i,j)
f  = a*b         % versus matrix multiplication a*b
```

Of course, we can solve the matrix equation $\mathbf{Ax} = \mathbf{b}$ easily. For example, we can try

```
A = [ 4  -2  -10;  2  10  -12;  -4  -6  16];
b = [-10;  32;  -16];
x = A\b          % Bingo!
```

---

[1] In computer science, this is referred to as polymorphism. The fact that mathematical operators can work on different data types is called overloading.

Let's check the solution by inverting the matrix[2] A:

```
C = inv(A);
x = C*b
```

We can find the eigenvalues and eigenvectors of A easily:

```
[X,D] = eig(A)
```

Finally, we do a simple polynomial fit illustration. Let's say we have a set of (x, y) data:

```
x = [ 0  1  2  4  6  10];
y = [ 1  7  23  109  307  1231];
```

To make a third-order polynomial fit of $y = y(x)$, all we need to enter is

```
c = polyfit(x,y,3)    % should obtain c = [1 2 3 1]
```

The returned vector c contains the coefficients of the polynomial. In this example, the result should be $y = x^3 + 2x^2 + 3x + 1$. We can check and see how good the fit is. In the following statements, we generate a vector xfit so that we can draw a curve. Then we calculate the corresponding yfit values and plot the data with a symbol and the fit as a line:

```
xfit=1:0.5:10;
yfit=xfit.^3 + 2*xfit.^2 + 3*xfit +1;
plot(x,y,'o', xfit,yfit)              % explanation on plotting
title('3rd order polynomial fit')     % is in the next subsection
legend('data','3rd order fit')
```

Speaking of plotting, this is what we get into next.

M1.2. Some Simple Plotting

The following features are covered in this session:

- Using the plot() function
- Adding titles, labels, and legends

Let's create a few vectors first:

```
x = 0:0.5:10;
y1= 2*x;
y2= sqrt(x);
```

Now we plot y1 versus x and y2 versus x together:

```
plot(x,y1,  x,y2)
```

___

[2] If you have taken a course on numerical methods, you would be pleased to know that MATLAB can do *LU* decomposition:

```
[L,U] = lu(A);
```

We have a limited selection of line patterns or symbols. For example, we can try[3]

```
plot(x,y1,'-.')
hold                % or use "hold on"
plot(x,y2,'--')
hold                % or use "hold off"
```

We can find the list of pattern selections with on-line help. The command `hold` allows us to add more plots to the same figure, and `hold` works as a toggle. That is why we do not have to state "on" and "off" explicitly.

We can add a title and axis labels too:

```
title('A boring plot')
xlabel('The x-axis label'), ylabel('The y-axis label')
```

We can issue multiple commands on the same line separated by commas. What makes MATLAB easy to learn is that we can add goodies one after another. We do not have to worry about complex command syntax. We can also do logarithmic plots. Try entering `help semilogx`, `semilogy`, or `loglog`. We'll skip them because they are not crucial for our immediate needs.

We can add a grid and a legend with

```
grid
legend('y1','y2')
```

A box with the figure legend will appear in the Graph Window. Use the mouse to drag the box to where you want it to be. We can also add text to annotate the plot with

```
text(1,9,'My two curves')     % starting at the point (1,9)
```

The text entry can be interactive with the use of

```
gtext('My two curves')
```

Now click on the Graph Window, and a crosshair will appear. Move it to where you want the legend to begin and click. Presto! Repeat for additional annotations.

In rare cases, we may not like the default axis scaling. To override what MATLAB does, we can define our own minimum and maximum of each axis with

```
axis([0 15 0 30])     % the syntax is [xmin xmax ymin ymax]
```

We need the brackets inside because the argument to the axis function is an array.

Plotting for Fun

We do not need to do three-dimensional (3-D) plots, but then it's too much fun not to do at least a couple of examples. However, we will need to use a few functions that we do not need otherwise, so do not worry about the details of these functions that we will not use

---

[3] To do multiple plots, we can also use

```
plot(x,y1,'-.', x,y2,'--')
```

again. We will get a pretty 3-D picture:

```
[x,y]=meshgrid(-10:0.5:10,   -10:0.5:10);

% meshgrid transforms the specified domain
% where -10 < x < 10,  and  -10 < y < 10
% into a grid of (x,y) values for evaluating z

r=sqrt(x.^2 + y.^2) + eps;   % We add the machine epsilon eps
z=sin(r)./r;                 % so 1/r won't blow up
mesh(z)
title('The Sinc Sombrero')
```

So you say wow! But MATLAB can do much more and fancier than that. We try one more example with Bessel functions, which you can come across in heat and mass transfer problems with cylindrical geometry:

```
% Here we do a 3-D mesh plot of Jo(sqrt(x^2+y^2))
% The x and y grids remain the same as in the previous plot

r=sqrt(x.^2+y.^2);
z=bessel(0,r);
mesh(z)
```

M1.3. Making M-files and Saving the Workspace

The following features are covered in this session:

- Executing repeated commands in a script, the so-called M-file
- Saving a session

For tasks that we have to repeat again and again, it makes sense to save them in some kind of a script and execute them. In MATLAB, these scripts are called M-files. The name came from the use of macros in the old days of computing. We can use M-files to write unstructured scripts or user-defined functions. MATLAB now refers to both as programs. You may want to keep in mind that a scripting interpretive language is not the same as a compiled language like C.

For our needs, a simple script suffices in most circumstances. To use an M-file[4]

(1) save all the repetitious MATLAB statements in a text file with the " .m" extension,
(2) execute the statements in that file by entering the file name *without* the " .m" extension.

---

[4] There is another easy way to "cheat." On UNIX/Linux workstations, open up a new text editor and enter your frequently used statements there. In Windows, you can use the really nice MATLAB Editor. You can copy and paste multiple commands back and forth between the text editor window and the MATLAB window easily. If you want to save the commands, you certainly can add comments and annotations. You can consider this text file as a "free-format notebook" without having to launch the Microsoft Word Notebook for MATLAB.

Here is one simple example. We need to plot x versus y repeatedly and want to automate the task of generating the plots. The necessary statements are

```
%  _____ M-file script: plotxy.m _____
% A very simple script to plot x vs y and add the labels
% ...the kind of things we don't want to repeat typing
%   again and again...

plot(x,y)
grid
xlabel('Time [min]')
ylabel('Step Response')
title('PID Controller Simulation')

% End of plotxy.m. An "end" statement is not needed.
```

Save these statements in a file named, say, `plotxy.m`. Anything after the "%" sign is regarded as a comment, which you do not have to enter if you just want to repeat this exercise. After we have defined or updated the values of x and y in the Command Window, all we need is to enter `"plotxy"` at the prompt and MATLAB will do the rest. The key is to note that the M-file has *no* "read" or "input" for x and y. All statements in an M-file are simply executed in the Command Window.

If you have an M-file, MATLAB may not find it unless it is located within its search path. Let's see where MATLAB looks first. On UNIX/Linux machines, MATLAB by default looks in the subdirectory from which it is launched. A good habit is to keep all your work in one subdirectory and change to that specific subdirectory before you launch MATLAB. On Windows machines, MATLAB looks for your files in the Work folder buried deep inside the Program Files folder. A good chance is that you want to put your files in more convenient locations. To coax MATLAB to find them, you need to change the directory or the search path. So the next question is how to do that, and the answer applies to both UNIX/Linux and Windows machines. The formal way is to learn to use the `"cd"` and `"path"` commands. The easy way is to use point-and-click features that can be found under pull-down menus, on toolbars, or in subwindows. Because these graphical interface features tend to change with each MATLAB upgrade, please refer to the *Web Support*, from which you can find updates of new changes and additional help.

If we want to take a coffee break and save all the current variables that we are working with, enter

```
save
```

before we `quit` MATLAB. When we launch MATLAB again, we type

```
load
```

and everything will be restored. Do not save the workspace if you are going to be away any longer because the old workspace is not very useful if you have all these variables floating around and you forget what they mean.

As a final comment, we can use `load` and `save` to import and export arrays of data. Because we do not really need this feature in what we do here, this explanation is deferred to the *Web Support*.

Session 2. Partial-Fraction and Transfer Functions

This tutorial is to complement our development in Chap. 2. You may want to go over the tutorial quickly before you read the text and come back later a second time for the details.

Partial-fraction and transfer functions

`poly`	Construct a polynomial from its roots
`residue`	Partial-fraction expansion
`roots`	Find the roots to a polynomial
`tf2zp`	Transfer function to zero-pole form conversion
`zp2tf`	Zero-pole form to transfer function conversion
Object-oriented functions:	
`tf`	Create a transfer function object
`get`	List the object properties
`pole`	Find the poles of a transfer function
`zpk`	Create a transfer function in pole-zero-gain form

M2.1. Partial Fractions

The following features are covered in this session:

- Finding the roots of a polynomial with `roots()`
- Generating a polynomial from its roots with `poly()`
- Doing partial fractions with `residue()`

Of secondary importance:

- Transfer function to zero-pole form, `tf2zp()`
- Zero-pole form to transfer function, `zp2tf()`

Let's first define a polynomial:

```
p = [1  5  4]      % makes p(s) = s^2 + 5*s + 4
```

We can find the roots of $p(s) = 0$ with the function `roots()`:

```
poles = roots(p)
```

MATLAB should return -4 and -1. That means the polynomial can be factored as $p(s) = (s + 4)(s + 1)$.[5]

---

[5] MATLAB has the function `fzero()` to find a root of a given function.

We can go backwards. Given the roots (or pole positions), we can get the polynomial with

```
p2 = poly(poles)
```

MATLAB returns the results in a column vector. Most functions in MATLAB take either row or column vectors, and we usually do not have to worry about transposing them.

We can do partial fractions with the `residue()` function. Say we have a transfer function

$$G(s) = \frac{q(s)}{p(s)} = \frac{1}{s^2 + 5s + 4},$$

where $q(s) = 1$ and $p(s)$ remains [1 5 4], as previously defined. We can enter

```
q = 1;
residue(q,p)
```

MATLAB returns the numbers -0.3333 and 0.3333. That is because the function can be factored as

$$\frac{1}{s^2 + 5s + 4} = \frac{-1/3}{s + 4} + \frac{1/3}{s + 1}.$$

How can we be sure that it is the -0.3333 coefficient that goes with the root at -4? We can use the syntax

```
[a,b,k]=residue(q,p)
```

MATLAB will return the coefficients in a, the corresponding poles in b, and whatever is left over in k, which should be nothing in this case. Note that [] denotes an empty matrix or vector.

Let's try another transfer function with poles at 0, -1, -2, and -3:

$$G(s) = \frac{1}{s(s + 1)(s + 2)(s + 3)}.$$

To find the partial fractions, this is what we can do[6]:

```
poles=[0  -1  -2  -3];
p=poly(poles);
q=1;
[a,b,k]=residue(q,p)
```

One more example. Find the partial fractions of the nasty-looking function

$$G(s) = \frac{s^2 + 4s + 3}{s^4 - 7s^3 + 11s^2 + 7s - 12}.$$

---

[6] If we need to write complex-conjugate roots, make sure there are no spaces within a complex number. For example, enter [-3+4*j   -3-4*j]. Either i or j can be used to denote $\sqrt{-1}$.

```
q=[1  4  3];
zeros=roots (q)          % should return -3, -1
p=[1  -7  11  7  -12];
poles=roots (p)          % should return 4, 3, 1, -1
[a,b,k]=residue (q,p)
```

See that MATLAB returns the expansion:

$$\frac{s^2 + 4s + 3}{s^4 - 7s^3 + 11s^2 + 7s - 12} = \frac{2.33}{s-4} - \frac{3}{s-3} + \frac{0.67}{s-1}.$$

Note that the coefficient associated with the pole at -1 is zero. That is because it is canceled by the zero at -1. In other words, the $(s + 1)$ terms cancel out. It is nice to know that the program can do this all by itself. We do not need to know the roots to use residue(), but it is a good habit to get a better idea of what we are working with.

A transfer function can be written in terms of its poles and zeros. For example,

$$F(s) = \frac{6s^2 - 12}{(s^3 + s^2 - 4s - 4)} = \frac{6(s - \sqrt{2})(s + \sqrt{2})}{(s + 1)(s + 2)(s - 2)}.$$

The RHS is called the pole-zero form (or zero-pole form). MATLAB provides two functions, tf2zp() and zp2tf(), to do the conversion. For instance,

```
q=[6  0  -12];
p=[1  1  -4  -4];
[zeros, poles, k]=tf2zp(q,p)
```

Of course, we can go backward with

```
[q,p]=zp2tf(zeros,poles,k)
```

Note: The factor k is 6 here, and in the MATLAB manual it is referred to as the "gain." This factor is really the ratio of the leading coefficients of the two polynomials $q(s)$ and $p(s)$. Make sure you understand that the k here is *not* the steady-state gain, which is the ratio of the *last* constant coefficients. (In this example, the steady-state gain is $-12/-4 = 3$.) MATLAB actually has a function called dcgain to do this.

One more simple example:

```
zero= -2;                % generate a transfer function
poles=[-4  -3  -1];      % with given poles and zeros
k=1;
[q,p]=zp2tf(zero,poles,k)
```

Double check that we can recover the poles and zeros with

```
[zero,poles,k]=tf2zp(q,p)
```

We can also check with

```
roots(q)
roots(p)
```

Try zp2tf or tf2zp on your car's license plate!

M2.2. **Object-Oriented Transfer Functions**

The following features are covered in this session:

- Defining a transfer function object with `tf()` or `zpk()`
- Determining the poles with `pole()`
- Using overloaded operators

MATLAB is object oriented. Linear-time-invariant (LTI) models are handled as objects. Functions use these objects as arguments. In classical control, LTI objects include transfer functions in polynomial form or in pole-zero form. The LTI-oriented syntax allows us to better organize our problem solving; we no longer have to work with individual polynomials that we can identify only as numerators and denominators.

We will use this syntax extensively starting in Session 3. Here, we see how the object-oriented syntax can make the functions `tf2zp()` and `zp2tf()` redundant and obsolete.

To define a transfer function object, we use `tf()`, which takes the numerator and denominator polynomials as arguments. For example, we define $G(s) = [s/(s^2 - 5s + 4)]$ with

```
G1 = tf([1 0], [1 -5 4])
```

We define $G(s) = [(6s^2 - 12)/(s^3 + s^2 - 4s - 4)]$ with

```
G2 = tf([6 0 -12], [1 1 -4 -4])
```

We can also use the zero-pole-gain function `zpk()` which takes as arguments the zeros, poles, and gain factor of a transfer function. Recall the comments after `zp2tf()`. This gain factor is not the steady-state (or dc) gain.

For example, we define $G(s) = \{4/[s(s + 1)(s + 2)(s + 3)]\}$ with

```
G3 = zpk([],[0 -1 -2 -3], 4)    % the [] means there is no zero
```

The `tf()` and `zpk()` functions also serve to perform model conversion from one form to another. We can find the polynomial form of G3 with

```
tf(G3)
```

and the pole-zero form of G2 with

```
zpk(G2)
```

The function `pole()` finds the poles of a transfer function. For example, try

```
pole(G1)
pole(G2)
```

You can check that the results are identical to the use of `roots()` on the denominator of a transfer function.

We may not need to use them, but it is good to know that there are functions that help us extract the polynomials or poles and zeros back from an object. For example,

```
[q,p]=tfdata(G1,'v')          % option 'v' for row vectors
[z,p,k]=zpkdata(G3,'v')
```

The addition and multiplication operators are overloaded, and we can use them to manipulate or synthesize transfer functions. This capability will come in handy when we analyze control systems. For now, let's consider one simple example. Say we are given

$$G_1 = \frac{1}{s+1}, \qquad G_2 = \frac{2}{s+2}.$$

We can find $G_1 + G_2$ and $G_1 G_2$ easily with

```
G1=tf(1,[1  1]);
G2=tf(2,[1  2]);
G1+G2                % or we can use zpk(G1+G2)
G1*G2                % or we can use zpk(G1*G2)
```

This example is simple enough to see that the answers returned by MATLAB are correct.

With object-oriented programming, an object can hold many properties. We find the associated properties with

```
get(G1)
```

Among the MATLAB result entries, we may find the properties `InputName`, `OutputName`, and `Notes`. We can set them with [7]

```
G1.InputName = 'Flow Rate';
G1.OutputName = 'Level';
G1.Notes = 'My first MATLAB function';
```

You will see the difference if you enter, from now on,

```
G1
get (G1)
```

MATLAB can use symbolic algebra to do the Laplace transform. Because this skill is not crucial to solving control problems, we skip it here. You can find a brief tutorial on the *Web Support*, and you are encouraged to work through it if you want to know what symbolic algebra means.

Session 3. Time-Response Simulation

This tutorial is to complement our development in Chap. 3. You may want to go over the tutorial quickly before you read the text and come back later a second time for the details.

---

[7] We are using the typical structure syntax, but MATLAB also supports the `set()` function to perform the same task.

Time-response simulation functions

---

damp	Find damping factor and natural frequency
impulse	Impulse response
lsim	Response to arbitrary inputs
step	Unit-step response
pade	Time-delay Padé approximation
ltiview	Launch the graphics viewer for LTI objects

M3.1. Step- and Impulse-Response Simulations

The following features are covered in this session:

- Using step() and impulse()
- Time response to any given input, lsim()
- Dead-time approximation, pade()

Instead of spacing out in the Laplace domain, we can (as we are taught) guess how the process behaves from the pole positions of the transfer function. However, wouldn't it be nice if we could actually trace the time profile without having to do the reverse Laplace transform ourselves? Especially the response with respect to step and impulse inputs? Plots of time-domain dynamic calculations are extremely instructive and a useful learning tool.[8]

The task of time-domain calculation is easy with MATLAB. Let's say we have

$$\frac{Y(s)}{X(s)} = \frac{1}{s^2 + 0.4s + 1},$$

and we want to plot $y(t)$ for a given input $x(t)$. We can easily do

```
q=1;
p=[1 0.4 1];   % poles at -0.2 ± 0.98j
G=tf(q,p)

step(G)        % plots y(t) for unit step input, X(s)=1/s

impulse(G)     % plots y(t) for impulse input, X(s)=1
```

What a piece of cake! Not only does MATLAB perform the calculation, but it automatically makes the plot with a properly chosen time axis. Nice![9] As a habit, find out more about a function with help as in

```
help step     % better yet, use helpwin or helpbrowser
```

---

[8] If you are interested, see the *Web Support* for using the Runge—Kutta integration of differential equations.

[9] How could we guess what the time axis should be? It is not that difficult if we understand how to identify the dominant pole, the significance behind doing partial fractions, and that the time to reach 99% of the final time response is approximately five time constants.

The functions also handle multiple transfer functions. Let's make a second transfer function in pole-zero form:

$$H(s) = \frac{2}{(s+2)(s^2 + 2s + 2)};$$

```
H=zpk([], [-2 -1+j -1-j], 2)
```

We can compare the unit-step responses of the two transfer functions with

```
step(G,H)
```

We can, of course, choose our own axis, or rather, time vector. Putting both the unit-step and impulse-response plots together may also help us understand their differences:

```
t=0:0.5:40;  % don't forget the semicolon!
ys=step(G,t);
yi=impulse(G,t);
plot(t,ys,t,yi)
```

Note: In the text, the importance of relating pole positions of a transfer function to the actual time-domain response was emphasized. We should get into the habit of finding what the poles are. The time-response plots are teaching tools that reaffirm our confidence in doing analysis in the Laplace domain. Therefore we should find the roots of the denominator. We can also use the damp() function to find the damping ratio and the natural frequency.

```
pole(G) % same result with roots(p)
damp(G) % same result with damp(p)
```

One more example. Consider the transfer function

$$\frac{Y(s)}{X(s)} = G(s) = \frac{2s+1}{(4s+1)(s+1)}.$$

We want to plot $y(t)$ if we have a sinusoidal input $x(t) = \sin(t)$. Here we need the function lsim(), a general simulation function that takes any given input vector:

```
q=[2 1];                  % a zero at -1/2
p=conv([4 1],[1 1]);      % poles at -1/4 and -1
G=tf(q,p)                 % (can use zpk instead)
t=0:0.5:30;
u=sin(t);
y=lsim(G,u,t);            % response to a sine function input
plot(t,y,t,u,'-.'), grid
```

Keep this exercise in mind. This result is very useful in understanding what is called frequency response in Chap. 8. We can repeat the simulation with higher frequencies. We can also add what we are familiar with:

```
hold
ys=step(G,t);
yi=impulse(G,t);
```

```
plot(t,ys,t,yi)
hold off
```

For fun, try one more calculation with the addition of random noise:

```
u=sin(t)+rand(size(t));
y=lsim(G,u,t);
plot(t,y,'r',t,u,'b'), grid    % Color lines red and blue
```

For useful applications, `lsim()` is what we need to simulate a response to, say, a **rectangular pulse**. This is one simple example that uses the same transfer function and time vector that we have just defined:

```
t=0:0.5:30;                % t = [0 .5 1 1.5 2 2.5 3 ... ]
u=zeros(size(t));          % make a vector with zeros
u(3:7)=1;                  % make a rectangular pulse from t=1
                           % to t=3
y=lsim(G,u,t);
yi=impulse(G,t);           % compare the result with impulse
                           % response
plot(t,u,  t,y,  t,yi,'-.');
```

Now we switch gears and look into the dead-time transfer function approximation. To do a Padé approximation, we can use the MATLAB function [10]

```
[q,p]=pade(Td,n)
```

where `Td` is the dead time, `n` is the order of the approximation, and the results are returned in `q(s)/p(s)`. For example, with `Td` = 0.2 and `n` = 1, entering

```
[q,p]=pade(0.2,1)          % first-order approximation
```

will return

```
q = -1 s + 10
p =  1 s + 10
```

We expected $q(s) = -0.1s + 1$ and $p(s) = 0.1s + 1$. Obviously MATLAB normalizes the polynomials with the leading coefficients. On second thought, the Padé approximation is so simple that there is no reason why we cannot do it ourselves as in a textbook. For the first-order approximation, we have

```
Td=0.2;
q = [-Td/2 1];
p = [ Td/2 1];
```

---

[10] When we use `pade()` without the left-hand argument `[q,p]`, the function automatically plots the step and the phase responses and compares them with the exact responses of the time delay. A Padé approximation has unit gain at all frequencies. These points will not make sense until we get to frequency-response analysis in Chap. 8. For now, keep the `[q,p]` on the LHS of the command.

We can write our own simple-minded M-file to do the approximation. You may now try

```
[q,p]=pade(0.2,2)      % second-order approximation
```

and compare the results of this second-order approximation with the textbook formula.

M3.2. LTI Viewer

The following feature is covered in this session:

- Graphics viewer for LTI objects, `ltiview`[11]

We can use the LTI Viewer to do all the plots, not only step and impulse responses, but also more general time-response and frequency-response plots in later chapters. If we know how to execute individual plot statements, it is arguable whether we really need the LTI Viewer. Nonetheless, that would be a personal choice. Here the basic idea and some simple instructions are provided.

To launch the LTI Viewer, enter in the MATLAB Command Window

```
ltiview
```

A blank LTI window will pop up. The first task would be to poke into features supported under the File and Tools pull-down menus and see what we can achieve by point and click. There is also a Help pull-down menu, which activates the Help Window.

The LTI Viewer runs in its own workspace, which is separate from the MATLAB workspace. The Viewer also works with only LTI objects generated by functions such as `tf()` and `zpk()`, and after Chap. 4, state-space objects, `ss()`. So let's generate a couple of objects in the MATLAB Command Window first:

```
G=tf(1,[1 0.4 1])
H=zpk([], [-2 -1+j -1-j], 2)
```

Now, go to the LTI Viewer window and select **Import** under the File pull-down menu. A dialog box will pop out to help import the transfer function objects. By default, a unit-step response will be generated. Click on the axis with the *right mouse button* to retrieve a pop-up menu that will provide options for other plot types, for toggling the object to be plotted, and for other features. With a step-response plot, the Characteristics feature of the pop-up menu can identify the peak time, rise time, and settling time of an underdamped response.

The LTI Viewer was designed to do comparative plots, either comparing different transfer functions or comparing the time-domain and (later in Chap. 8) frequency-response properties of a transfer function. Therefore a more likely (and quicker) case is to enter, for example,

```
ltiview('step',G,H)
```

The transfer functions G and H will be imported automatically when the LTI Viewer is launched, and the unit-step response plots of the two functions will be generated.

[11] The description is based on Version 5.1 of the MATLAB control toolbox. If changes are introduced in newer versions, they will be presented on the *Web Support*.

Another useful case is, for example,

```
ltiview({'step';'bode'},G)
```

In this case, the LTI Viewer will display both the unit-step response plot and the Bode plot for the transfer function G. We will learn about Bode plots in Chap. 8, so don't panic yet. Just keep this possibility in mind until we get there.

Session 4. State-Space Functions

This tutorial is to complement our development in Chap. 4. You may want to go over the tutorial quickly before you read the text and come back later a second time for the details.

State-space functions

canon	Canonical state-space realization
eig	Eigenvalues and eigenvectors
ss2ss	Transformation of state-space systems
ss2tf	Conversion from state-space to transfer function
tf2ss	Conversion from transfer function to state-space
printsys	Slightly prettier looking display of model equations
ltiview	Launch the graphics viewer for LTI objects
ss	Create state-space object

M4.1. Conversion between Transfer Function and State-Space

The following features are covered in this session:

- Using ss2tf() and tf2ss()
- Generating object-oriented models with ss()

We need to revisit Example 4.1 with a numerical calculation. Let's use the values $\zeta = 0.5$ and $\omega_n = 1.5$ Hz to establish the transfer function and find the poles:

```
z=0.5;
wn=1.5;                    % Should find
q=wn*wn;                   % q=2.25
p=[1 2*z*wn wn*wn]         % p=[1  1.5  2.25]
roots(p)                   % -0.75 ± 1.3j
```

From the results in Example 4.1, we expect to find

$$\mathbf{A} = \begin{bmatrix} 0 & 1 \\ -2.25 & -1.5 \end{bmatrix}, \quad \mathbf{B} = \begin{bmatrix} 0 \\ 2.25 \end{bmatrix}, \quad \mathbf{C} = [1 \quad 0], \quad \mathbf{D} = 0.$$

Now let's try our hands with MATLAB by using its transfer function to state-space conversion function:

```
[a,b,c,d]=tf2ss(q,p)
```

MATLAB returns with

$$a = \begin{bmatrix} -1.5 & -2.25 \\ 1 & 0 \end{bmatrix}, \quad b = \begin{bmatrix} 1 \\ 0 \end{bmatrix}, \quad c = [0 \quad 2.25], \quad d = 0,$$

which are not the same as those in Example 4.1. You wonder what's going on? Before you kick the computer, a closer look should reveal that MATLAB probably uses a slightly different convention. Indeed, MATLAB first "split" the transfer function into product form:

$$\frac{Y}{U} = \frac{X_2}{U}\frac{Y}{X_2} = \frac{1}{\left(s^2 + 2\zeta\omega_n s + \omega_n^2\right)}\omega_n^2 = \frac{1}{(s^2 + 1.5s + 2.25)}2.25.$$

From $X_2/U = 1/(s^2 + 2\zeta\omega_n s + \omega_n^2)$ and with the state variables defined as

$$x_1 = \frac{dx_2}{dt}, \qquad x_2 = x_2 \text{ (i.e., same)},$$

we should obtain the matrices a and b that MATLAB returns. From $Y/X_2 = \omega_n^2$, it should be immediately obvious how MATLAB obtains the array c.

In addition, we should be aware that the indexing of state variables in MATLAB is in *reverse order* of textbook examples. Despite these differences, the inherent properties of the model remain identical. The most important of all is to check the eigenvalues:

```
eig(a)     % should be identical to the poles
```

A conversion from state-space back to a transfer function should recover the transfer function:

```
[q2,p2]=ss2tf(a,b,c,d,1)     % same as q/p as defined earlier
```

The last argument in `ss2tf()` denotes the ith input, which must be 1 for our SISO model. To make sure we cover all bases, we can set up our own state-space model as in Example 4.1,

```
a=[0 1;  -2.25 -1.5]; b=[0; 2.25]; c=[1 0]; d=0;
```

and check the results with

```
eig(a)                       % still the same!
[qs,ps]=ss2tf(a,b,c,d,1)
```

The important message is that there is no unique state-space representation, but all model matrices should have the same eigenvalues. In addition, the number of state variables is the same as the order of the process or system.

The fact that the algorithm used by MATLAB does not return a normalized output matrix **C** can create problems when we do feedback calculations in Chap. 9. The easy solution is to rescale the model equations. The output equation can be written as

$$y = [\alpha \quad 0]x = [1 \quad 0]\bar{x},$$

where $\bar{x} = \alpha x$. Substitution for **x** by $\bar{x}$ in $dx/dt = \mathbf{Ax} + \mathbf{B}u$ will lead to

$$\frac{d\bar{x}}{dt} = \mathbf{A}\bar{x} + \alpha\mathbf{B}u = \mathbf{A}\bar{x} + \bar{\mathbf{B}}u,$$

where $\bar{\mathbf{B}} = \alpha \mathbf{B}$. In other words, we just need to change $\mathbf{C}$ to the normalized vector and multiply $\mathbf{B}$ by the scaling factor. We can see that this is correct from the numerical results of Example 4.1. (Again, keep in mind that the MATLAB indexing is in reverse order of textbook examples.) We will use this idea in Chap. 9.

We now repeat the same exercise to show how we can create object-oriented state-space LTI models. In later chapters, all control toolbox functions take these objects as arguments. We first repeat the statements above to regenerate the state matrices a, b, c, and d. Then we use ss() to generate the equivalent LTI object.

```
q=2.25;
p=[1  1.5  2.25];
[a,b,c,d]=tf2ss(q,p);
sys_obj=ss(a,b,c,d)
```

We should see that the LTI object is identical to the state-space model. We can retrieve and operate on individual properties of an object. For example, to find the eigenvalues of the matrix a inside sys_obj, we use

```
eig(sys_obj.a)     % find eigenvalue of state matrix a
```

We can obtain the transfer function, as analogous to using ss2tf(), with

```
tf(sys_obj)
```

Now you may wonder if we can generate the state-space model directly from a transfer function. The answer is, of course, yes. We can use

```
sys2=ss(tf(q,p))
eig(sys2.a)            % should be identical to the poles
```

MATLAB will return with matrices that look different from those given previously:

$$a = \begin{bmatrix} -1.5 & -1.125 \\ 2 & 0 \end{bmatrix}, \quad b = \begin{bmatrix} 1 \\ 0 \end{bmatrix}, \quad c = [\,0 \quad 1.125\,], \quad d = 0.$$

With what we know now, we bet ss() uses a different scaling in its algorithm. This time, MATLAB factors the transfer function into this product form:

$$\frac{Y}{U} = \frac{X_2}{U}\frac{Y}{X_2} = \frac{2}{(s^2 + 1.5s + 2.25)}1.125.$$

From $X_2/U = 2/(s^2 + 1.5s + 2.25)$ and with the state variables defined as

$$x_1 = \frac{1}{2}\frac{dx_2}{dt}\left(\text{i.e., } \frac{dx_2}{dt} = 2x_1\right), \qquad x_2 = x_2,$$

we should obtain the new state matrices. Again, the key result is that the state matrix a has the same eigenvalue.

This exercise underscores one more time that there is no unique way to define state variables. Because our objective here is to understand the association between transfer function and state-space models, the introduction continues with the ss2tf() and the tf2ss() functions.

Two minor tidbits before we move on. First, the `printsys()` function displays the model matrices or polynomials in a slightly more readable format. Sample usage:

```
printsys(a,b,c,d)
printsys(q,p,'s')
```

Second, with a second-order transfer function, we can generate the textbook state-space matrices, given a natural frequency wn and damping ratio z:

```
[a,b,c,d]=ord2(wn,z)     % good for only q=1
```

If we examine the values of b and c, the result is restricted to a unity numerator in the transfer function.

M4.2. Time-Response Simulation

To begin with, we can launch the LTI Viewer with

```
ltiview
```

as explained in MATLAB Session 3. The graphics interface is designed well enough so that no further explanation is needed.

The use of `step()` and `impulse()` on state-space models is straightforward as well. Here just a simple example is provided. Let's go back to the numbers that we have chosen for Example 4.1 and define

```
a=[0 1; -2.25 -1.5]; b=[0; 2.25]; c=[1 0]; d=0;
sys=ss(a,b,c,d);
```

The `step()` function also accepts state-space representation, and generating the unit-step response is no more difficult than using a transfer function:

```
step(sys)
```

Now we repeat the calculation in the transfer function form and overlay the plot on top of the last one:

```
G=tf(2.25,[1 1.5 2.25]);
hold
step(G,'x')
hold off
```

Sure enough, the results are identical. We would be in big trouble if it were not! In fact, we should get the identical result with other state-space representations of the model. (You may try this yourself with the other set of a,b,c,d returned by `tf2ss()` when we first went through Example 4.1.)

Many other MATLAB functions, for example, `impulse()`, `lsim()`, etc., take both transfer function and state-space arguments (what can be called polymorphic). There is very little reason to do the conversion back to the transfer function once you can live in state-space with peace.

M4.3. Transformations

The following features are covered in this session:

- Similarity and canonical transforms
- Using functions `canon()` and `ss2ss()`

First a similarity transform is demonstrated. For a nonsingular matrix **A** with distinct eigenvalues, we can find a nonsingular (modal) matrix **P** such that the matrix **A** can be transformed into a diagonal made up of its eigenvalues. This is one useful technique in decoupling a set of differential equations.

Consider the matrix **A** from Example 4.6. We check to see if the rank is indeed 3, and compute the eigenvalues for reference later:

```
A=[0 1 0;  0 -1 -2;  1 0 -10];
rank(A)
eig(A)      % -0.29,  -0.69,  -10.02
```

We now enter

```
[P,L] = eig(A)      % L is a diagonal matrix of eigenvalues
                    % P is the modal matrix whose columns are the
                    % corresponding eigenvectors
a = inv(P)*A*P      % Check that the results are correct
```

Indeed, we should find a to be the diagonal matrix with the eigenvalues.

The second route is to diagonalize the entire system. With Example 4.6, we further define

```
B=[0;  2;  0];
C=[1  0  0];
D=[0];
S=ss(A,B,C,D);      % Generates the system object
SD=canon(S)
```

The `canon()` function by default will return the diagonalized system and, in this case, in the system object SD. For example, we should find SD.a to be identical to the matrix L that we obtained a few steps back.

The third alternative to generate the diagonalized form is to use the state-space to state-space transform function. The transform is based on the modal matrix that we obtained earlier:

```
SD=ss2ss(S,inv(P))
```

To find the observable canonical form of Example 4.6, we use

```
SO=canon (S,'companion')
```

In the returned system SO, we should find SO.a and SO.b to be

$$
\mathbf{A}_{ob} = \begin{bmatrix} 0 & 0 & -2 \\ 1 & 0 & -10 \\ 0 & 1 & -11 \end{bmatrix}, \qquad \mathbf{B}_{ob} = \begin{bmatrix} 1 \\ 0 \\ 0 \end{bmatrix}.
$$

Optional reading:

The rest of this section requires material on the *Web Support* and is better read together with Chap. 9. Using the supplementary notes on canonical transformation, we find that the observable canonical form is the transpose of the controllable canonical form. In the observable canonical form, the coefficients of the characteristic polynomial (in reverse sign) are in the last column. The characteristic polynomial is, in this case,

$$P(s) = s^3 + 11s^2 + 10s + 2.$$

We can check that with

```
roots([1  11  10  2])     % Check the roots
poly(A)                   % Check the characteristic polynomial of A
```

We can find the canonical forms ourselves. To evaluate the observable canonical form $\mathbf{A}_{ob}$, we define a new transformation matrix based on the controllability matrix:

```
P=[B  A*B  A^2*B];
inv(P)*A*P                % Should be A_ob as found by canon()
inv(P)*B                  % Shoud be B_ob (Bob!)
```

To find the controllable canonical form,

$$\mathbf{A}_{ctr} = \begin{bmatrix} 0 & 1 & 0 \\ 0 & 0 & 1 \\ -2 & -10 & -11 \end{bmatrix}, \quad \mathbf{B}_{ctr} = \begin{bmatrix} 0 \\ 0 \\ 1 \end{bmatrix},$$

we use the following statements based on the *Web Support* supplementary notes. Be very careful when constructing the matrix **M**:

```
poly(A);                  %To confirm that it is [1 11 10 2]
M=[10 11 1; 11 1 0; 1 0 0];
T=P*M;
inv(T)*A*T
inv(T)*B
```

We now repeat the same ideas one more time with Example 4.9. We first make the transfer function and the state-space objects:

```
G=zpk([],[-1 -2 -3],1);
S=ss(G);
```

As a habit, we check the eigenvalues:

```
eig(S)                    % Should be identical to eig(G)
```

To find the modal matrix, we use

```
[P,L]=eig(S.a)
inv(P)*S.a*P              % Just a check of L
```

The observable canonical form is

```
SD=canon(S)
```

The component `SD.a` is, of course, the diagonalized matrix `L` with eigenvalues. We can check that `SD.b` and `SD.c` are respectively computed from

```
inv(P)*S.b        % Identical to SD.b
S.c*P             % Identical to SD.c
```

Finally, the observable canonical form is

```
SO=canon(S, 'companion')
```

The matrix `SO.a` is

$$\mathbf{A}_{ob} = \begin{bmatrix} 0 & 0 & -6 \\ 1 & 0 & -11 \\ 0 & 1 & -6 \end{bmatrix},$$

meaning that

$$P(s) = s^3 + 6s^2 + 11s + 6,$$

which is the characteristic polynomial

```
poly([-1 -2 -3])
```

as expected from the original transfer function.

Session 5. Feedback Simulation Functions

This tutorial is to complement our development in Chaps. 5 and 6. You may want to go over the tutorial quickly before you read the text and come back later a second time for the details.

Feedback simulation functions

feedback	Generate feedback-system transfer function object
simulink	Launch Simulink

M5.1. Simulink

Comments with Respect to Launching Simulink

Simulink is a user-friendly simulation tool with an icon-driven graphics interface that runs within MATLAB. The introduction here is more conceptual than functional for two reasons. One, the Simulink interface design is very intuitive and you may not need help at all! Second, for a thorough introduction, we need to reproduce many of the graphics windows. To conserve

Figure M5.1.

paper (and trees), these print-intensive and detailed explanations have been moved to the *Web Support*. Furthermore, the Helpbrowser of MATLAB is extremely thorough and should serve as our main guide for further applications.

To launch Simulink, enter in the Command Window

```
simulink
```

and MATLAB will launch the Simulink Block Library window with pull-down menus. A few sample block library icons are shown in Fig. M5.1. Each icon represents a toolbox and contains within it a set of models, which will make themselves available if we double-click on the toolbox icons. For example, we can find within the Sources toolbox (Fig. M5.1) a model for generating a step input function and within the Sinks toolbox a model for graphing results. Within the Continuous toolbox are the important models for generating transfer functions and state-space models (Fig. M5.2).

All we need is to drag and drop the icons that we need from the toolboxes into a blank model window. If this window is not there, open a new one with the File pull-down menu. From here on, putting a feedback loop together to do a simulation is largely a point-and-click activity. An example of what Simulink can generate is shown in Fig. M5.3.

Simulink is easy to learn, fun, and instructive, especially with more complex MIMO systems. For systems with time delays, Simulink can handle the problem much better than the classical control toolbox. Simulink also has ready-made objects to simulate a PID controller.

A few quick pointers:

- These are some of the features that we use most often within the Simulink Block Library:

  Sources: Step input; clock for simulation time
  Sinks: Plotting tools; output to MATLAB workspace or a file
  Continuous: Transfer functions in polynomial or pole-zero form; state-space models; transport delay
  Math: Sum; gain or gain slider
  Nonlinear: Saturation; dead zone
  Blocksets: From the Blocksets and Toolboxes, choose "Simulink Extras," and then "Additional Linear." In there are the PID and the PID with approximate derivative controllers.

Figure M5.2.

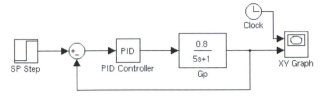

Figure M5.3.

- All Simulink simulation block diagrams are saved as ASCII files with the "mdl" extension.
- Before we start a simulation, choose Parameters under the Simulation pull-down menu to select the time of simulation. If we are using the XY Graph, we need to double-click its icon and edit its parameters to make the time information consistent.
- Simulink shares the main MATLAB workspace. When we enter information into, say, the transfer function block, we can use a variable symbol instead of a number. We then define the variable and assign values to it in the MATLAB Command Window. This allows for a much quicker route for doing parametric studies than does changing the numbers within the Simulink icons and dialog boxes.
- We can build our own controllers, but two simple ones are available: an ideal PID and a PID with approximate derivative action.

For curious minds: The first time you use the PID controllers, drag the icon onto a new simulation window, select the icon, and then *Look under mask* under the Edit pull-down menu. You will see how the controllers are put together. The simple PID controller is

$$G_c(s) = K_c + \frac{K_I}{s} + K_D s,$$

and the PID with approximate derivative controller is

$$G_c(s) = K_c + \frac{K_I}{s} + \frac{K_D s + 1}{s/N + 1}.$$

We also see the transfer functions used by each icon when we double-click on it and open up the parameter entry dialog window. Therefore, in terms of notation, we have $K_I = K_c/\tau_I$, $K_D = K_c \tau_D$, and $N = 1/\alpha \tau_D$.

M5.2. Control Toolbox Functions

The following feature is covered in this session:

- Synthesizing a closed-loop transfer function with `feedback()`

The closed-loop transfer function of a servo problem with proper handling of units is Eq. (5.11) in text:

$$\frac{C}{R} = \frac{K_m G_c G_p}{1 + G_m G_c G_p}.$$

It can be synthesized with the MATLAB function `feedback()`. As an illustration, we use a simple first-order function for G_p and G_m and a PI controller for G_c. When all is done, we test

251

the dynamic response with a unit-step change in the reference. To make the reading easier, we break the task up into steps. Generally, we would put the transfer function statements inside an M-file and define the values of the gains and time constants outside in the workspace.

Step 1: Define the transfer functions in the forward path. The values of all gains and time constants are arbitrarily selected:

```
km=2;                            % Gc is a PI controller
kc=10;
taui=100;
Gc=tf(km*kc*[taui 1], [taui 0]);

kp=1;
taup=5;
Gp=tf(kp, [taup 1]);             % Gp is the process function
```

In the definition of the controller G_c, we have included the measurement gain K_m, which usually is in the feedback path and the reference (Fig. 5.4). This is a strategy that helps to eliminate the mistake of forgetting about K_m in the reference. One way to spot whether you have made a mistake is if the system calculation has an offset when in theory you know that it should not.

Step 2: Define the feedback path function. Let's presume that our measurement function is first order too. The measurement gain has been taken out and implemented in Step 1:

```
taum=1;                          % Gm is the measurement function
Gm=tf(1, [taum 1]);              % Its s.s. gain km is in Gc
```

Step 3: Define the closed-loop function:

```
Gcl=feedback(Gc*Gp,Gm);    % Gcl is the closed-loop function C/R
```

Comments:

- By default, `feedback()` uses negative feedback.
- With unity feedback, i.e., $G_m = K_m = 1$, we would simply use

```
    Gcl=feedback(Gc*Gp,1)
```

  to generate the closed-loop function.
- We could generate a closed-loop function with, for example, `Gc*Gp/(1 + Gc*Gp)`, but this is not recommended. In this case, MATLAB simply multiplies everything together with no reduction and the resulting function is very unclean.

Step 4: We can now check (if we want to) the closed-loop poles and do the dynamic simulation for a unit-step change in R:

```
disp('The closed-loop poles & s.s. gain:')
pole(Gcl)
dcgain(Gcl)

step(Gcl)    % Of course, we can customize the plotting
```

This is the general idea. You can now put it to use by writing M-files for different kinds of processes and controllers.

When we have a really simple problem, we should not even need to use `feedback()`. Yes, we can derive the closed-loop transfer functions ourselves. For example, if we have a proportional controller with $G_c = K_c$ and a first-order process, all we need are the following statements, which follow Example 5.1 and Eq. (E5.1) in text:

```
kc=1;
kp=0.8;
taup=10;
Gcl=tf(kc*kp,[taup 1+kc*kp]);
pole(Gcl)
step(Gcl);    % Again for unit-step change in R
```

Try a proportional controller with a second-order process as derived in Example 5.2 in text. This is another simple problem for which we do not really need `feedback()`.

We now finish up with what we left behind in Session 4. Let's revisit Example 4.6. For checking our results later, we first find the poles of the closed-loop transfer function with

```
q=2*[1 10];
p=[1 11 10 2];
roots(p)      % -0.29, -0.69, and -10.02
```

Next, we define each of the transfer functions in the example:

```
G1=tf(1,[1 0]);
G2=tf(2,[1 1]);
H=tf(1,[1 10]);
```

Note that the numbering and notation are entirely arbitrary. We now generate the closed-loop transfer function and check that it has the same closed-loop poles:

```
Gcl=feedback(G1*G2,H);
pole(Gcl)
```

We can also easily obtain a state-space representation and see that the eigenvalues of the state matrix are identical to the closed-loop poles:

```
ssm=ss(Gcl);
eig(ssm.a)
```

For fun, we can recover the closed-loop transfer function Gcl with:

```
tf(ssm)
```

One final check with our own derivation. We define the coefficient matrices with Eqs. (E4.23) and (E4.24) and then do the conversion:

```
a=[0 1 0; 0 -1 -2; 1 0 -10];
b=[0; 2; 0];
c=[1 0 0];
d=0;
eig(a)                         % should return the same
[q3,p3]=ss2tf(a,b,c,d,1)       % eigenvalues and transfer
                               % function
```

253

If this is not enough to convince you that everything is consistent, try `step()` on the transfer function and different forms of the state-space model. You should see the same unit-step response.

Session 6. Root-Locus Functions

This tutorial is to complement our development in Chap. 7. You may want to go over the tutorial quickly before you read the text and come back later a second time for the details.

Root-locus functions

`rlocus`	Root-locus plot
`rlocfind`	Find the closed-loop gain graphically
`sgrid`	Draw the damping and natural frequency lines
`sisotool`	Launch the SISO system design graphics interface

M6.1. Root-Locus Plots

The following features are covered in this session:

- Root-locus calculation and plots, `rlocus()`
- Frequency and damping factor grid, `sgrid()`
- Obtaining gain of chosen closed-loop pole, `rlocfind()`

In simple terms, we want to solve for s in the closed-loop equation

$$1 + G_0(s) = 1 + kG(s) = 0,$$

where we further write $G_0 = kG(s)$ and $G(s)$ is the ratio of two polynomials, $G(s) = q(s)/p(s)$. In the simplest case, we can think of the equation as a unity feedback system with only a proportional controller (i.e., $k = K_c$) and $G(s)$ as the process function. We are interested in finding the roots for different values of the parameter k. We can either tabulate the results or we can plot the solutions s in the complex plane – the result is the root-locus plot.

Let's pick an arbitrary function such that $q(s) = 1$ and $p(s) = s^3 + 6s^2 + 11s + 6$. We can generate the root-locus plot of the system with:

```
p=[1 6 11 6];
roots(p)                    % Check the poles
G=tf(1,p);
rlocus(G)                   % Bingo!
```

For the case in which $q(s) = s + 1$, we use

```
G=tf([1 1],p);             % Try an open-loop zero at -1
rlocus(G)                  % to cancel the open-loop pole at -1
```

MATLAB automatically selects a reasonable vector for k, calculates the roots, and plots them. The function `rlocus()` also adds the open-loop zeros and poles of $G(s)$ to the plot.

Let's try two more examples with the following two closed-loop characteristic equations:

$$1 + K\frac{1}{(s+1)(s+3)} = 0, \qquad 1 + K\frac{1}{(s+1)(s+2)(s+3)} = 0;$$

```
G=zpk([],[-1 -3],1)         % The second-order example
rlocus(G)

G=zpk([],[-1 -2 -3],1)      % The third-order example
rlocus(G)
```

The point of the last two calculations is that a simple second-order system may become extremely underdamped, but it never becomes unstable.

Reminder: We supply the polynomials $q(s)$ and $p(s)$ in $G(s)$, but do not lose sight that MATLAB really solves for s in the equation $1 + kq(s)/p(s) = 0$.

In the initial learning stage, it can be a bad habit to rely on MATLAB too much. Hence the following two exercises take the slow way in making root-locus plots, which, it is hoped, may make us more aware of how the loci relate to pole and zero positions. The first thing, of course, is to identify the open-loop poles:

```
q=[2/3 1];                  % Redefine q(s) and p(s)
p=[1 6 11 6];
poles=roots(p)'             % display poles and zeros as row vectors
zeros=roots(q)'
G=tf(q,p);
k=0:0.5:100;                % define our own gains; may need
                            % 0:0.1:10 to better see the break-off point
rlocus(G,k);                % MATLAB will plot the roots with '+'
```

Until we have more experience, it will take some trial and error to pick a good range and increment for k, but then that is the whole idea of trying it ourselves. This manual approach makes us better appreciate the placements and changes of closed-loop poles as we vary the proportional gain.[12]

We may also want to override the MATLAB default format and use little dots:

```
r=rlocus(G,k);              % Save loci to array "r" first
plot(r,'.')                 % Now use plot() to do the dots
hold                        % hold the plot to add goodies
pzmap(G)                    % pzmap() draws the open-loop poles
hold off                    % and zeros
```

Be careful to read where the loci are on the real axis because pzmap() also traces the axis with little dots, which can be confusing.

We may want to find the ultimate gain when the loci cross the imaginary axis. Again there are many ways to do it. The easiest method is to estimate with the MATLAB function rlocfind(), which is introduced next.

---

[12] The gain vector generated automatically by MATLAB is not always instructive if we want to observe the region close to the imaginary axis. We can use "tricks," like making two gain vectors with different increments, concatenating them, and using the result in rlocus(). However, we should not get bogged down with fine details here. Certainly for day-to-day routine calculations, we can omit the gain vector and let MATLAB generate it for us.

There are two very useful MATLAB features. First, we can overlay onto the root-locus plot lines of the constant damping factor and the natural frequency. These lines help us pick the controller gain if the design specification is in terms of the frequency or the damping ratio,

```
sgrid                    % use the default grid
```

or, better yet,

```
sgrid(zeta,wn)           % plot only lines with given damping ratio
                         % and natural frequency
```

Example:

```
sgrid(0.7,1)             % add the approx. 45° line for zeta=0.7 and
                         % the unit circle (frequency=1)
```

The second feature is the function `rlocfind()`, which allows us to find the gain associated with a closed-loop pole. We can enter

```
[ck,cpole]=rlocfind(G)
```

or

```
rlocfind(G)              % this simple form will not return
                         % the values of the closed-loop poles
```

MATLAB will wait for us to click on a point (the chosen closed-loop pole) in the root-locus plot and then return the closed-loop gain (`ck`) and the corresponding closed-loop poles (`cpole`). MATLAB does the calculation with the root-locus magnitude rule, which is explained on the *Web Support*.

What if we click a point not exactly on a root locus? When we select a point s^*, MATLAB calculates the value $k^* = -p(s^*)/q(s^*)$, which will be a real positive number only if s^* satisfies the closed-loop equation. Otherwise, k^* is either complex or negative if the pole is a real number. In this case, MATLAB calculates the magnitude of k^*, uses it as the gain, and computes the corresponding closed-loop poles. Thus we find that the chosen points are always right on the root loci, no matter where we click.

We may also want to use the `zoom` feature of MATLAB to zoom in and out of a plot to get a better picture of, say, the break-off point of two loci. Make sure you enter "`zoom off`" when you are done.

M6.2. SISO System Design Graphics Interface

The following feature is covered in this session:

- Graphics user interface for designing SISO systems, `sisotool` [13]

The control toolbox supports an extremely nice SISO system design tool that is ideal for experimentation. This graphics interface is even more intuitive and self-explanatory than that of Simulink. The same approach is taken as that of our introduction to Simulink, and

---

[13] The description is based on Version 5.1 of the MATLAB control toolbox. If changes are introduced in newer versions, they will be presented on the *Web Support*.

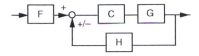

Figure M6.1.

the not-so-necessary and print-intensive window display and instructions have been moved to the *Web Support*. Only a very brief conceptual introduction is provided here.

To launch the SISO system design tool, enter in the MATLAB Command Window

```
sisotool              % default view
```

or

```
sisotool('rlocus')    % root-locus view only
```

A graphics window with pull-down menus and tool buttons will pop out, slowly. The default view displays both the root-locus and the Bode editors. Because we have not learned Bode plots yet, the second option with `rlocus` is less intimidating for the moment. Here are some pointers on the usage of the tool:

- The SISO design tool supports a flexible block diagram, as shown in Fig. M6.1. The feedback can be either positive or negative. Similar to the LTI Viewer, the tool runs in its own functional space. We have to import the transfer functions under the File pull-down menu. By default, the transfer functions F, C, G, and H are all assigned the value "1," so we have to import at least a transfer function for G to do meaningful calculations.
- The default compensator C in Fig. M6.1 is a proportional controller, but it can be changed to become a PI, PD, or PID controller. The change can be accomplished many ways. One is to retrieve the compensator-editing window by clicking on the C block or by using the Compensator pull-down menu. We can also use the set of button on the toolbar to add or move open-loop poles and zeros associated with the controller.
- Once a root-locus plot is generated, we can interactively change the locations of the closed-loop poles and the tool will compute the corresponding controller gain for us.
- For a given system and chosen closed-loop poles displayed in the root-locus plot, we can generate its corresponding time-response and frequency-response plots with features under the Tools pull-down menu.

In the next section, you can use the SISO design tool if you prefer, but the explanation is given with commands. It is easier to get the message across with commands, and in the beginner's learning stage, entering your own command can give you a better mental imprint of the purpose of the exercise.

M6.3. Root-Locus Plots of PID Control Systems

The following feature is covered in this session:

- Making root-locus plots that model situations of PID control systems

Here are some useful suggestions regarding root-locus plots of control systems. In the following exercises, we consider only the simple unity feedback closed-loop characteristic equation:

$$1 + G_c G_p = 0.$$

We ignore the values of any gains. We focus on only the probable open-loop pole and zero positions introduced by a process or by a controller, or, in other words, the shape of the root-locus plots.

Let's begin with a first-order process $G_p = 1/(s+1)$. The root-locus plot of a system with this simple process and a proportional controller, $G_c = K_c$, is generated as follows:

```
Gp=tf(1,[1 1]);                    % open-loop pole at -1
subplot(221), rlocus(Gp)           % Gc = Kc
```

To implement an ideal PD controller, we will have an additional open-loop zero. Two (of infinite) possibilities are

```
taud=2;                            % open-loop zero at -1/2
Gc=tf([taud 1],1);
subplot(222), rlocus(Gc*Gp)
```

and

```
taud=1/2;                          % open-loop zero at -2
Gc=tf([taud 1],1);
subplot(223), rlocus(Gc*Gp)
```

What are the corresponding derivative time constants? Which one would you prefer?

We next turn to a PI controller. We first make a new figure and repeat proportional control for comparison:

```
figure(2)
subplot(221), rlocus(Gp)      % Gc = Kc
```

Integral control will add an open-loop pole at the origin. Again, we have two regions where we can put the open-loop zero:

```
taui=2;                            % open-loop zero at -1/2
Gc=tf([taui 1],[taui 0]);
subplot(222), rlocus(Gc*Gp)
```

and

```
taui=1/2;                          % open-loop zero at -2
Gc=tf([taui 1],[taui 0]);
subplot(223), rlocus(Gc*Gp)
```

Once again, what are the corresponding integral time constants? Which one would you prefer?

Finally, let's take a look at the probable root loci of a system with an ideal PID controller, which introduces one open-loop pole at the origin and two open-loop zeros. For illustration, we will not use the integral and derivative time constants explicitly, but refer to only the two zeros that the controller may introduce. We will also use zpk() to generate the transfer functions:

```
figure(3)
subplot(221), rlocus(Gp)        % redo Gc = Kc

op_pole=[0];                    % open-loop pole at 0

op_zero=[-0.3 -0.8];            % both zeros larger than -1
Gc=zpk(op_zero,op_pole,1);
subplot(222),rlocus(Gc*Gp)

op_zero=[-1.5 -3];              % both zeros less than -1
Gc=zpk(op_zero,op_pole,1);
subplot(223),rlocus(Gc*Gp)

op_zero=[-0.5 -1.8];           % one zero in each region
Gc=zpk(op_zero,op_pole,1);
subplot(224),rlocus(Gc*Gp)
```

Yes, you know the question is coming. Which case would you prefer? We can use the rule of thumb that the derivative time constant is usually approximately one fourth the value of the integral time constant, meaning that the zero farther away from the origin is the one associated with the derivative time constant.

Note that the system remains stable in all cases, as it should for a simple first- or second-order system. One final question: Based on the design guidelines by which the system should respond faster than the process and the system should be slightly underdamped, what are the ranges of derivative and integral time constants that you would select for the PD, PI, and PID controllers? And in what region are the desired closed-loop poles?

We'll finish with implementing the P, PI, and PD controllers on a second-order over-damped process. As in the previous exercise, try to calculate the derivative or integral time constants and take a minute to observe the plots and see what may lead to better controller designs.

Let's consider an overdamped process with two open-loop poles at -1 and -2 (time constants at 1 and 0.5 time units). A system with a proportional controller would have a root-locus plot as follows. We stay with tf(), but you can always use zpk().

```
figure(1)
p=poly([-1 -2]);               % open-loop poles -1, -2
Gp=tf(1,p);
subplot(221),rlocus(Gp)        % proportional control
```

To implement an ideal PD controller, we now have three possible regions in which to put the zero:

```
taud=2;                        % open-loop zero at -1/2
Gc=tf([taud 1],1);
subplot(222), rlocus(Gc*Gp)

taud=2/3;                      % open-loop zero at -1.5
Gc=tf([taud 1],1);
subplot(223), rlocus(Gc*Gp)
```

```
taud=1/3;                          % open-loop zero at -3
Gc=tf([taud 1],1);
subplot(224), rlocus(Gc*Gp)
```

We will put the PI controller plots on a new figure:

```
figure(2)
subplot(221),rlocus(Gp)            % redo proportional control
```

The major regions in which to place the zero are the same, but the interpretation as to the choice of the integral time constant is very different. We now repeat, adding the open-loop zeros:

```
taui=2;                            % open-loop zero at -1/2
Gc=tf([taui 1],[taui 0]);
subplot(222), rlocus(Gc*Gp)

taui=2/3;                          % open-loop zero at -1.5
Gc=tf([taui 1],[taui 0]);
subplot(223), rlocus(Gc*Gp)

taui=1/3;                          % open-loop zero at -3
Gc=tf([taui 1],[taui 0]);
subplot(224), rlocus(Gc*Gp)
```

You may want to try some sample calculations using a PID controller. One way of thinking: We need to add a second open-loop zero. We can limit the number of cases if we assume that the value of the derivative time constant is usually smaller than the integral time constant.

Session 7. Frequency-Response Functions

This tutorial is to complement our development in Chap. 8. You may want to go over the tutorial quickly before you read the text and come back later a second time for the details.

Frequency-response functions

bode	Bode plots
freqresp	Frequency response of a transfer function
logspace	Logarithmically spaced vector
margin	Gain margin and crossover frequency interpolation
nichols, ngrid	Nichols plots
nyquist	Nyquist plots
sisotool	Launch the SISO system design graphics interface

M7.1. Nyquist and Nichols Plots

The following feature is covered in this session:

- Nyquist plots, `nyquist()`

The SISO system design tool `sisotool`, as explained in Session 6, can be used to do frequency-response plots. Now we want to use the default view, so we just need to enter

```
sisotool
```

Hints to make better use of the design tool are on the *Web Support*. We use commands here because they give us a better idea behind the calculations. This section is brief as our main tool will be Bode plots, which will be explained in the next section.

Let's say we have a simple open-loop transfer function G_0 of the closed-loop characteristic equation,

$$1 + G_0 = 0,$$

and we want to find the proportional gain that will give us an unstable system. For this simple exercise, we take $G_0(s) = K G(s)$:

```
p=poly([-1; -2; -3]);      % Open-loop poles at -1, -2, -3
G=tf(10,p);                % Arbitrary value K=10
nyquist(G);                % Bingo!
```

We'll see two curves. By default, MATLAB also maps and plots the image of the negative imaginary axis. That can make the plot too busy and confusing, at least for a beginner. So we'll stay away from the default in the following exercises:

```
[re,im]=nyquist(G);
plot(re(1,:),im(1,:))      % Only the positive Im-axis image
```

Of course, we can define our own frequency vector:

```
w=logspace(-1,1);          % Generate numbers between [10^-1, 10^1]
[re,im]=nyquist(G,w);
plot(re(1,:),im(1,:))
```

The function `logspace()` generates a vector with numbers nicely spaced on the logarithmic scale. Its use is optional. The default of the function gives 50 points and is usually adequate. For a smoother curve, use more points. For example, this command will use 150 points: `logspace(-1,1,150)`.

```
hold                          % to add the (-1,0) point and the axes
                              % on the plot
x=-1; y=0;
xh=[-2 2]; yh=[0 0];          % the axes
xv=[0 0];   yv=[-2 1];
plot(x,y,'o',xh,yh,'-',xv,yv,'-')
```

We can increase the gain K and repeat the calculation with, for example, two more trials[14]:

```
G=tf(50,p);                    % try again
[re,im]=nyquist(G,w);
plot(re(1,:),im(1,:))

G=tf(60,p);                    % and again
[re,im]=nyquist(G,w);
plot(re(1,:),im(1,:))
hold off
```

We do not use the Nichols plot (log magnitude versus phase) much, but it is nice to know that we can do it just as easily:

```
p=poly([-1; -2; -3]);
G=tf(10,p);
nichols(G)
ngrid
zoom                           % need to zoom into the meaningful region
```

The plot with default settings is quite useless unless we use `ngrid` to superimpose the closed-loop gain and phase grid lines. Instead of zooming in, we can reset the axes with

```
axis([-360 0 -40 20])
```

M7.2. Magnitude and Phase-Angle (Bode) Plots

The following features are covered in this session:

- Bode plot calculation, `bode()`
- Finding the gain and phase margins, `margin()`
- Bode plots for transfer functions with dead time

We begin with one simple example. Let's say we want to analyze the closed-loop characteristic equation

$$1 + \frac{1}{s^2 + 0.4s + 1} = 0.$$

We generate the Bode plot with

```
G=tf(1,[1 0.4 1]);
bode(G)                        % Done!
```

The MATLAB default plot is perfect! That is, except when we may not want decibels as the unit for the magnitude. We have two options. One, learn to live with decibels, the

---

[14] All functions like `nyquist()`, `bode()`, etc., can take on multiple LTI objects, as in

```
nyquist(G1,G2,G3)
```

but only when we do not use LHS arguments.

convention in the control industry, or two, we do our own plots. This is a task that we need to know when we analyze systems with dead time. This is how we can generate our own plots:

```
w=logspace(-1,1);
[mag,phase]=bode(G,w);
mag=mag(1,:);                        % required since MATLAB v.5
phase=phase(1,:);

subplot(211), loglog(w,mag)
               ylabel('Magnitude'), grid
subplot(212), semilogx(w,phase)
               ylabel('Phase, deg'), grid
               xlabel('Frequency (rad/time)')
```

As an option, we can omit the subplot command and put the magnitude and phase plots in individual figures.

This is how we can make a Bode plot with decibels as the scale for the magnitude.

```
dB=20*log10(mag);     % converts magnitude to decibels
```

Now we do the plotting. Note that the decibel unit is already a logarithmic scale:

```
subplot(211),      semilogx(w,dB)    % Use semilogx for dB
                   ylabel('Magnitude(dB)')
                   grid
subplot(212),      semilogx(w,phase)
                   ylabel('Phase angle (degree)')
                   xlabel('Frequency, w')
                   grid
```

We most often use radians per second as the unit for frequency. In the case in which cycles per second or hertz are needed, the conversion is

```
f=w/(2*pi);               % Converts w [rad/s] to [Hz]
```

After using the `subplot()` command and before doing any other plots, we should make it a habit to reset the window with

```
clf                   % clear figure
```

We now find the gain margin with its crossover frequency (Gm, Wcg) and phase margin with its crossover frequency (Pm, Wcp) with either one of the following options:

```
[Gm,Pm, Wcg,Wcp]=margin(mag,phase,w)     % option 1
```

where mag and phase are calculated with the function bode() beforehand. We can skip the bode() step and use the transfer function directly as the argument,

```
[Gm,Pm, Wcg,Wcp]=margin(G)     % option 2
```

or simply

```
margin(G)            % option 3, Gm in dB
```

In the last option without any LHS arguments, MATLAB will do the Bode plot, and display the margin calculations on the plot.
Two important comments:

(1) With G=tf(1,[1 0.4 1]), i.e., a simple second-order system, it is always stable. The gain margin calculation is meaningless. Nevertheless, MATLAB returns a set of results anyway. Again, a computer is not foolproof. All margin() does is an interpolation calculation.

(2) If you use option 1 or 2 above, margin() returns the linear-scale gain margin in the variable Gm. With option 3, however, the gain margin displayed in the plot is in decibels, you need to convert it back with $10^{dB/20}$.

To handle dead time, all we need is a simple modification using the fact that the time delay transfer function has magnitude 1 and phase angle $-t_d\omega$. We need one single statement to "tag on" the lag that is due to dead time, and we do it after the bode() function call.
So let's start with the second-order function, which is always stable:

```
G=tf(1,[1 0.4 1]);
freq=logspace(-1,1);            % freq is in rad/time
[mag,phase]=bode(G,freq);
mag=mag(1,:);
phase=phase(1,:);
```

Now let's say we also have dead time:

```
tdead=0.2;      % [time unit]
```

The following statement is the only addition needed to introduce the phase lag that is due to dead time:

```
phase = phase - ((180/pi)*tdead*freq);      % phase is in degrees
```

We can now proceed with the plotting and phase/gain margin interpolation:

```
subplot(211),      loglog(freq,mag)
                   ylabel('Magnitude'),title('Bode Plot')
                   grid
subplot(212),      semilogx(freq,phase)
                   ylabel('Phase(degree)'),xlabel('Frequency')
                   grid

% now using new phase variable that includes dead-time phase lag
[Gm,Pm,Wcg,Wcp]=margin(mag,phase,freq)
```

The whole idea of handling dead time applies to other types of frequency-domain plots, but the Bode plot is the easiest to learn from.

There is no magic in the functions `nyquist()` or `bode()`. We could have done all our calculations by using the more basic `freqresp()` function. What it does essentially is make the $s = j\omega$ substitution numerically in a given transfer function $G(s)$. A sample usage is

```
w=logspace(-1,1);
gjw=freqresp(G,w);      % does the s=jw calculation for each
                        % value in w
```

After that, we can use the result to do frequency-response analysis. If you are interested in the details, they are provided in the Session 7 Supplement on the *Web Support*.

Homework Problems

Part I. Basic Problems

Where appropriate, use MATLAB to solve the problem or to check your derivation.

(1) For the given transfer function

$$\frac{Y(s)}{X(s)} = \frac{2}{(s+2)(s^2+9)}.$$

  (a) Derive $y(t)$ with respect to a unit-step input and an impulse input.
  (b) What are the values of $y(t)$ at $t = 0$ and $t = \infty$ with respect to a unit-step input?

(2) For the given ODE,

$$y'' + y' + y = \sin \omega t,$$

with initial condition $y(0) = y'(0) = 0$. What are the characteristic features of $y(t)$?
What is $y(t)$ when $t \to \infty$?

(3) For the given ODE,

$$\frac{d^2 y}{dt^2} + 4\frac{dy}{dt} + 5y = f(t), \quad \text{with } y(0) = y'(0) = 0,$$

sketch the probable time-domain response if $f(t)$ is a unit-step function.

(4) Derive time-domain function $y(t)$ of the following Laplace transforms

  (a) $Y(s) = \dfrac{10}{(s+1)^2(s+3)}$

  (b) $Y(s) = \dfrac{s+3}{(s^2+2s+5)}$

  (c) $Y(s) = \dfrac{e^{-4s}}{s(2s^2+3s+2)}$

  (d) $Y(s) = \dfrac{e^{-2s}}{s(s^2+9)}$

(5) Find the partial fractions of the following transfer function:

$$G(s) = \frac{s+1}{s^2(10s+1)}.$$

(6) For the following transfer function,

$$\frac{Y(s)}{X(s)} = \frac{3s(s+2)(s-2)}{5s^4 + 6s^3 + 2s^2 + 3s},$$

and given that the input $x(t)$ is a unit-step function, what is $y(t)$ as $t \to \infty$? Under what condition, as related to the property of the transfer function, is this result valid?

(7) For the following transfer function and a unit-step input, sketch *qualitatively* the time response $y(t)$:

$$\frac{Y(s)}{X(s)} = \frac{(s+1)e^{-0.5s}}{(s+2)(s^2 - 2s + 5)}.$$

Explain how each "term" in the transfer function may contribution to your sketch.

(8) Plot the zeros and poles of the following transfer function on the complex plane:

$$\frac{Y(s)}{X(s)} = \frac{10s(s+1)}{(s+2)(s^2 + 2s + 2)}.$$

Now, ignore the zeros, i.e., consider only

$$\frac{Y(s)}{X(s)} = \frac{10}{(s+2)(s^2 + 2s + 2)}.$$

If $X(s)$ represents a unit-step function, what is qualitatively the expected time response $y(t)$? What is the steady-state gain and what are some of the characteristic parameters that we can associate with $(s^2 + 2s + 2)$?

What about the case for the following function?

$$\frac{Y(s)}{X(s)} = \frac{10}{(s+2)(s^2 + 2)}.$$

(9) The dynamic behavior of a process is modeled by the transfer function

$$\frac{Y(s)}{F(s)} = \frac{18}{(s^2 + 3s + 9)}.$$

(a) With a given step change input of $f(t) = 3u(t)$, where $u(t)$ is the unit-step function, what is the new steady-state value of the step response?

(b) If it is required that the output must never exceed an upper limit of 10, what is the largest step input that we can tolerate without exceeding this limit?

(c) Consider the following transfer function:

$$\frac{Y(s)}{F(s)} = \frac{5}{s^2 + s + 9}.$$

Sketch qualitatively the output $y(t)$ in response to when $F(s)$ represents a unit-step input and an impulse input. In each case, label clearly the final value at large times. With the step response, label the value of one characteristic feature (of your choice) of the response.

(10) (a) Consider two stirred-tanks-in-series. The first one has a volume of 4 m^3 and the second is 3 m^3. The flow rate and the inlet concentration of an inert species into the first stirred-tank under normal circumstances are 0.02 m^3/s and 1 gmol/m^3, respectively. Consider that the inlet (deviation) concentration is being perturbed and is described by an impulse function as

$$c(t) = 6\delta(t).$$

What is the time response, i.e., $c_2(t)$ in the second tank? What is $c_2(t)$ as time is sufficiently large?

(b) Consider five stirred-tanks-in-series. They are identical with a volume of 4 m^3, and the steady inlet flow rate is 2 m^3/min. Consider now the change in inlet concentration as an impulse function; plot the responses in each of the tanks in the same figure. Repeat with a unit-step input change.

(c) On what basis would you choose the range of the time axis, i.e., how long should the simulation be in order for you to observe the final steady state?

(11) (a) From standard step-response measurements, the poles of a process are determined to be located at $-4.5 \pm 2.5j$. The same process is subjected to a rectangular-pulse input. The pulse has a value of 1.3 between the period $t = 0$ and $t = 3$ (time unit). What are the pole positions of the transfer function now?

(b) We are given the following transfer function:

$$\frac{Y(s)}{F(s)} = \frac{12}{(6s^2 + 3s + 9)}.$$

We are asked to find the steady-state gain and time constant. If someone asks "Is that for the case when the input is a unit-step function," what is your response?

(c) We are asked to find the differential equation that represents the transfer function in part (b). If someone asks "Can I assume zero initial condition," what is your response?

(12) From a standard step-input experiment, you determined that a process under investigation was first order with a steady-state gain of 8.5 (unit) and a process time constant of 3.5 (time unit). Your colleague did a different step-response experiment with the same equipment and determined that the steady-state gain was 5.5 and the process time constant was 2.5. The units being correct and the discrepancies being statistically significant, who is right? Your colleague is a very careful experimentalist, and so are you.

(13) For the given transfer function

$$\frac{Y(s)}{X(s)} = \left[\frac{K_1}{\tau_1 s + 1} + \frac{K_2}{\tau_2 s + 1} \right],$$

what is the steady-state gain? What are the poles and zeros?

(14) For the given Laplace transform $Y(s)$, derive $y(t)$ in terms of constants and sine function:

$$Y(s) = \frac{1}{s(s^2 + 2s + 3)}.$$

269

What is the value of $y(t)$ as time $\to \infty$?

(15) Consider the transfer function

$$\frac{Y}{F} = \frac{3}{s(s^2 + 2s + 4)}.$$

(a) Qualitatively, what is the time response $y(t)$ if $f(t)$ represents a unit-step input? What is the value of $y(t)$ when time is sufficiently large? What is the time constant that we may use to evaluate the "speed" of response?

(b) Repeat step (a) if $f(t)$ represents an impulse input. What is $y(t)$ when time is sufficiently large?

(16) We have two identical isothermal CSTRs-in-series. They both have space time $\tau = 4$ min, a first-order reaction rate constant $k = 1.5 \text{ min}^{-1}$, and both parameters are constants.

(a) Write the mass balances of the reactant concentration in the two CSTRs. What is the transfer function relating the concentration of the second CSTR, C_2, to the concentration entering the first one, C_0? What are the steady-state gain and the process time constant of the transfer function?

(b) Let's say the reaction rate constant is zero. We now subject the two stirred-tanks to a unit-step response experiment by increasing C_0 and measuring C_2. Someone suggested that we can measure the space time by the 63.2% time response for C_2 to reach the new steady state. We, of course, know better, and say that it is plain wrong. What is the percent response when time is equal to the process time constant in this case?

(17) We have a first-order process with the transfer function

$$\frac{Y}{F} = \frac{K_p}{(\tau_p s + 1)}.$$

Because the use of a temperature ramp is common in gas chromatographs, it is suggested that an experiment with a ramp input be used. If the ramp input is $f(t) = \alpha t$, show how K_p and τ_p can be determined from $y(t)$ when time is sufficiently large.

(18) Linearize the following two functions. Put the results in terms of deviation variables.

(a) Vapor–liquid equilibrium: $y(x) = \frac{\alpha x}{1 + (\alpha - 1)x}$ about the steady state x_s.

(b) Antoine equation: $P_i^s(T) = P_c \exp(A_1 - \frac{A_2}{T + A_3})$ about the steady state T_s.

(19) Consider the balances of the cell density x and food concentration S in a simple fermentor model:

$$\frac{dx}{dt} = -Dx + \mu(S)x,$$

$$\frac{dS}{dt} = D(S_{in} - S) - \frac{1}{Y}\mu(S)x,$$

where

$$\mu(S) = \frac{\mu_m S}{K_m + S}.$$

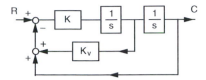

Figure PI.21.

Coefficients Y, μ_m, and K_m are constants. Quantities that can vary with time are inlet concentration S_{in} and dilution rate D, which really is q/V, the reciprocal of space time. Derive the transfer functions that relate changes in S_{in} and D to changes in cell density x.

(20) In heterogeneous catalysis, the reaction kinetics takes on the so-called Langmuir–Hinshelwood form. In a catalytic reactor, the mass balance for reactant A may take the following form:

$$\frac{dC_A}{dt} = \frac{1}{\tau}(C_{A0} - C_A) - r(C_A, C_B),$$

where $r(C_A, C_B) = [(kC_AC_B)/(1 + K_AC_A + K_BC_B)]$, and $C_A(0) = C_{AS}$. In the balance, C_{A0} is the inlet concentration, C_B is the concentration of reactant B, and both are functions of time. The rate constant is k. The adsorption equilibrium constants are K_A and K_B, and the space time is τ. All these quantities are constants. Linearize the reactant A mass balance.

(21) Feedback compensation can be achieved with what is called *rate feedback*. One example is shown in Fig. PI.21.

(a) Derive the transfer function relating changes in R to those in C, i.e., C/R.

(b) If $K = 4$ and if we desire the system response $c(t)$ to have a damping ratio of 0.7, what should be the value of K_v?

(22) We want to relate the temperature in the center of an oven to what we can measure on the outside. The oven has two insulating layers as shown in Fig. PI.22. The thermal resistance in both layers is modeled with heat transfer coefficients, h_1 and h_2, and the heat balance equations are

$$C_{p,1}M_1\frac{dT_m}{dt} = h_1 A_1(T_i - T_m),$$

$$C_{p,2}M_2\frac{dT_i}{dt} = h_2 A_2(T_o - T_i),$$

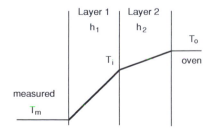

Figure PI.22.

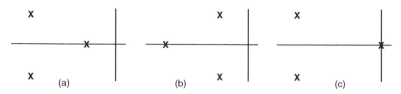

Figure PI.23.

where the heat capacity $C_{p,i}$, mass M_i, transfer area A_i, and heat transfer coefficients h_i are all constants.

(a) Derive the transfer function relating changes in the oven temperature T_o to the measured temperature T_m.

(b) If the change in oven temperature T_o is a unit step, what is the response we would expect (i.e., qualitatively) in T_m? What is (are) the time constant(s)?

(c) Will the measured temperature T_m ever be identical to the oven temperature?

(23) We have a third-order process. Three possible sets of pole positions are depicted in Fig. PI.23. Sketch the time-response characteristics of the three cases with respect to an impulse input and a step input.

(24) Consider the Lotka–Volterra equations that are used to model predator–prey problems in ecology:

$$\frac{dy_1}{dt} = (1 - \alpha y_2)y_1,$$

$$\frac{dy_2}{dt} = (-1 + \beta y_1)y_2.$$

The initial conditions are $y_1(0) = y_{1s}$ and $y_2(0) = y_{2s}$. Derive the linearized equations in terms of deviation variables using y_{1s} and y_{2s} as the references. What is the characteristic polynomial of the linearized model?

(25) The poles and zeros of a process are represented by the $\times$ and $\circ$ symbols, respectively, in Fig. PI.25. It was determined that the steady-state gain of the process is 2 (units).

(a) First, refer to Fig. PI.25(a). What is the order of the transfer function? What is the transfer function of the process?

(b) What is (are) the time constant(s) of the process?

(c) Sketch qualitatively the expected time response of the process with respect to a unit-step input. What is a reasonable (approximation of) settling time?

(d) If the pole-zero plot is changed to that shown in Fig. PI.25(b), what does the time response look like? Does the steady-state gain remain the same as before?

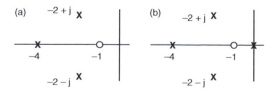

Figure PI.25.

Part I. Basic Problems

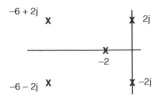

Figure PI.26.

(26) For a given transfer function with the pole positions shown in Fig. PI.26, reconstruct what the time-domain terms should be.

(27) For the given transfer function,

$$\frac{Q_2}{Q_0} = \frac{1}{(6s + 1)^2}.$$

If $q_0(t)$ is an impulse function (and without doing partial fractions), what is the most probable expression for $q_2(t)$? What is the value of q_2 at large times? What is the value of q_2 at large times if we have instead

$$\frac{Q_2}{Q_0} = \frac{1}{s(6s + 1)}?$$

(28) A task expected of us in the Chap. 3 Review Problems: Derive $y(t)$ for the case in which $\tau_1 \neq \tau_2$ and

$$Y(s) = \frac{1}{s(\tau_1 s + 1)(\tau_2 s + 1)}.$$

(29) Find the transfer function C/R for the block diagram in Fig. PI.29.

(30) Consider the closed-loop system in Fig. PI.30.
  (a) Derive the closed-loop transfer function C/R.
  (b) What is the steady-state error if we impose a unit-step change in R?
  (c) What should be the value of K if we desire the system response (C/R) to have a 10% overshoot?

(31) In process applications, there are different ways in which we can ramp up a variable such as temperature. As shown in Fig. PI.31(a), the variable is ramped up and kept constant. In Fig. PI.31(b), it is ramped down immediately to form a triangular pulse, and in Fig. PI.31(c), we do something in between.

   Derive the Laplace transform of these three input functions. Make use of the principle of superposition as to how we generate a rectangular pulse. Do not try to apply the integral definition.

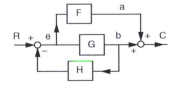

Figure PI.29.

Homework Problems

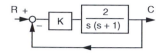

Figure PI.30.

(32) Consider the block diagram in Fig. PI.32(a) that represents a system used to handle dead time. Find C/R for a servo problem and also the equivalent controller transfer function G_c in Fig. PI.32(b).

(33) The model for the change in the liquid level of a gravity flow vessel in a textbook,

$$A \frac{dh}{dt} = q_0 - q,$$

is an oversimplification. The dependence of q on the level is more than just simply $\sqrt{h}$. Moreover, the flow rate also depends on the valve coefficient and characteristic. For an *equal-percentage valve*, the proper equation for the outlet flow rate is a function of the liquid level h and the valve opening (the so-called lift) l, where $0 \le l \le 1$, such that $q = q(l, h)$ is defined as

$$q = C_v R^{l-1} \sqrt{\frac{\Delta P(h)}{\rho_s}}, \quad \Delta P(h) = \frac{(P_0 + \rho g h) - P_1}{g_c}.$$

The valve coefficient C_v, rangeability R, fluid specific gravity ρ_s, density ρ, gravity g, gravitational constant g_c, ambient pressure P_0, and valve outlet pressure P_1 are all constants. The inlet flow is a function of time, $q_0 = q_0(t)$. Derive the transfer functions that describe this liquid-level model. Identify the steady-state gain(s) and time constant(s).

(34) A typical exercise is to linearize the heat and mass balances of a CSTR. Let's say we have a chemical reaction $2A \rightarrow B$ and the reaction is second order in A. The CSTR balances (i.e., the model equations) are

$$\tau \frac{dC_A}{dt} = C_{A,i} - C_A - \tau 2 k_0 e^{-E/RT} C_A^2,$$

$$\tau \frac{dT}{dt} = T_i - T + \frac{(-\Delta H)\tau}{\rho C_p} 2 k_0 e^{-E/RT} C_A^2 - \frac{U A_c}{\rho C_p Q}(T - T_c),$$

where τ is the space time, Q is the volumetric flow rate, U is the overall heat transfer coefficient, A_c is the heat transfer area, and the other notations carry their usual meanings in chemical kinetics. Derive the transfer functions that account for

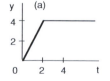
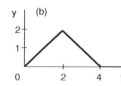
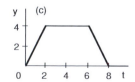

Figure PI.31.

Part II. Intermediate Problems

(a)

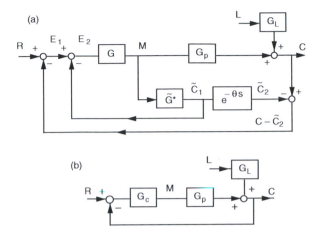

(b)

Figure PI.32.

the effects of changes in the inlet temperature T_i and the cooling jacket temperature T_c on the reactor concentration C_A and temperature T. You can consider the inlet concentration $C_{A,i}$, space time, and all other parameters as constants.

Part II. Intermediate Problems

(1) After adding a real PD controller as in Eq. (5.7) to a first-order process function, we essentially have a second-order system.
   (a) Derive the system steady-state gain, natural time period, and damping ratio.
   (b) How does the choice of α affect the natural time period?
   (c) If we now have a regulator problem and the load change is a unit step, what is the final steady-state value of the controlled variable?
   (d) For a servo system, hand-sketch the expected time responses of the systems with an ideal PD controller and the real PD controller with respect to a unit-step change in set point.
   (e) Will this system with the real PD controller become underdamped? (*Hint*: With the proper technique, you can arrive at the answer without doing any algebra.)
(2) For the following given process transfer function,

$$G_p(s) = \frac{(-s + 1)}{(s + 1)(s + 2)},$$

and the closed-loop system characteristic polynomial $1 + G_c G_p = 0$, if we want to use a simple proportional controller, answer the following questions:
   (a) What is the range of K_c that provides stable closed-loop responses?
   (b) From this exercise, can you tell the rationale behind, for example, the use of the 1/1 Padé approximation, and the general effect of time delay in a given system?
   (c) Sketch the Bode plot of the system with proportional control. Identify the contribution of magnitude and phase angle from each of the terms.

(3) Consider the process transfer function

$$G_p = \frac{2(s+1)}{s(s+2)(s^2+1)}$$

and the closed-loop system characteristic polynomial $1 + G_c G_p = 0$:
  (a) Show, with an analytical technique of your choice, that the closed-loop system cannot be stabilized with a proportional controller at all.
  (b) Suppose we can use a PI or a PD controller with a positive proportional gain; write the appropriate system characteristic equations for each controller. Without further analysis, can you make a judgment whether a PI or a PD controller is more feasible in stabilizing the system?
  (c) Use root-locus plots to illustrate your answer.
(4) Consider a unity feedback system with characteristic polynomial $1 + G_c G_p = 0$. The process function is

$$G_p = \frac{K}{s(\tau_p s + 1)},$$

where the steady-state gain K is 0.6 (unit) and the time constant τ_p is 3 (time unit).
  (a) If we use a proportional controller, what is the offset? What statement can you make regarding the selection of controllers?
  (b) With the proportional controller, sketch the root locus of the system.
  (c) With a proportional controller, what is the proportional gain if the system is to have a damping ratio of $1/\sqrt{2}$? What are the locations of the closed-loop poles?
  (d) What is the time constant of the system in part (c)?
  (e) What is the time constant of the system when the damping ratio is to be 0.2?
(5) Consider a system with either an ideal PD [Fig. PII.5(a)] or a PI controller [Fig. PII.5(b)].
  (a) What is the steady-state error of the system in each case?
  (b) With the PD controller, derive the stability criteria with respect to K_c with an appropriate method of your choice. What statement can you make regarding the proportional gain and the stability of the system?
  (c) What are the different probable ranges of derivative time constants that we may have with the PD controller? Explain with sketches of root-locus plots.
  (d) From part (c), which case would allow for an underdamped-type response? How would you evaluate the damping ratio? Which one would you choose as the basis of your controller design?

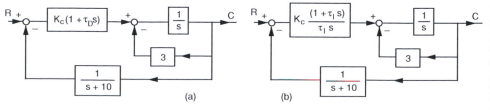

(a)   (b)

Figure PII.5.

Part II. Intermediate Problems

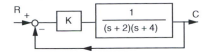

Figure PII.6.

    (e) With the PI controller, what are the different probable ranges of integral time constants that we may have?

    (f) Let's say we have $\tau_I = 0.3$ min. What is the proportional gain such that the dominant poles of the system are equivalent to a damping ratio of 0.9? Can the system behave with a damping ratio of 0.2?

    (g) What happens to the system when we choose $\tau_I = 1/3$ min? Explain with a hand-sketch.

(6) Consider the simple unity feedback loop with a proportional gain K in Fig. PII.6:

    (a) Sketch the root-locus plot of the system.

    (b) What is the value of K when the system exhibits critically damped behavior?

    (c) What is the value of K when the desired system damping ratio is 0.707? Identify the corresponding closed-loop pole(s) on the root-locus plot.

    (d) What is the steady-state error with the value of K in part (c)?

    (e) What is the settling time with the value of K in part (c)?

    (f) What is the value of K if the required system overshoot with respect to a unit set-point change is 10%?

    (g) If there is to be a load function to the system, what property must the load function have?

(7) Consider a simple unity feedback system with the closed-loop characteristic equation:

$$1 + \frac{K_c e^{-0.35s}}{(5.1s + 1)(1.2s + 1)} = 0.$$

    (a) Calculate analytically the exact values of the magnitude and the phase lag on the Bode plot for the specific case in which $K_c = 7.5$ and $\omega = 0.8$ rad/min.

    (b) What is the critical gain of the system? What would it be if there is no dead time?

    (c) What is the proportional gain when the desired gain margin is 1.7?

    (d) What is the proportional gain when the desired phase margin is 30°?

(8) Consider a simple unity feedback system with the closed-loop characteristic equation $1 + G_c G_p = 0$ and in which G_p is a first-order function with a steady-state gain of 2 (unit) and a time constant of 2 min. We now want to implement a PI controller. We are also given the choices of using integral time constants with values of 1, 2, or 4 min. Our task is to design a system with a damping ratio of 3/4.

    (a) Make a rational choice of the integral time constant. Provide a brief explanation.

    (b) With your chosen integral time constant, what is the proportional gain that you would use?

(9) Consider the closed-loop system in Fig. PII.9. What is the restriction on the value of β (a positive number) if we want an underdamped system? What is the restriction if the system must be stable? What is the offset of this system?

277

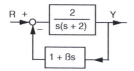

Figure PII.9.

(10) Feedback compensation can be achieved with what is called rate feedback, as shown in Fig. PII.10.
   (a) Derive the closed-loop transfer function of the block diagram.
   (b) For simplicity, let's take $K = 1$ and $\tau = 1$. If we desire the system response to have a damping factor of 0.7, what is a proper choice of the value of b?
   (c) Based on the transfer function $(1 + bs)$, can you say why the term "rate feedback" is used?
   (d) In terms of the damping ratio, what is the general effect of b on the system response? What if b is zero?

(11) We have learned that it helps to avoid derivative kick if the derivative control is on the measured variable only. Let's do a simple exercise to understand why. Consider the two situations with ideal PD control; Fig. PII.11(a) is the textbook ideal PD control of a first-order process, and Fig. PII.11(b) is the so-called rate control in the feedback loop.
   (a) Sketch the root-locus plots when we use an ideal PD controller, as in Fig. PII.11(a). Should you choose $\tau_p > \tau_D$ or $\tau_p < \tau_D$? Use the proper choice for parts (b)–(f).
   (b) Redo the root-locus plot in which now the controller is now a real PD controller.
   (c) Sketch the root-locus plot for Fig. PII.11(b).
   (d) What is the difference in the characteristic polynomial between Figs. PII.11(a) and PII.11(b)?
   (e) Which system configuration can eliminate offset?
   (f) What are the closed-loop transfer functions of the two cases in a servo problem? With what we have learned about the behavior of lead–lag elements, what is your expected time-domain responses of these two cases with respect to a unit-step change in the reference?

(12) We have a system with the closed-loop characteristic equation

$$1 + K_c \left(\frac{\tau_I s + 1}{\tau_I s} \right) \frac{1}{(s + 1)(2s + 1)} = 0,$$

where the time constants are in minutes. We want to design a slightly underdamped system. We face the possibilities of using integral time constants of 0.5, 1.5, or 2.5 min. Any oscillation of the system should exhibit behavior that resembles a

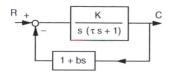

Figure PII.10.

Part II. Intermediate Problems

(a)   (b)

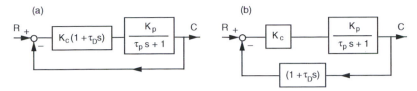

Figure PII.11.

damping ratio of 0.7, i.e., the oscillation may be due to complex poles that are not the dominant closed-loop poles.

(a) For each choice of the integral time constant, find the proportional gain that may meet the specification with the use of root-locus plots.

(b) Among the three choices, which case would you use in your controller design? State your reason.

(13) Let's consider a unity feedback system with an ideal PD controller and the closed-loop characteristic equation $1 + G_c G_v G_p = 0$, where

$$G_c = K_c(1 + \tau_D s), \quad G_v = \frac{1}{(12s + 1)}, \quad G_p = \frac{1}{(s + 1)}.$$

The problem seeks to find a system with a damping ratio of $\zeta = 0.7$. We have three probable regions to select our derivative time constant: (1) $\tau_D < 1$, (2) $1 < \tau_D < 12$, or (3) $\tau_D > 12$.

(a) Which region must we choose for the derivative time constant?

(b) What is the stability region for different ranges of the derivative time constant and the proportional gain in this problem?

We now consider the same problem with a PI controller, $G_c = K_c(1 + \frac{1}{\tau_I s})$.

(a) If we seek a system that can provide underdamped behavior, which are the probable region(s) that we should select for the integral time constant? (1) $\tau_I < 1$, (2) $1 < \tau_I < 12$, or (3) $\tau_I > 12$? You must give a proper explanation to your answer.

(b) Let's say we have selected $\tau_I = 0.5$. Use the Routh–Hurwitz criterion to determine the range of K_c that provides a stable system.

(c) Find the ultimate gain and the ultimate frequency for the case in (b).

(d) What is the probable offset in this system?

(14) Let's consider a system with inherent dead time. We will use only a proportional controller. The closed-loop characteristic equation is $1 + G_c G_p = 0$. The process transfer function is

$$G_p = \frac{3e^{-s}}{(4s + 1)}.$$

(a) If you approximate the dead-time function with a first-order Padé formula, use any analytical method of your choice to find the approximate range of proportional gain such that the system is stable.

(b) Show how you would find the exact ultimate gain with the direct substitution method. (*Hint for an initial guess*: The nontrivial ultimate frequency lies between $\pi/2$ and 1.8.)

(c) Repeat, using a Bode plot.

279

(15) Consider a simple unity feedback system with the equation $1 + G_cG_p = 0$. The process transfer function is $G_p = K/s^2$.

    (a) Before we can really design the controller, we need to measure the value of K. Why is it difficult to evaluate K with a common method such as the open-loop step test?

    (b) We can measure K in a closed-loop by using a proportional controller with a known value of K_c. Show how we may measure K with a step change in the reference.

    (c) Between a PI and a PD controller, which one should we use? Show, with root-locus plots (can be hand-sketched), how you would proceed to design the controller of your choice.

(16) Controller design based on direct synthesis is not unique. The result depends on our specification of the closed-loop response. The result is also dependent on the assumptions that we make along the way. Take the case of a process function with dead time,

$$G_p(s) = \frac{K_p e^{-t_d s}}{(\tau_p s + 1)},$$

and the desired closed-loop function $(C/R) = [e^{-\theta s}/(\tau_c s + 1)]$. In Chap. 6, we used a first-order Taylor expansion to approximate the exponential term in G_c, which we know is not very good unless the dead time θ is small.

    (a) If we use the first-order Padé approximation instead, show that the resulting G_c function resembles a real PID controller. *Hint:* Do not multiply the terms in the numerator. One of them will become the integral and the other the real derivative controller terms. The choice of the integral time constant is similar to other case studies in direct synthesis.

    (b) The same can be said with the use of IMC in Example 6.5. If we choose the low-pass filter such that G_c^* is a semiproper function (i.e., $r = 2$), the resulting G_c again resembles a real PID controller. Derive the parameters of the real PID controller.

    (c) In both parts (a) and (b), how are the parameters associated with the real derivative control term different from what we are familiar with?

(17) Consider a unity feedback system with the closed-loop characteristic equation $1 + G_cG_vG_p = 0$. Two of the functions are

$$G_p = \frac{2}{(20s + 1)}, \qquad G_v = \frac{0.5}{(s + 1)}.$$

A root-locus plot of the system has loci that originate from four points, 0, -0.05, -1, and -5, on the real axis, and two of the loci terminate at points -0.1 and -0.5.

    (a) What is the specific controller being used? Provide proper explanations by sketching the probable root-locus plot of the system. What can you say about the values, where possible, of the controller parameters?

    (b) What is the order of this system?

    (c) Illustrate with a root-locus plot the probable locations of the dominant closed-loop poles in this system.

Part II. Intermediate Problems

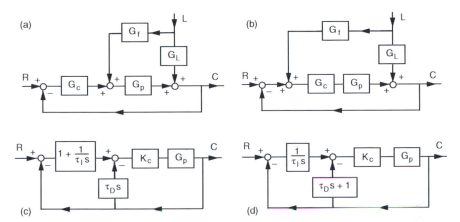

Figure PII.19.

(18) Consider a unity feedback system with the closed-loop characteristic equation $1 + G_c G_p = 0$. The process and controller functions are

$$G_p = \frac{10}{s(s+5)}, \quad G_c = K\frac{(s+5)}{(s+20)}.$$

(a) Determine the ranges of the gain K that will provide overdamped, critically damped, and underdamped behavior of the closed-loop dynamic response.

(b) Based on the closed-loop characteristic polynomial, sketch the probable shape of the root-locus plot.

(c) What is the value of the gain K that provides a closed-loop response that has a damping factor ζ of 0.7? What is the steady-state error of this system with respect to changes in the set point?

(19) To implement feedforward control, two schemes in Figs. PII.19(a) and PII.19(b) are proposed to eliminate the effect of disturbances with the transfer function G_f:

(a) Derive the closed-loop transfer functions relating changes in R and L to changes in C in both cases.

(b) What should G_f be if we want perfect elimination of any load changes, again in both cases?

(c) Which case is a more sensible or probable design? And why?

(d) We now look at two alternative schemes of implementing PID control: the so-called PI-D and I-PD controllers as shown in Figs. PII.19(c) and PII.19(d), respectively. Derive the closed-loop servo function C/R in each case.

(e) How have these two systems changed the characteristic equation as compared with when we use a standard textbook ideal PID controller?

(20) A classical example in stability analysis is that a control system can stabilize an open-loop unstable process. Consider

$$G_p = \frac{0.5}{s - 2}$$

and a PI controller with an integral time constant $\tau_I = 1$. What is the range of proportional gain for which the closed-loop system is stable? What is the frequency of oscillation when the system is marginally stable? What is the associated proportional gain?

(21) Consider the unstable process transfer function

$$G_p = \frac{K}{s - 2}$$

and the characteristic equation of a simple control system: $1 + G_c G_p = 0$.

(a) What is the stability criterion if we use a PD controller? How is it different from a proportional controller?

(b) What is the stability criterion if we use a PI controller?

(c) Sketch the root-locus plots for parts (a) and (b).

(d) Derive the ultimate gain and the ultimate frequency in parts (a) and (b).

(22) For the given open-loop unstable process and load functions,

$$G_p(s) = \frac{2}{s - 4}, \qquad G_d(s) = \frac{0.5}{s - 4},$$

we know that a proportional controller can stabilize the process in a servo problem. Would you expect the result to be different with respect to load changes? Consider a unity feedback system with the closed-loop characteristic equation $1 + G_c G_p = 0$, but now the regulating problem.

(a) For a simple proportional controller, find the range of K_c that yields stable system responses with respect to load changes.

(b) Draw the root-locus plot of the system.

(c) What is the ultimate gain of the system? At this position, will the response be stable with respect to a step input and an impulse load input?

(d) If we apply the direct synthesis method to design the controller, what is the analytical result and what is your recommendation?

(23) Consider a closed-loop system with only a proportional controller; the characteristic equation is given as

$$1 + K_c \frac{1.2 e^{-0.7s}}{(0.2s + 1)(4s + 1)} = 0.$$

(a) What is the ultimate gain of this system? The unit of the time constants is in minutes.

(b) If you want to use a PID controller, choose the controller settings based on an appropriate empirical tuning relation of your choice. With this setting, what is the gain margin of the system with the PID controller?

(c) If you want a gain margin of 1.7 instead, what proportional gain should you use (presumably you keep the same τ_I and τ_D)?

(24) Consider the system with cascade control in Fig. PII.24. In this case, we implement only proportional control in both the primary and the secondary loops. We are given

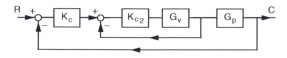

Figure PII.24.

that

$$G_v = \frac{1}{5s + 1},$$

$$G_p = \frac{1}{(s + 1)(0.1s + 1)}.$$

The cascade controller is designed to speed up the time constant of the actuating valve 10 times. What is the ultimate gain of this cascade system? Explain the stability with and without cascade control with Bode plots. What is the proportional gain that we should use if we desire a gain margin of 2 in the system?

(25) We have the following transfer function:

$$G(s) = \frac{18(2s + 1)}{(s^2 + 3s + 9)(s + 4)}.$$

(a) Identify all the corner frequencies and frequency asymptotes on the magnitude and the phase-angle plot of $G(j\omega)$.
(b) Write the analytical equations with which we can calculate the magnitude and the phase angle of $G(s)$.
(c) If $G(s) = Y(s)/X(s)$ and $X(s)$ denotes a unit-step input, what is the steady-state value of $y(t)$? Is there a dominant pole in $G(s)$?
(d) If the transfer function is in a closed-loop system with the characteristic equation $1 + K_cG(s) = 0$, what should K_c be if we want the gain margin to be 2?

(26) Consider a system with the closed-loop characteristic equation $1 + G_cG_p = 0$. We are given that the open-loop poles are located at -0.5, $-3 \pm j\sqrt{3}$, $-6 \pm j\sqrt{5}$, and -9, and there is an open-loop zero at -0.5. The steady-state gain of the process function G_p is 2.

(a) What kind of controller could G_c be? What is a probable justification for the zero at -0.5?
(b) Suppose we want to design the system such that it has a decay ratio of 0.25. What dominant pole(s) must we use as the basis of our calculation? What are the controller settings?
(c) Suppose the preceding design is too underdamped and we want a decay ratio of 0.1. What are the controller settings?
(d) Sketch the root-locus plot of this system.
(e) With respect to system stability, do we have enough information to design a controller that can handle disturbances?

(27) A chemical reactor can be cooled by either a cooling jacket or a condenser. The transfer functions are measured as

$$\frac{T}{Q_c} = \frac{-1}{s + 1}\left(\frac{^\circ C}{gpm}\right), \qquad \frac{T}{Q_j} = \frac{-5}{10s + 1}\left(\frac{^\circ C}{gpm}\right),$$

where Q_c and Q_j are the cooling water flow rates to the condenser and the jacket, respectively, and the time unit is in minutes. The range of the temperature transmitter is 100–200 °C, with a corresponding output of 0–10 mV. Control valves have linear trim and constant-pressure drop and are half-open under normal operating conditions. Normal condenser flow is 30 gpm. Normal jacket cooling water flow is

20 gpm. The temperature measurement through a thermowell is a first-order function with a time constant of 12 s. Our objective is to determine which is the better method to manipulate the temperature.

(a) Which cooling method do you prefer? Consider first the case of a proportional controller. What are the proportional gains for a closed-loop damping coefficient of 0.707?

(b) With your chosen manipulated variable, design a PID controller with your own design objectives and choice of methods.

(28) Consider a system with the closed-loop characteristic equation $1 + G_c G_v G_p G_m = 0$ and the following transfer functions:

$$G_v(s) = \frac{1}{10s + 1}, \quad G_p(s) = \frac{1}{20s + 1}, \quad G_m(s) = \frac{1}{0.5s + 1}.$$

The controller was selected to be

$$G_c(s) = K_c \left[1 + \frac{1}{3s} + \frac{2}{3}s \right].$$

(a) Sketch the root-locus plot for this system. Find the range of K_c that provides stable closed-loop responses. *Hint:* This problem is tricky. The system is stable for very small and very large proportional gains, but not in between. This is called *conditional stable*.

Being a good control engineer, you, of course, dislike the arbitrary choice of the controller. So let's design our own controller. We also want to take advantage of empirical tuning relations.

(b) First, find a first-order with dead-time function that may serve as the process reaction curve function. Compare the time response (with respect to a step input) of the approximation with the original functions.

(c) With only a proportional controller, what is the approximate range of the proportional gain that provides us with a stable system?

(d) Now we can design a more complete controller for the system. Use either the Cohen–Coon or the ITAE tuning relations. State your reason. For your controller design, provide the controller settings and a time-response sketch with respect to a unit set-point change.

(e) How does the addition of derivative control affect the controller settings and the overall system response?

(29) A process was pulsed tested and, by use of a fast Fourier transform, the open-loop frequency-response data in Fig. PII.29 were obtained: Identify the open-loop transfer function. If we had performed the frequency-response experiment on a closed-loop system by a pulse input into the set point, what would we be measuring?

(30) Consider the model of the liquid level in a stirred-tank,

$$A \frac{dh}{dt} = q_0 - q,$$

where $h(0) = 0$. The outlet flow rate is kept constant. We'll look into a couple of problems with this integrating process.

Part II. Intermediate Problems

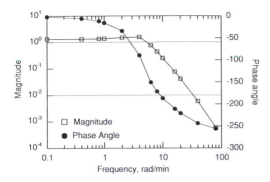

Figure PII.29.

(a) A simple rate feedback scheme is shown in Fig. PII.30. What is the value of K_v that provides a critically damped behavior? The cross-sectional area of the tank, A, is 2 (area units).

(b) If we increase the value of K_v, will the system be underdamped or overdamped?

(c) If the set point h_{sp} is increased by 10 (height units), what is the value of K_v such that the liquid level h will also change exactly by 10 (height units)?

(31) Consider the transfer function that contains integrating action:

$$\frac{Y(s)}{U(s)} = \frac{K}{s(\tau s + 1)}.$$

We take a sinusoidal input $u(t) = \sin \omega t$, with frequency $\omega = 0.5$ rad/s. The input and the output are shown in Fig. PII.31. Estimate the values of K and τ from the data. Why is it that the mean of the sinusoidal output is 0.5 and not zero? *Hint:* There is only one term in $Y(s)$ that we really need for this last part.

(32) Consider two stirred-tanks with electrical heaters that are connected in series. The system block diagram is shown in Fig. PII.32, where we manipulate the voltage V_1 of the heater in the first stirred vessel. The controlled variable is the temperature of the second tank. The load variables are inlet flow rate Q_0 and voltage to the second tank heater V_2. The transfer functions that we found from previous experiments are

$$K_m = 0.15 \text{ V/°C}, \quad t_d = 0.5 \text{ min}, \quad G_2 = \frac{4 \text{ °C/V}}{s+1}, \quad G_3 - \frac{-5e^{-0.5s} \text{ °C/gpm}}{(10s+1)(s+1)},$$

$$G_4 = \frac{2.5 \text{ °C/V}}{10s+1}, \quad G_5 = \frac{1 \text{ °C/°C}}{10s+1}.$$

All the time constants have units in minutes.

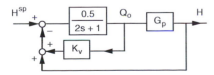

Figure PII.30.

285

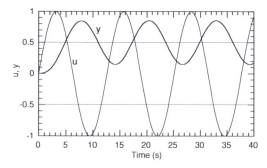

Figure PII.31.

(a) We first use a proportional controller. What is the ultimate gain of the system? What would it be if there were no transport lag between G_2 and G_5?

(b) What is the proportional gain when the desired gain margin is 2?

(c) What is the proportional gain when the desired phase margin is 40°?

(d) If we want a system that exhibits "slight overshoot," what are the PID controller settings that we may select based on the Ziegler–Nichols ultimate-gain tuning relation?

(e) Provide the PID settings based on the process reaction curve and the Ziegler–Nichols tuning relation. How different is this result compared with that of part (d)?

(f) Provide on the same plot the time-domain responses of the system subject to a unit-step change in the set point when we use PID controller settings based on the Cohen–Coon empirical tuning relation, the settings from part (d), and the settings from part (e).

(g) If you have to design a feedforward controller to handle fluctuations in flow rate Q_0, what is an appropriate function for G_f?

(h) Repeat part (g) with a feedforward controller to handle fluctuations in the variable V_2.

(33) Consider the process function

$$G_p = \frac{0.5}{0.01s^2 + 0.04s + 1},$$

where time is in the unit of seconds. We need to design a control system that cannot permit steady-state error and excessive underdamped behavior. On this note, we aim for a slightly underdamped system that, if it exhibits oscillations, is at

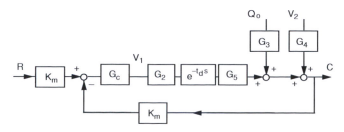

Figure PII.32.

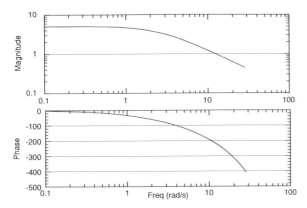

Figure PII.34.

most equivalent to a damping ratio of 0.707. We use a simple unity feedback system with the characteristic equation $1 + G_c G_p = 0$. Also, for the sake of this exercise, we have our choice of using $\tau_I = 1/8$ or $1/4$ s and, similarly, $\tau_D = 1/8$ or $1/4$ s.

(a) With respect to the function G_p, what is the time constant?

(b) With root-locus plots, illustrate how the choices of P, PI, and ideal PD controllers may influence the system time response. (Identify clearly the open-loop poles and zeros and the direction of root loci.) Provide reasons for controllers that you want to eliminate for this system design.

(c) Demonstrate your choice of an ideal PID controller design with a root-locus plot. What are the settings of the controller? What is (are) the dominant pole(s) in this system?

(34) A transfer function is used to generate the Bode plot in Fig. PII.34. Identify the transfer function and evaluate all parameters that are associated with this transfer function. It is also given that the corner frequency is 2.5 rad/s and, at a frequency of 1 rad/s, the magnitude is 4.9 and the phase lag is $-33°$.

(35) With respect to Fig. PII.35, which represents a unity feedback system with the closed-loop equation $1 + G_c G_p = 0$, derive the state-space model given that $G_c = K_c(1 + 1/\tau_I s)$ and $G_p = K_p/(\tau_p s + 1)$. Find the characteristic polynomial of the closed-loop system matrix $\mathbf{A}$. The PI controller function can be rearranged in the system block diagram, as shown in Fig. PII.35. The locations for the two state variables are also identified as shown.

(36) Derive the state-space model of the unity feedback system in Fig. PII.36, where now $G_c = K_c(s + z/s + p)$ is a lead–lag compensator and $G_p = [K/(s + a)]$.

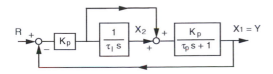

Figure PII.35.

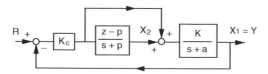

Figure PII.36.

The lead–lag element can be rearranged as

$$\frac{s+z}{s+p} = 1 + \frac{z-p}{s+p},$$

as shown in Fig. PII.36, such that the function is represented by proper fractions.

(37) Derive the state-space model for a system with a second-order process and a PID controller, as shown in Fig. PII.37.

(38) In Example 8.10, we discussed how the integral time constant must be chosen such that it is larger (slower) than the process time constants if we are to stabilize the system. This is an exercise to test this hypothesis. Consider a simple unity feedback system with the closed-loop characteristic equation $1 + G_c G_p = 0$, where

$$G_p = \frac{2}{(4s+1)(5s+1)}$$

and G_c is a PI controller with either (1) $\tau_I = 1$ or (2) $\tau_I = 10$. All time constants have the unit of minutes.

(a) Derive the stability criterion with the Routh array and using general notation for τ_I.

(b) Provide the stability criterion with two specific values of the integral time constant: $\tau_I = 1$ and $\tau_I = 10$.

(c) Provide the root-locus plots for both cases of integral time constants. Indicate clearly if the system may become unstable or not.

(d) When $\tau_I = 1$, what is the ultimate gain? If we choose a proportional gain such that the gain margin is 2, what is the phase margin under this controller setting?

(e) Repeat part (c) with $\tau_I = 10$.

(f) When $\tau_I = 1$, what is the proportional gain that we must choose such that the system has a damping ratio of $1/\sqrt{2}$? What is the dominant closed-loop pole?

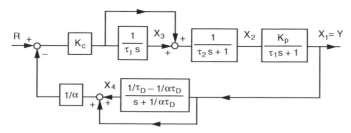

Figure PII.37.

(g) Repeat part (e) with $\tau_I = 10$.
(h) Provide the Bode plots for both cases of $\tau_I = 1$ and $\tau_I = 10$. Make use of corner frequencies and the high- and low-frequency asymptotes to explain the stability features of this problem.

(39) In Subsection 3.4.3, we gave up further analysis of the model of two interacting tanks-in-series, saying that state-space representation is better for handling such problems. This is our chance to give that a try now. Let's say we have two tanks such that $A_1 = 5$ m^2, $A_2 = 2$ m^2, and $R_1 = R_2 = 1$ min/m^2.
(a) Compute state-space model matrices in which h_2 is the only output.
(b) Confirm that the eigenvalues of $\mathbf{A}$ are the same as the poles of characteristic equation (3.46).
(c) Compute the transfer function and use that to find for a classical proportional control system such that the damping ratio is 0.7.
(d) With the closed-loop poles taken from part (c), find the state feedback gain. You should find it to be identical to classical control.
(e) With classical PI control, use root-locus plots to show that $\tau_I = 0.5$ min is a reasonable choice for the control system. Find the proportional gain such that the damping ratio is 0.7.
(f) Find the corresponding state feedback gain by using state-space integral control.
(g) Do the time-domain simulation for the state-space system with and without integral control.

(40) A classic example of MIMO problem is a distillation column. From open-loop step tests, the following transfer functions are obtained:

$$\begin{bmatrix} X_D(s) \\ X_B(s) \end{bmatrix} = \begin{bmatrix} \dfrac{0.6}{(7s+1)^2} & \dfrac{-0.5e^{-0.5s}}{(7s+1)^2} \\ \dfrac{0.3e^{-0.5s}}{(16s+1)(0.5s+1)} & \dfrac{-0.4}{(14s+1)(0.4s+1)} \end{bmatrix} \begin{bmatrix} L(s) \\ V(s) \end{bmatrix}.$$

In this model, x_D and x_B are the distillate and the bottom compositions, respectively, L is the reflux flow rate, and V is the boil-up rate. We want to design a 2 × 2 MIMO system with only PI controllers.
(a) From the relative gain array, determine the proper pairing of variables. Draw the block diagram and set up the system with Simulink.
(b) To begin the design, we need the controllers based on a simple SISO system first. For each of the two controllers, what are the proportional gains if the gain margin is to be 2?
(c) What is the proportional gain if the phase margin is to be 45°?
(d) What are the PI controller settings based on the Ziegler–Nichols tuning relations?
(e) How should we detune the controllers when we implement them in the MIMO system?
(f) Determine how effective the use of decouplers is, as in Fig. 10.14, in reducing loop interactions.

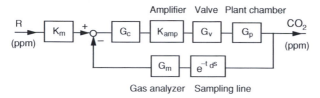

Figure PIII.1.

Part III. Extensive Integrated Problems

(1) To address one aspect of the accumulation of greenhouse gases, we want to build a chamber that can be used to study the level of CO_2 on the growth of plants on a space shuttle. In the system design, there is a time delay of 0.75 min in the sampling line. The sensor is a gas analyzer that can measure CO_2 content from 0 to 600 ppm linearly and the output is 0 to 5 V. The gas analyzer can be modeled as a first-order function with a time constant of 0.1 min. The plant chamber itself can be modeled as a well-mixed reactor (CSTR) with a pseudo-first-order reaction. This process function has a steady-state gain of 0.23 (ppm min)/ml and a time constant of 5 min.

Under normal circumstances, we would use a solenoid valve. For the present problem, we take a simpler approach. We use instead a regulating valve that can be modeled as a first-order function with a time constant of 0.02 min and a steady-state gain of 0.2 ml/(min mV). An amplifier is needed to drive the valve, and it has a gain of 10 mV/mV. The block diagram is shown in Fig. PIII.1.

(a) Design a proportional controller with a gain margin of 1.7.

(b) What if we want a PID controller? Provide a system design that gives us an acceptable underdamped behavior, say an apparent (because the system is not second order) damping ratio of 0.45.

(2) (a) We are assigned to identify a process that is believed to be second order. Let's presume that the magnitude and the phase plots in Fig. PIII.2(a) are obtained from a pulse experiment.

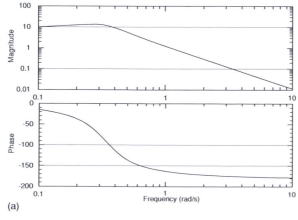

(a)

Figure PIII.2a.

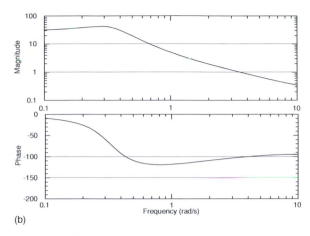

(b)

Figure PIII.2b.

We are also given the information that from a unit-step response experiment, the process exhibited a 25% overshoot. Estimate the parameters K_p, τ, and ζ of the transfer function:

$$G_p(s) = \frac{K_p}{(\tau^2 s + 2\zeta \tau s + 1)}.$$

(b) The second-order process in part (a) is put in a closed-loop system with a controller. The closed-loop characteristic equation is a simply $1 + G_c G_p = 0$, and the Bode plot of the open-loop transfer function of this system is given in Fig. PIII.2(b). What is the controller (P, PI, or PD) being used? Explain your decision with hand-sketched Bode plots with features such as low- and high-frequency asymptotes and corner frequencies. It would be difficult to read the controller time constants from the figure, but we should be able to estimate the order of magnitude of the time constants, especially if they are smaller or larger than that in the process transfer function.

(c) Repeat part (b) with Fig. PIII.2(c).

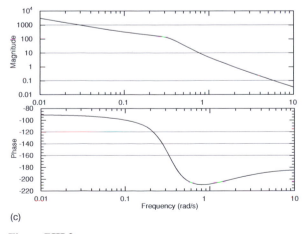

(c)

Figure PIII.2c.

(3) We want to design a control system for a gas absorber to reduce SO_2 emission by manipulating the water flow rate. The gas absorber can be modeled by the transfer function

$$\frac{C(s)}{F(s)} = \frac{-0.05}{(2s + 1)},$$

where, in the time domain, $c(t)$ is the effluent SO_2 concentration as deviated from its nominal operating value of 100 ppm, and $f(t)$ is the water flow rate as deviated from its nominal operating value of 250 gpm. Time is measured in minutes. The system has the closed-loop equation $1 + G_c G_a G_p G_m = 0$.

(a) For the moment, we are told to ignore the dynamics of the actuator and measurement transfer functions, hence taking $G_m = K_m = 1$ mV/ppm. However, we have the choice of $G_a = \pm 1$ gpm/mV at this moment. We are using a controller with positive proportional gain. Should G_a be $+1$ or -1? Explain your choice.

(b) We will stay with the choice of G_a in part (a) and $G_m = 1$ mV/ppm, but now a PI controller, still with positive proportional gain, is to be used. The integral time constant is 0.5 min. What must be the choices of the proportional gain such that the closed-loop system is stable?

(c) Repeat part (b), but now we have an ideal PD controller instead. The derivative time constant is also 0.5 min. What must be the choices of the proportional gain such that the closed-loop system is stable?

(d) We want a closed-loop system response that is underdamped. The system should have a dynamic response equivalent to a damping ratio of 0.7. Select the proper controller from either part (b) or part (c). Estimate the proportional gain that can conform to the damping ratio specification. If you have more than one choice, select one and explain why. For the given closed-loop poles that you have selected, what is the corresponding natural frequency?

(e) Now reality strikes! We can no longer ignore the measurement function. The voltage output of the gas analyzer was previously calibrated, and the result in actual measurement variables is

$$v = 0.45 + 1.2 [SO_2],$$

where the voltage output is in millivolts and the concentration of SO_2 is in parts per million. The gas analyzer is extremely fast, but the sampling line has a dead time of 0.2 min. The actuator, which really is a valve, has a steady-state gain of ± 0.9 gpm/mV and exhibits a first-order response with a time constant of 0.3 min. [The sign should of course be the same as the one determined in part (a).]

Now the EPA is tightening up its regulation, and we need to lower the outlet SO_2 concentration. Use the controller settings that we have obtained in part (d) and provide a time-response simulation plot representative of the system subject to a step change of -10 ppm in the set point.

(4) Let's demonstrate our understanding and command of our control theory with the following experiment, which is on a torsion disk, but may just as well be a system

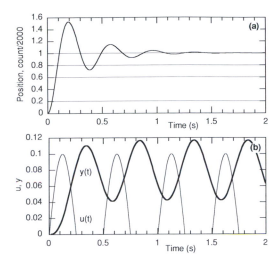

Figure PIII.4.

with a flow vessel with fixed-rate pumps. The position (angle rotation) of the disk is measured in the arbitrary unit of counts. If you need an introductory explanation of mechanical models, see the *Web Support*.

- An open-loop step test resulted in a response that was ramplike. A closed-loop servo experiment with only a proportional controller and $K_c = 0.02$ and a set-point change of 2000 counts resulted in an underdamped system response, and there is no offset. The response is shown in Fig. PIII.4(a).

(a) What is your interpretation at this point with respect to the two experiments?

- Based on the first two experiments, a suggestion was made that an open-loop frequency-response experiment be performed to identify the process function. We applied a sinusoidal input with an amplitude of 0.1 V and different frequencies to the actuator (a dc motor), and the application software returned the response in counts. We gathered from the responses that, at very high frequencies, the phase lag appeared to be approximately −180°. There appeared to have been a phase lag even at low frequencies. A simulated plot in which the input frequency was 2 Hz is shown in Fig. PIII.4(b). The output of the computation is scaled arbitrarily in this plot. Note that the mean of the large time response is not zero, as opposed to the input sine wave.

(b) Now, our immediate task is to identify the open-loop function. Write the probable form of the transfer function G_p (really the lumped $G_a G_p G_m$). Show that the mean of the sinusoidal time response is not zero and that it can be used to estimate the process gain.

(c) Sketch a block diagram of the system. We can assume very fast dynamics in the dc motor (the actuator) and the encoder (the position sensor). Derive the closed-loop transfer function for servo control.

(d) Because of limitations in the application software, we cannot estimate the measurement gain and actuator gain easily. Rearrange the block diagram to a form

293

such that we can lump the K_m together with the process and actuator steady-state gains.

(e) For your proposed (lumped) G_p, find the time constant of the process based on the data of the open-loop frequency-response experiment in Fig. PIII.4(b).

(f) Show how you could have used the magnitude of the proposed (lumped) G_p in helping to estimate the (lumped) steady-state gains from the open-loop frequency-response experiment.

(g) Now with the closed-loop step-response data [Fig. PIII.4(a)], estimate the lumped steady-state gain ($K_m K_a K_p$).

(h) Show how, in the closed-loop step-response data, the overshoot, settling time, and time to peak are consistent with the time period, damping ratio, and lumped steady-state gain of your proposed closed-loop transfer function.

• At this point, it was apparent that, because of the difficulty in analyzing data accurately, the estimated values of the open-loop and the closed-loop experiments were not in perfect agreement. Two more experiments were proposed and performed to resolve the discrepancies.

First, we used a PD controller in a closed-loop experiment. We picked a reasonable proportional gain and varied the derivative time constant. What we observed was that when τ_D was roughly 0.2 s or larger, the servo system did not exhibit any oscillations even as we increased the proportional gain further. When τ_D was roughly 0.1 s or less, the servo system exhibited underdamped behavior. (Yes, you could have guessed that the "answer" lies somewhere in between.)

Second, we did a closed-loop frequency response experiment – we imposed a sinusoidal wave (with an amplitude of 1000 counts) on the set point and observed the response of the controlled variable. We used a proportional controller with $K_c = 0.02$.

When the input sine wave had a frequency as low as 0.2 Hz, the amplitude of the closed-loop response (the amplitude ratio) relative to the input amplitude was approximately one and the phase lag was virtually zero. When the input sine wave had a frequency of 4 Hz, the amplitude ratio was only approximately 0.7. However, when we used an input frequency of 2 Hz, the ratio was almost 2!

(i) With root-locus plots, show how varying τ_D with a PD controller could have helped to resolve the discrepancies. What time constant could we have been able to identify and estimate in this experiment?

(j) What is the probable explanation of the three closed-loop frequency-response experiments? With the numbers that you have estimated regarding the system with a proportional controller, show that, at a frequency of 2 Hz, the observed result is indeed probable.

(k) If we had also measured the phase lag in the closed-loop frequency-response experiments, how could we have made used of the data? Write the appropriate magnitude and phase-angle equations and explain how we could apply them.

(l) Finally, we are ready to design the controller. What controller should we use? Can we use direct synthesis?

Part III. Extensive Integrated Problems

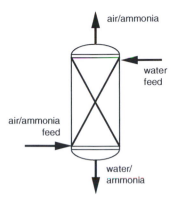

air/ammonia

water
feed

air/ammonia
feed

water/
ammonia

Figure PIII.5.

(m) For the proper choice of controller, sketch Bode plots of the open-loop transfer function of your closed-loop system. How important is the application of
the Nyquist stability criterion in your design? Of course we do not know the
values of τ_I and/or τ_D yet. We can consider only the relative value with respect
to τ_p.
(n) Consider a design specification of a system damping ratio of 0.7; provide an
acceptable controller design, assuming saturation is not a problem.
(o) I took my result of K_c and $\tau_D = 0.05$ s and tried it for real. The data showed
approximately a 10% overshoot or the equivalent of a damping ratio of 0.6. How
should I field tune the controller to achieve $\zeta = 0.7$?
(5) Consider a gas absorber that is designed to remove NH_3 from an air stream
(Fig. PIII.5). The gas stream entering the absorber contains air and NH_3. The design
specification is such that the outlet NH_3 concentration in the vapor phase must be
50 ppm or less.
   Available to us is an electronic concentration transducer that is calibrated to be
linear over the range of 0–200 ppm NH_3 and with an output range of 4–20 mA. This
fancy device has a negligible time lag. There is also an air-actuated valve that, when
fully opened and for the 10-psi pressure change that is available, will allow for a
throughput of 500 gpm of water. You have your choice of an air-to-open or an air-to-
close valve. The time constant of the valve is 5 s. Also available is an I/P transducer
with which, for a change of 4 to 20 mA, the output changes from 3 to 15 psi. If you
need more instrumentation to complete the design, you have permission from your
manager to purchase whatever you need.
   (a) Draw the instrument diagram of a control loop that can maintain the outlet NH_3
concentration at a set point of 50 ppm.
   (b) Draw the block diagram for the closed-loop system. Identify your manipulated, controlled, measured, and disturbance variables and their units. With
the exception of the absorber itself, provide the transfer function for each
block.
      We now need to find the process transfer function for the absorber. Several
open-loop step tests were performed, and after averaging the data, we obtained
following table:

295

Data for the NH_3 concentration in the outlet air stream in response to a step change of the water flow rate that dropped from 250 to 200 gpm

Time (s)	NH_3 (ppm)	Time (s)	NH_3 (ppm)
0	50.00	90	51.20
20	50.00	100	51.26
30	50.12	110	51.35
40	50.30	130	51.55
50	50.60	160	51.70
60	50.77	220	51.77
70	50.90		
80	51.05		

(c) Identify the absorber process transfer function with an appropriate model of your choice.

(d) Specify the action of the control valve and the controller (i.e., positive or negative gains).

(e) With proper assumptions or justifications, find the PID controller settings, using an empirical tuning relation of your choice.

(f) From a time-domain simulation of the control system (using a unit-step change in the set point), find the percentage of overshoot of the system.

(g) If the design specification is such that no offset is acceptable and the percentage of overshoot should not exceed 10%, provide one new controller setting that can meet the requirement. Provide justification of your design and a confirmation with a time-response simulation.

(6) The temperature of a chemical reactor with a very exothermic organic synthesis is monitored by a thermocouple with an amplifier. The reactor can be cooled by a water jacket and/or a condenser that traps the vapor and returns the condensed liquid reflux (Fig. PIII.6). Our task is to design a temperature controller system for the reactor. This is some of the information that we have.

- From an experiment involving the chemical reactor and the cooling jacket, we found that changes in the reactor temperature with respect to changes in the cooling

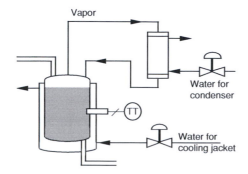

Figure PIII.6.

water flow rate is a first-order function. The steady-state gain is $-5\,°C/gpm$. The process can reach a new steady state with respect to a step change in flow rate in ~ 10 min.

- From another experiment involving the chemical reactor and the condenser, we found the change in reactor temperature with respect to changes in the condenser water flow rate to be first order as well. The steady-state gain is $-0.8\,°C/gpm$. The time constant is ~ 2 min.

- The span of the temperature transmitter is calibrated within the range $70–120\,°C$ for a full-scale output of $0–5$ V. The thermowell introduces a first-order lag of 15 s.

- The regulating valves are similar in design. They have a linear trim and constant-pressure drop. They can be described by a first order function with a time constant of 30 s. They are half open under normal operation conditions: The normal water jacket flow rate is 10 gpm, and the normal condenser water flow rate is 5 gpm.

- The controller has a full-scale output of $0–5$ V, which is designed to drive the entire range of the regulating valve. The bias signal is 2.5 V.

(a) Make a proper decision as to how you need to control the reactor temperature with a SISO system. Draw the instrumental diagram and the corresponding block diagram.

(b) With the understanding that the reaction is extremely exothermic, what are your choices of instrumentation (controller and regulating valve) in terms of their actions?

(c) Make sure you remember to identify the units on the block diagram. Determine the necessary transfer functions. Identify the probable disturbances.

(d) If you use only a proportional controller, use an analytical method of your choice to find the stability limit on the proportional gain.

(e) With a root-locus plot, confirm your result in part (d). In addition, find the proportional gain that you need to use if you desire the system to have a dynamic response with a 5% overshoot.

(f) If you want to use an empirical tuning relationship for a PID controller setting, provide a reasonable approximation of the process reaction curve function. Provide the controller settings based on the tuning relations of Cohen–Coon, Ziegler–Nichols, and ITAE.

(g) Can we make use of the results of IMC? If so, what is the corresponding PID controller setting?

(h) Back to part (d). We know there are other methods to arrive at the same answer. Confirm your value of ultimate gain with a proportional controller by using frequency-response techniques.

(i) With the ultimate gain known, what would be probable PID controller settings for our system if we desire a system response with a "slight" overshoot?

(j) Let's use an ideal PID controller and the integral and derivative time constants that we have chosen in part (i). What proportional gain should we use if the design specification is to have a gain margin of 1.7? (*Hint:* This is a trick question.)

(k) Provide the time-response simulations of the servo system with a unit-step change in the set point. Use three different, but in your opinion most reasonable,

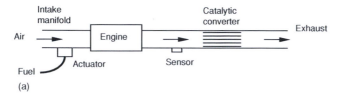

(a)

Figure PIII.7a.

PID controller settings based on previous results. Put all simulations on the same plot. Select the most "desirable" closed-loop system response with which you may do the final field tuning. State your reason.

(7) To meet emission standards, the fuel-to-air (F/A) ratio of a combustion engine must be metered precisely. Fuel-injection systems are designed not only to accommodate engine combustion performance, but also to achieve a desired F/A ratio for the catalytic converter to function properly. The catalyst is ineffective when the F/A ratio is more than 1% different from the stoichiometric ratio of 1:14.7. A feedback control loop must be implemented. The F/A ratio is monitored by a sensor before the catalytic converter, and, based on the measurement, the fuel flow rate is adjusted accordingly [Fig. PIII.7(a)].

The most difficult task in designing the control system is due to the nonlinearity of the sensor. Made of zirconium oxide, almost all the change in output voltage occurs near the F/A value at which the feedback system must operate. The gain of the sensor falls off quickly for F/A excursions just slightly away from 1:14.7 [Fig. PIII.7(b)].

In addition to the very nonlinear gain, the sensor exhibits a first-order lag with a time constant of 0.1 s. There is also a time delay in the system from the air intake manifold to where the sensor is located. The value depends on driving conditions, but on average, we can take the transport time delay to be 0.2 s. Physically, the engine could be driven by a fast fuel flow in the form of vapor or droplets and a slow fuel flow in the form of a liquid film on the manifold wall. The air manifold and the engine can be approximated altogether with the linear model in Fig. PIII.7(c).

The time constants can change considerably as a function of engine load and speed, but for this exercise, we will take $\tau_1 = 0.02$ s and $\tau_2 = 1$ s.

(a) Draw the block diagram of the control system with proper labels.

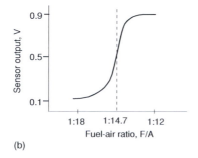

(b)

Figure PIII.7b.

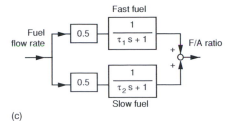

(c)

Figure PIII.7c.

    (b) Based on the information provided, select a proper mode of controller and state your reasons.

    (c) Design the controller. Use at least two different methods in your analysis. State your assumptions and design specifications.

(8) Consider the temperature control of a furnace in which we manipulate the fuel-gas flow rate. The temperature transmitter with a 0–1 V analog output is calibrated for 300–500 °C. The open-loop step-test data of the furnace temperature (Fig. PIII.8) were measured with a step change on the manipulated variable that is equivalent to a 5% increase of the biased output of the controller.

    (a) Draw the block diagram of the system with proper labels and units. Based on the information given, provide probable transfer functions of each block or of blocks with transfer functions lumped together.

    (b) Design an appropriate controller using IMC. If in theory we dictate a first-order response, why is the system response underdamped?

    (c) If the design criterion is a phase margin of 30°, what controller settings would you adjust and what is the final result?

    (d) Finally, select a different controller setting using an empirical tuning relation.

    (e) Perform time-response simulations. Which design do you prefer? What is your criterion?

(9) An exothermic chemical reactor (CSTR) can be open-loop unstable. The chemical reaction rate increases as the temperature shoots up, and subsequently more heat is given off. This heats the reactor to an even higher temperature such that the reaction rate is still faster and even more heat is evolved. If this process continues, what we have is an explosion. Thus open-loop instability means that the reactor temperature will take off when there is no feedback control of the cooling rate on the reactor.

       We do a simplified analysis on this problem. We must first assume that any changes in the reactant concentration can be neglected, i.e., we take the reactant

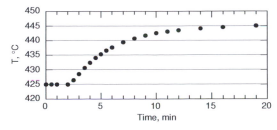

Figure PIII.8.

concentration C_A as a constant. Hence we consider only the energy balance in this simplified reactor model:

$$\frac{dT}{dt} = \frac{F}{V}(T_i - T) + \frac{(-\Delta H_R)}{\rho C_p} k_0 e^{-E/RT} C_A - \frac{UA_t}{\rho C_p V}(T - T_c),$$

where T_c is the cooling jacket temperature that we can manipulate. Other notations take on their usual meanings.

(a) Derive the process transfer function that relates the reactor temperature to the cooling jacket temperature, assuming that we have no fluctuations in the inlet temperature T_i.

(b) Derive the criterion that provides the condition under which the chemical reactor is open-loop stable. What is the physical significance (interpretation) of the result?

(c) For the set of conditions given below, is the reactor open-loop stable?

(d) Open-loop stable or not, with a proportional controller, is there a criterion on the controller gain that you must satisfy to stabilize the system? If yes, what is it?

(e) Discuss qualitatively how you would choose a controller (P, PI, PD, or PID) for the reactor. What would you consider as important dynamic criteria for the control system?

(f) Suppose we want to raise the temperature of the reactor by 5°. Provide the design and time response of a feedback control system that meets your design criteria stated in part (d).

Reactor parameters

$F = 0.1$ m^3/min,	$U = 0.062$ Kcal/m^2 K s
$V = 1.4$ m^3, $A_t = 230$ m^2	
$k_0 = 7 \times 10^{10}$ s^{-1}	$(-\Delta H_R) = 16{,}000$ Kcal/kgmol
$E = 32$ Kcal/gmol	$\rho = 800$ kg/m^3
$R = 2$ cal/gmol K	$C_p = 0.95$ Kcal kg^{-1} K^{-1}

Reactor steady-state values

$T = 600$ K
$C_A = 0.8$ kgmol/m^3

---

(10) We consider here a simplified fermentor classical control system (as opposed to that of Example 4.8).

We all agree that it would be best to use the sugar concentration S as the controlled variable, and use the volumetric flow rate Q as the manipulated variable. Any change in inlet concentration S_i is taken as disturbance. To make this problem a SISO system, we have to assume that the cell mass is a constant. This is possible only if we somehow trap the cells in the reactor and if they somehow stop growing. The simplified fermentor model is now

$$\frac{dS}{dt} = D(S_i - S) - \frac{\mu_m C}{Y_{c/s}} \frac{S}{K_m + S},$$

with the initial condition at $t = 0$ and $S = S_0$. The dilution rate D (defined as Q/V) is the reciprocal of space time. Because we are manipulating the flow rate, it actually helps to clean up our algebra by defining a variable that is directly proportional to Q. The fermentor volume is constant.

Our job is to design a controller that can maintain the sugar concentration at a specified value without steady-state errors. We can tolerate only slight oscillations. After any change in operating conditions, we want the fermentor to reach a new steady-state in less than 7 h. (Fermentation is a slow process, but still we want the job done before the plant operator has to go home.) The regulating valve is rather slow. The sugar sensor is worse. Their transfer functions with proper units for their steady-state gains are

$$G_v = \frac{3}{0.06s + 1}, \qquad G_m = \frac{e^{-0.15s}}{0.24s + 1}.$$

As for the fermentor, we have the following data:

S_i (steady-state) $= 10$ g sugar/L
$Y_s = 0.4$ g cell/g sugar
$\mu_m = 0.4$ h^{-1}
$K_m = 0.05$ g/L

In developing the process model, we must pick some nominal steady-state value about which the fermentor is being operated. A good choice is to specify 95% conversion of the inlet sugar concentration. Thus we can calculate the steady-state sugar concentration. At steady state, D and C are given by the steady-state solutions

$$D = \frac{\mu_m S}{K_m + S},$$

as it is also true that $C = Y_{c/s}(S_i - S)$. (We could not have derived these equations without the cell mass balance. If you really want to know how they came about, revisit Example 4.8.)

(11) Consider a rectilinear system that is naturally underdamped. In simple terms, it is a mass and spring system with a fraction. Using an input of 0.2 V to the actuator, we obtained the open-loop data in Fig. PIII.11. We also performed a closed-loop

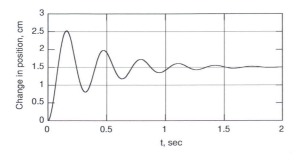

Figure PIII.11.

experiment with a proportional controller and $K_c = 1$ and found the offset to be 40%.

(a) Estimate the approximate $G_a G_p G_m$ as a second-order function. The sensor is extremely fast and we can take $G_m = K_m$.

(b) The function in part (a) is only an estimate because the actuator dynamics is so fast that it is masked by the slower process in the open-loop step test. One strategy to find the actuator function is to find the ultimate gain and frequency with a closed-loop experiment and from that back out what the actuator should be. Here we may assume that the actuator (a dc motor) has the function

$$G_a(s) = \frac{1}{\tau_a s + 1}.$$

The steady-state gain in G_a is written as 1, but its actual value is really contained in the data that we used to obtain the result in part (a). We did the closed-loop ultimate-gain experiment and found $K_{cu} = 1.65$ and $w_{cg} = 36.6$ rad/s. Find what τ_a should be.

(c) Do an open-loop step-test simulation to show that indeed the dynamics of the actuator was masked by the process in Fig. PIII.11.

(d) Use a method of your choice and design an ideal PID controller. Can the open-loop zeros be complex numbers? State your design specification and confirm the design with time-response simulations.

(12) We have three CSTRs-in-series (Fig. PIII.12), and we want to control the concentration of the third CSTR. In terms of reaction engineering, a mole fraction is the more sensible variable, but we stay with concentration so that the model equations are easier to interpret in this problem. In general, the balance of a reactant in a CSTR with a second order reaction is

$$V \frac{dC}{dt} = Q(C_0 - C) - V k C^2.$$

The rate constant k is 1.5 m^3 gmol^{-1} min^{-1} and volume V at 2 m^3 is the same in all three reactors. The inlet steady-state flow rate Q^s is 1 m^3/min. In your work, use the notation $\tau = V/Q^s$.

(a) Consider only one CSTR and derive the transfer functions associated with Q and C_0. What are the time constants and steady-state gains?

(b) Derive the transfer functions of all three CSTRs and put them in numerical form.

(c) You want to design a unity feedback servo system with which you can control the outlet concentration in the third CSTR, $c_3(t)$. The characteristic equation is

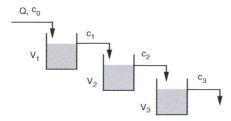

Figure PIII.12.

simply $1 + G_c G_p = 0$. Make your proper choice of manipulated, controlled, and load variables and derive the process transfer function G_p.

(d) If you use only a proportional controller, find the range of proportional gain that is necessary for a stable system. Find the proportional gain if the gain margin is 2.

(e) You want to design a PI controller, and you are given the choice of $\tau_I = 0.5$ min or $\tau_I = 2$ min. Which choice of integral time constant is more likely to lead to a stable system? Provide proper explanation. You use either root-locus plots or frequency-response analysis.

(f) Choose the integral time constant ($\tau_I = 0.5$ or 2 min) that can give us a more stable system. Find the proportional gain of this particular system if the gain margin is 2.

(g) If we want this system with a PI controller to have a damping ratio of 0.7, what is the proportional gain?

(h) Show on a plot that the process function may be approximated with a first-order with dead-time function that has a dead time of 0.5 min and a time constant of 2 min.

(i) Calculate the PI controller settings using the Ziegler–Nichols, ITAE, and Ciancone–Marlin tuning relations. Compare the time responses with a unit-step input for the set point.

We look into the rationale of the choice of integral and derivative time constants in this problem. Use the following so-called interacting PID controller function for the remainder of the problem:

$$G_c(s) = K_c \left(1 + \frac{1}{\tau_I s}\right)(1 + \tau_D s).$$

To begin, use the Ziegler–Nichols ultimate-gain tuning relation to find the PID controller settings for the case of one-quarter decay ratio system response.

(j) You want to find out if it is desirable to use this integral time constant as opposed to using an integral time constant of 0.5 min. Use a derivative time constant of 0.3 min in both cases. Use Bode plots to explain why one of the two choices can stabilize the system. State clearly and explain the crucial corner frequencies and, for the function $G_c G_p$, the overall low- and high-frequency asymptotes, and the one singularly most important feature that distinguishes the stability of the two cases.

(k) Sketch the root-locus plots for the two cases. In terms of dynamic response, identify what is considered as the dominant pole(s) of the *system*. Identify the ultimate gain where appropriate.

(l) The one-quarter decay ratio setting is of course too underdamped. If we insist on using the same integral and derivative time constants, what is the proportional gain that provides a system with a damping ratio of $1/\sqrt{2}$?

(m) Confirm your calculations with a time-response simulation of a unit-step change in the set point.

(n) Put the three CSTR's in a state-space model. Consider the case of only one input, c_0.

(o) Show that the step responses of c_3 with respect to a step change in c_0 are identical whether we use a transfer function or state-space representation calculations.

(p) Show that this model is both controllable and observable.

(q) Compute the state feedback gains if the closed-loop poles are identical to what we obtained in part (d). Do a time-response simulation to show that the design is identical to that of classical control.

(r) Design the state feedback controller you obtain if you add integral action.

(s) Find the estimator gains if you want to implement a full-state estimator.

References and Suggested Reading

Only a handful of introductory control textbooks in chemical processes is given here. A more thorough reference section with annotation is available on the *Web Support*.

1. A. B. Corripio. *Design and Application of Process Control Systems*, 2nd ed. Research Triangle Park, NC: ISA, 1998.
2. D. R. Coughanowr. *Process Systems Analysis and Control*, 2nd ed. New York: McGraw-Hill, 1991.
3. M. L. Luyben and W. L. Luyben. *Essentials of Process Control*. New York: McGraw-Hill, 1997.
4. T. E. Marlin. *Process Control: Designing Processes and Control Systems for Dynamic Performance*. New York: McGraw-Hill, 1995.
5. B. A. Ogunnaike and W. H. Ray. *Process Dynamics, Modeling and Control*. New York: Oxford University Press, 1994.
6. J. B. Riggs. *Chemical Process Control*, Lubbock, TX: Ferret Publishing, 1999.
7. D. E. Seborg, T. F. Edgar, and D. A. Mellichamp. *Process Dynamics and Control*. New York: Wiley, 1989.
8. C. A. Smith and A. B. Corripio. *Principles and Practice of Automatic Process Control*, 2nd ed. New York: Wiley, 1997.
9. G. Stephanopoulos. *Chemical Process Control. An Introduction to Theory and Practice*. Englewood Cliffs, NJ: Prentice-Hall, 1984.

Index

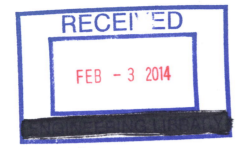

Made in the USA
Lexington, KY
27 January 2014